SYMBIOTIC ASSOCIATIONS

Other Publications of the
Society for General Microbiology

SYMBIOTIC ASSOCIATIONS

THIRTEENTH SYMPOSIUM OF THE
SOCIETY FOR GENERAL MICROBIOLOGY
HELD AT THE
ROYAL INSTITUTION, LONDON
APRIL 1963

CAMBRIDGE
Published for the Society for General Microbiology
AT THE UNIVERSITY PRESS
1963

PUBLISHED BY
THE SYNDICS OF THE CAMBRIDGE UNIVERSITY PRESS

Bentley House, 200 Euston Road, London, N.W. I
American Branch: 32 East 57th Street, New York, 22, N.Y.
West African Office: P.O. Box 33, Ibadan, Nigeria

Printed in Great Britain at the University Press, Cambridge
(Brooke Crutchley, University Printer)

CONTRIBUTORS

ARBER, W., Laboratoire de Biophysique, Université de Génève, Switzerland.

BAKER, J. M., Forest Products Research Laboratory, Princes Risborough, Buckinghamshire.

BOND, G., Department of Botany, University of Glasgow.

BROOKS, MARION A., Department of Entomology and Economic Zoology, University of Minnesota, St Paul, Minnesota, U.S.A.

COLEMAN, G. S., Biochemistry Department, Institute of Animal Physiology, Babraham, Cambridge.

DROOP, M. R., The Marine Station, Millport, Scotland.

DUBOS, RENÉ, The Rockefeller Institute, New York, U.S.A.

HUNGATE, R. E., University of California, Davis, U.S.A.

KESSLER, A., The Rockefeller Institute, New York, U.S.A.

LEV, M., National Institute for Research in Dairying, Shinfield, Reading.

MELIN, E., Institute of Physiological Botany, Uppsala, Sweden.

MOSSE, BARBARA, Soil Microbiology Department, Rothamsted Experimental Station, Harpenden.

NICHOLAS, D. J. D., Research Station, Long Ashton, Bristol.

NÜESCH, J., Eidgenössische Technische Hochschule, Zürich, Switzerland.

NUTMAN, P. S., Soil Microbiology Department, Rothamsted Experimental Station, Harpenden.

SMITH, D. C., University Department of Agriculture, Oxford.

CONTENTS

EDITORS' PREFACE

The pressing problems of co-existence in world affairs may have in-fluenced the Committee in their choice of subject for this years' Symposium. If so, it is to be hoped that the more bizarre examples of symbiosis illustrated in this volume will not be followed in the world at large; there are many other ways of escaping the Hobbesian predicament that without 'commonwealth' life must be 'nasty, brutish and short'.

It has often been maintained that the subject of symbiosis has no independent standing, and that its name tends to disguise a fundamentally parasitic relationship—surrounding its admitted complexity with an unscientific aura of teleology. A study of the pathological aspects alone however would be incomplete at best, and it is questionable whether such an approach would contribute much to the understanding of those associations which have no obvious pathological symptoms, but only beneficial effects. Indeed it could more cogently be urged that the study of symbiosis may illuminate some facets of the host-parasite relationship.

In this volume the origin, organization and functioning of symbiotic systems are discussed. In the field of microbiology these include associations between bacteria and their temperate phages, between bacteria, fungi and higher plants, and between micro-organisms and animals. Principally considered are ecology and taxonomy, methods of transmission and metabolic interactions between symbionts, especially at the nutritional level. The difficulty and frequent failure to grow many of the micro- and some of the macro-symbionts in pure culture are measures of our present incomplete understanding of symbiosis. There are also many non-microbial symbiotic associations, some comprising several components, and of even greater complexity in structure and behaviour than the microbial systems discussed in this symposium.

Unfortunately some of the papers originally invited were withdrawn at a late stage and substitutes could not easily be found. It is for this reason that Dr Nüesch's article, which was invited late, appears at the end of the volume.

The editors would like to thank the contributors and the Cambridge University Press for their help and co-operation which have made possible publication well in advance of the meeting at which the papers are to be discussed.

P. S. NUTMAN
BARBARA MOSSE

Soil Microbiology Department
Rothamsted Experimental Station
Harpenden, Herts
December 1962

INTEGRATIVE AND DISINTEGRATIVE FACTORS IN SYMBIOTIC ASSOCIATIONS

RENÉ DUBOS AND ALEX KESSLER

The Rockefeller Institute, New York, U.S.A.

While other contributors in this symposium deal with the description, evolutionary history and mechanisms of particular cases of 'mutualistic associations', we shall attempt in contrast to present some problems which pertain to symbiosis considered as a general biological phenomenon. Three aspects will be considered here: the biological specificity of symbiotic associations, the dependence of symbiotic processes on environmental factors, and the creative manifestations of symbiosis.

SPECIFICITY IN SYMBIOTIC RELATIONSHIPS

The organizing committee of this symposium have directed us to focus attention on those '*symbiotic relations involving micro-organisms in which both members derive ecological advantage*'. It is obvious, however, that there occur frequently in nature various types of biological associations from which the partners derive ecological advantage, but which do not come within the purview of this symposium because they are accidental and transient. As the word symbiosis has come to be used, it generally implies a certain degree of specificity and of permanency. Specificity refers to that 'property of two interacting systems which permits the one to interact with the other with some degree of selectivity' (Weiss, 1955). The concept is most easily visualized in the biochemical fields where, for example, antigen–antibody interactions, and substrate–enzyme systems can be formulated on a molecular basis in stereo-chemical terms.

In a more purely biological context, specificity denotes the pattern of attributes of the two organisms constituting a symbiotic association which allows them to interact with some degree of selectivity—the interaction taking the form of a relationship where each organism forms a more or less critical portion of the environment of its partner. In other words, specificity in our discussion is concerned with the overall pattern of adjustment between two components of a symbiotic pair, it is an expression of the complementariness of all their dynamically interacting attributes. The greater the number of factors which interact and the more selective any one of them becomes, the smaller is the probability of

I

achieving the required combination for complementariness and hence the higher the degree of symbiotic specificity. Symbiotic specificity is therefore an expression of the mechanisms which determine why and how certain organisms are frequently found in association with others, and why they are restricted in their use of living environment to these organisms—even to particular regions of the latter. A converse aspect of this problem is the knowledge of the factors that can break down established symbiotic associations.

It is taken for granted that the dynamic equilibrium that exists between two members of any symbiotic pair is the result of the evolutionary development of both. But while any one or several of the characters of each symbiont may play a role in this process, and direct its partner along certain lines of development, few are the cases in which the relevant genetic changes have been recognized. The relation of orchids to their mycorrhizal fungi constitutes one of the first examples studied in some detail. In the primitive orchid *Bletilla hyacinthina*, the seeds can germinate aseptically and develop into slender seedlings with distinct leaves, but fail to develop a protocorm; in this species, therefore, the effect of the fungus is morphogenetic. Moreover, the rhizome of the infected *Bletilla* is free from mycorrhizal fungi; the new roots must be infected every year. In the less primitive orchids, such as *Cattleya*, *Cypripedium* and *Ophridium*, aseptic germination can take place, but the embryo does not develop at all although it becomes green. Here also, symbiosis is transient with infection taking place anew each year as the new roots develop. In the Sarcanthinae, which are more specialized, germination occurs only in association with *Rhizoctonia mucoroides*, and the fungus persists throughout the year in the plant. The most complete type of association has been observed in *Neottia nidus-avis* in which the same fungal infection persists throughout the life of the plant extending from the protocorm via the rhizome into the roots and into the base of the inflorescence instead of being limited to the roots (see review, Caullery, 1952).

These examples illustrate that selective processes can progressively increase the complexity of structural and physiological adaptation of each symbiont to its new habitat, finally reaching complete integration. The specializations evolved by each symbiont are generally related to the particular functions of environment performed for it by its partner— whether it be providing for space, protection, food or reproduction. It is the evolution of such specializations by symbiotic organisms within the living environment which lays the groundwork for symbiotic specificity. Such specialization may revolve around the function of surface to surface adhesion which permits the initiation of the partnership; increasing adjustment to that surface will then serve to restrict adaptability of this function to other environments, i.e. adhesion to other surfaces. In other cases highly selective factors of the internal environment are involved—including nutritional complementariness, ionic and gaseous composition, resistance to cellular defence mechanisms, etc. The increased degree of specificity of a single function may help to restrict a particular organism to a single or a few closely related partners.

Specialization within a living framework may have additional influences on the subsequent evolution of the organisms concerned, for example by tending to encourage retrogressive evolution. The interaction of two organisms may also involve an increasing number of activities. What starts out as a place to hold can become a protective site as well, and eventually may become a source of readily available food provided by the partner—either by its food-getting activities, or more frequently in the form of its own tissues. Each specialized adjustment evolved for each added activity performed by the partner entails decreasing adaptability to different conditions, i.e. raises the degree of specificity. As more and more of a symbiont's needs are satisfied by its partner, selective pressure on it decreases. Structures and functions no longer required are lost or degenerate, since natural selection does not operate against genetic changes which damage useless somatic attributes.

In general, the less self-sufficient a symbiont becomes, the more it depends on its partner and the less adaptable it becomes to other partners. Increasing mutual dependence leads to increasingly complex forms of integration between the two symbionts. Such integration results in decreasing flexibility, or conversely, in increasing fastidiousness or specificity for either or both partners.

All of these evolutionary processes may of course, and usually do, affect the two symbionts differentially. Some symbionts are less and others more dependent on their partners. There are thus degrees of self-sufficiency, from the symbiont which unconcernedly 'hangs his hat' on its equally unconcerned partner, to the totally interdependent symbiotic pair forming an integrated mechanism, little different from that which binds together the different parts of a single organism.

The fact that profound morphological, physiological and even behavioral differences can result from the interaction of a symbiont with different species or even with different individuals is well exemplified by the so-called insect parasitoids (Salt, 1941). In a given insect parasitoid the gross size, the proportion of body parts, the presence and absence of wings, the length and nature of the developmental period, the fecundity and the vigour, may be markedly influenced by the different species or individuals of hosts in which they develop. Even the behaviour is differentially affected, different hosts influencing the behaviour of the larvae, or of the emergent adults, or of the free adult's later activities. Although parasitoids represent a very specialized type of symbiosis, the same kind of phenomenon has been described for entirely different organisms such as trematodes. The polymorphism displayed by trypanosomes depending on the tissues or fluid within which they develop represents basically the same thing. This dynamic aspect of symbiosis is underscored by repeated observations of the effect of a change in one symbiont altering the specificity of the entire association; this has been noted for changes involving age, sex, hormonal state and nutritional needs as well as social behaviour.

ENVIRONMENT AND SYMBIOTIC RELATIONSHIPS

The manifestations of the interplay between two components of any symbiotic pair are of course conditioned by the factors of the environment which operate at the time of observation. For this reason, the terms most commonly used to denote biological relationships—such as symbiotic, commensal, mutualistic, parasitic or even pathogenic—are rather misleading if they imply a state of permanency of the relationship which does not take into account its dynamic character. Many of yesterday's parasites have now become mutualistic symbionts, and perhaps more importantly today's commensals and symbionts can behave as destructive parasites under certain conditions.

This dependence of the performance of biological associations upon environmental factors certainly accounts in part for the many controversies concerning the nature of symbiosis. Whereas the word symbiosis usually connotes an association from which both partners derive benefit, many authors emphasize instead the fact that a struggle is constantly going on between the two members of the association. In this respect, it seems appropriate to quote here the statement made by Caullery (1952) in his book on *Parasitism and Symbiosis*. Speaking of the presence of the mycorrhizal fungus in orchids he wrote: 'It is a phenomenon of parasitism, an infection, a state of disease which has become essential to the development of the plant, but can also stunt and even kill the plant.' Or again speaking of lichens: 'It is necessary to abandon the idea of a purely mutualistic association with equivalent reciprocal benefits. It is a conflict between alga and fungus.'

There is no doubt that the mycorrhizal fungi can on the one hand invade their hosts and kill them, while on the other hand much of the mycorrhizal mycelium is destroyed within the phagocytic cells of all orchids.

Even in the case of lichens—for which de Bary (1879) invented the word symbiosis to symbolize his belief that the association was beneficial to both the alga and the fungus—there are many biologists who hold the view that one of the organisms of the pair in reality behaves as a parasite. Consistent with this view is the well-known fact that the association between alga and fungus can be readily disturbed by changing the composition of the atmosphere or the supporting medium. It is for this reason that lichens disappear from the trees and rocks of urban areas where the air is polluted, or when abundant nutrients are added to the medium. Conversely, it is because many ill-defined factors are involved in achieving and maintaining the proper equilibrium between alga and fungus that it has proved so difficult to synthesize lichens from their two separate constituents. Interestingly enough, the synthesis of lichens can be realized by placing the fungus and the alga in a medium so deficient that it is unable to support the growth of either one of them separately (Ahmadjian, 1962)!

The change from mutually beneficial to pathological relationships can be illustrated also by examples taken from the behaviour of the intestinal flora in mammals. There is no doubt that certain members of the intestinal flora which are usually regarded as innocuous can exert a pathological role under certain circumstances—for example, following total body radiation. On the other hand, it has long been thought that the intestinal flora plays a useful role in animal economy by synthesizing certain growth factors and perhaps also by stimulating certain natural defence mechanisms—a view which has recently received support from observations on germ-free animals. These animals have proved to be more exacting in their nutritional requirements than are conventional animals, and furthermore they differ from the latter in certain physiological and anatomical features, as will be reported on p. 8 in this chapter.

We shall now deal with other aspects of the complex interplay between intestinal flora and host, quoting briefly some results obtained with a new colony of mice developed four years ago at the Rockefeller Institute (Schaedler & Dubos, 1962; Dubos & Schaedler, 1962a, b; Dubos, Schaedler & Costello, 1963; Dubos, Schaedler & Stephens, 1963).

This colony, designated NCS, has been derived from the standard colony of so-called Swiss mice (SS) and is maintained in such a manner that it is now free of many of the common mouse pathogens. Of special interest is the fact that no *Escherichia coli*, *Proteus*, or *Pseudomonas* can be isolated from the stools of NCS mice as long as they are maintained under proper conditions. In contrast, the faecal flora of NCS mice contains very large numbers of viable lactobacilli, belonging to at least three different morphological types.

It must be emphasized at this point that the diet fed the mice has a profound effect on the composition of their faecal flora. The lactobacilli decrease in numbers (100-fold), and one type (Rhizoid) disappears completely within a few days after the animals have been placed on a semi-synthetic diet containing purified casein as sole source of protein—this despite the fact that the casein diet contains all the known growth factors and supports rapid weight gain. In contrast, a very high lactobacillus population including the Rhizoid type is found in the stools of mice fed diets containing certain unidentified natural substances. A particular brand of commercial pellets (D & G) proved very effective in enriching the lactobacillus population, both quantitatively and qualitatively.

When penicillin or tetracycline is administered to the mice in their drinking-water, all viable lactobacilli immediately disappear from the faecal flora. Within one or a very few days the stools are then found to contain enormous numbers of enterococci and Gram-negative bacilli (including *Escherichia coli*). Concomitant with this change in intestinal flora there occurs a marked loss of body weight, amounting to 10 % or more of the initial weight of the animal.

The faecal flora, and the body weight, progressively return to their original level after discontinuance of the drug. However, while this reversal is relatively rapid in animals fed the D & G pellets, it takes place much more slowly with the semi-synthetic casein diet. The differences in rate are illustrated in Fig. 1.

Mice fed the D & G pellets are also more resistant to certain experimental bacterial infections than are those fed the casein diet. Thus it appears as if some nutritional factors can favour the establishment of the lactobacillus flora while at the same time increasing the ability of the animal to overcome other bacterial species (in the intestinal tract as well as in the inner organs).

Ever since Metchnikoff and Tissier, it has been considered that lactobacilli constitute the dominant microbial species in breast-fed human infants as well as in other animals. As suggested by the findings with mice, such lactobacilli may play a useful role with regard to weight gain and resistance to infection. Other findings, not to be discussed here, also suggest that the presence of these organisms in large numbers is correlated with higher resistance to certain toxic agencies—see also the experience with specific pathogen free (SPF) mice recently reported by Paget (1962).

It must be acknowledged that the information on the relation of intestinal flora to biological characteristics is far too incomplete to warrant definite conclusions. It is obvious, for example, that penicillin, tetra-cycline and diet affect not only the lactobacillus population but also other components of the digestive flora (such as the diphtheroid and bacteroides types) and influence certain physiological activities of the animal. Never-theless, granted our present lack of knowledge of the interplay between the host and its digestive flora, it is probable that disturbance of their

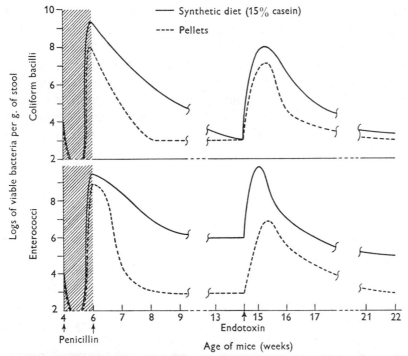

Fig. 1. The numbers of enterococci and coliform bacteria in NCS mice which received penicillin in the drinking-water for 1 week, and were fed pellets, or a synthetic casein diet, and were later given a single injection of endotoxin (0·34 mg).

equilibrium has unfavourable effects on the host, the lactobacilli or most commonly both. We have emphasized above the role of the nutritional state in the maintenance of this equilibrium. Had space permitted we could have mentioned other factors which may be of equal importance—ranging from the weather to the type of litter on which the animals are kept and the effect of crowding in the mouse colony.

Thus it is clear that in symbiotic associations between micro-organisms and animals, just as in other symbioses, the details of the interrelationship depend not only upon the intrinsic properties of the symbionts but also upon their joint environment.

THE CREATIVE MANIFESTATIONS OF
SYMBIOSIS

The most extensively studied, because most obvious, manifestations of symbiosis are those concerned with nutrition. In lichens, the alga supplies carbohydrates obtained by photosynthesis while the fungus secures moisture and minerals. In the digestive tract of mammals, various types of bacteria synthesize certain vitamins and probably amino acids from materials present in the excretions of the animal. Such examples of advantageous nutritional relationship have been or could be worked out for most symbiotic systems. But important as they are, the nutritional effects of symbiosis are not its most interesting manifestation. More remarkable is the fact that many symbiotic systems produce substances and structures that neither one of the two components produces when growing alone. These creative effects are so numerous and so varied that it would be impossible to catalogue them here, and a few examples will have to suffice, namely: the production of peculiar organic acids, pigments, and complex reproductive bodies by lichens; the production of haemoglobin, and of special structures in the bacterial root nodules of leguminous plants; the morphogenetic effects associated with the presence of unidentified micro-organisms in the leaf nodules of certain tropical plants; the synthesis of diphtheria toxin or of new somatic antigens by bacteria when they are lysogenized by the proper strains of bacteriophage, etc. Description of these and other examples will be found in Allen & Allen (1954), Humm (1944), Zinder (1959), Jacob & Wollman (1961) and Dubos (1960). To this list we should like to add the very recent finding that the intestinal flora also exerts morphogenetic effects.

Germ-free animals present anatomical features which differentiate them from conventional animals, particularly in the poor development of the lymphatic system and the large size of the caecum. In general such abnormalities are rapidly corrected when germ-free animals are recontaminated. Recently it has been found in guinea-pigs that the germ-free state results in an incomplete histological development of the intestinal mucosa (Sprinz *et al.* 1961; Sprinz, 1962). The intestinal mucosa of the germ-free guinea-pig resembles that of the prenatal pig but differs from that of the conventionally reared animal by a near absence of inflammatory cells in the lamina propria, distinctly shallow crypt glands lined by numerous markedly distended goblet cells, absence of degenerative changes in the epithelium lining the villi, taller and more delicately shaped villi in the small intestine, and a villous pattern in the caecum. Oral introduction of bacteria rapidly brings about a change in the architecture and histology of the bowel; within a matter of a few weeks the pattern approaches that of conventionally raised animals.*

There is of course no reason to assume that a common mechanism will be found for such varied effects. They are probably unrelated and each particular case will be found to involve highly specialized reactions. Nevertheless, a few general remarks seem in order to suggest that some of these phenomena may be related more closely than appears at first sight.

There now exists convincing evidence that single enzymes can catalyse several types of reactions; one of the examples most carefully studied is that of crystalline glutamic dehydrogenase which can not only oxidize glutamic acid but can also catalyse the oxidative deamination of L-alanine (Tomkins & Yielding, 1961; Tomkins, Yielding & Curran, 1961). This is relevant to our discussion in that certain steroid hormones, 1:10 phenanthroline and alkaline pH, all promote disaggregation of the enzyme protein into subunits, stimulate alanine deamination and inhibit glutamic dehydrogenase. In contrast, ADP, DPN and TPN, which cause association of the enzyme protein, inhibit alanine deamination and stimulate glutamic dehydrogenase. These findings suggest a means by which steroid hormones and other biologically active substances and conditions can alter both the kinetic properties and the substrate specificity of an enzyme by changing its physical structure. It is known, furthermore, that the affinity of antigen for antibody, and more generally of one large molecule for another, is determined not only by molecular configuration, but also by the environment in which the two come into contact. Taken together, these facts may help in interpreting some of the problems posed by the specificity and creative effects of symbiotic associations.

In the first section of this chapter we have considered specificity as operating at the level of the interacting organisms. In reality, however, the mechanisms of specificity may not operate through the organism as a whole, but rather through particular structures and processes. As in the interaction of molecules, the interactions between organisms may take place between functional groups. We have seen, moreover, that the reactivity of biologically active molecules can be modified by the environment. In the light of these facts it becomes clear that specificity cannot be defined completely on the basis of chemical structures at a given time because living things change when their environment changes. In other words, the concept of biological specificity involves factors other than those considered in orthodox chemistry, because it must include the total environment.

* It is of interest that SPF rats, which can be assumed to have an intestinal flora somewhat different from that of conventional animals, also exhibit differences in the structure of their intestinal mucosa; particularly the intestinal villi are more 'delicate' structures than in conventional animals (Paget, 1962).

The same considerations are also relevant to the new structures and biochemical characteristics which often emerge when two organisms become associated. The fact that a given enzyme can catalyse a different chemical reaction when placed in a new environment might be of importance in the phenomena of differentiation which occur in symbiotic systems. Indeed, it might help us to understand how the integration of two organisms, genetically unrelated, can give rise to a new entity which may then deserve to be regarded as a new organism.

In a very speculative mood, we shall make bold to suggest here the possible relevance of these considerations to the cell theory. As is well known, it is possible by several different techniques to remove from certain cells various organelles which have long been considered essential components of cellular structure; for example, *Euglena* cells can be rid of chloroplasts by treatment with streptomycin. Organelles can be thus removed without killing the cell, and more interestingly an indeterminate number of cell divisions can occur without production of new units of these organelles, provided the cell can call into play alternate metabolic pathways. This has been clearly shown with yeast and *Euglena*. Examples and discussions bearing on these statements will be found in Lederberg (1952), Salser (1961) and Dubos (1960).

The meaning of these facts is that several types of cells can and do exist with several mechanisms of information storage which are and remain independent of each other. In other words, it might be useful to think of the cell, not as a genetic unit but rather as an assembly of several independent genetic organelles which have become thoroughly integrated. Whether it is legitimate to extend the concept of symbiosis to this unorthodox view of cell structure is a problem which might be worth considering during the discussions of the symposium.

It seems, in summary, that the kind of knowledge required to understand the mechanisms of symbiotic relationships can be grouped around the following questions which are derived from discussions in a recent symposium on information theory in biology (Quastler, 1958).

(1) Which of the attributes of each organism are involved in their interaction, and which are not?

(2) In what way does the process of interacting itself alter or modify these or other attributes? Information obtained in answer to these questions might when tabulated distinguish the crucial from the incidental mechanisms involved in a particular symbiotic partnership. With frequent sampling, such information would define the characteristics of the symbiotic complex during its entire life span.

(3) What is there about a pattern of interacting attributes which

differentiates a particular symbiotic state from any other? The answer to this question would help to explain how certain mechanisms can evolve, and thereby alter symbiotic specificity as well as phenotypic expression.

REFERENCES

AHMADJIAN, V. (1962). Investigations of lichen synthesis. *Amer. J. Bot.* **49**, 3.
ALLEN, O. N. & ALLEN, E. K. (1954). Morphogenesis of the leguminous root nodule. In *Abnormal and Pathological Plant Growth. Brookhaven Symp. in Biology*, no. 6, 209.
DE BARY, A. (1879). *Die Erscheinung der Symbiose*. Strassburg: Trübner.
CAULLERY, M. (1952). *Parasitism and Symbiosis*. London: Sidgwick and Jackson Ltd.
DUBOS, R. J. (1960). Integrative and creative aspects of infection. In *Perspectives in Virology*. Ed. M. Pollard. New York: John Wiley and Sons Inc.
DUBOS, R. J. & SCHAEDLER, R. W. (1962a). The effect of diet on the fecal bacterial flora of mice and on their resistance to infection. *J. exp. Med.* **115**, 1161.
DUBOS, R. J. & SCHAEDLER, R. W. (1962b). Le rôle biologique de la flore digestive. *J. Immunol.* (in the Press).
DUBOS, R. J., SCHAEDLER, R. W. & COSTELLO, R. L. (1963). The effect of anti-bacterial drugs on the weight of mice. *J. exp. Med.* (in the Press).
DUBOS, R. J., SCHAEDLER, R. W. & STEPHENS, M. (1963). The effect of anti-bacterial drugs on the fecal flora of mice. *J. exp. Med.* (in the Press).
HUMM, H. J. (1944). Bacterial leaf nodules. *J. N.Y. bot. Gdn*, **45**, 193.
JACOB, F. & WOLLMAN, E. L. (1961). *Sexuality and the Genetics of Bacteria*. New York: Academic Press.
LEDERBERG, J. (1952). Cell genetics and hereditary symbiosis. *Physiol. Rev.* **32**, 403.
PAGET, G. E. (1962). The pathological state of specific pathogen-free animals. *Proc. R. Soc. Med.* **55**, 262.
QUASTLER, H. (1958). *Information Theory in Biology*. University of Illinois.
SALSER, W. (1961). Non-genetic biological information mechanisms. *Perspectives in Biol. and Med.* **4**, 177.
SALT, G. (1941). The effect of hosts upon their insect parasites. *Biol. Rev.* **16**, 239.
SCHAEDLER, R. W. & DUBOS, R. J. (1962). The fecal flora of various strains of mice. Its bearing on their susceptibility to endotoxin. *J. exp. Med.* **115**, 1149.
SPRINZ, H. (1962). Morphological response of intestinal mucosa to enteric bacteria and its implication for sprue and Asiatic cholera. *Fed. Proc.* **21**, 57.
SPRINZ, H., KUNDEL, D. W., DAMMIN, G. J., HOROWITZ, R. E., SCHNEIDER, H. & FORMAL, S. B. (1961). The response of the germ free guinea pig to oral bacterial challenge with *Escherichia coli* and *Shigella flexneri*. *Amer. J. Path.* **39**, 6.
TOMKINS, G. M. & YIELDING, K. L. (1961). Regulation of the enzymic activity of glutamic dehydrogenase mediated by changes in its structure. *Cold Spr. Harb. Symp. quant. Biol.* **26**, 331.
TOMKINS, G. M., YIELDING, K. L. & CURRAN, J. (1961). Steroid hormone activation of L-alanine oxidation catalysed by a sub-unit of crystalline glutamic dehydrogenase. *Proc. nat. Acad. Sci., Wash.* **47**, 270.
WEISS, F. (1955). Specificity in growth control. In *Biological Specificity and Growth*. Ed. E. G. Butler. Princeton University Press.
ZINDER, N. D. (1959). Genetic interaction between bacteriophage and bacteria. In *Perspectives in Virology*. Ed. M. Pollard. New York: John Wiley and Sons Inc.

BACTERIOPHAGE LYSOGENY

WERNER ARBER

Laboratoire de Biophysique, Université de Genève, Switzerland

INTRODUCTION

Many bacterial strains are known to produce and liberate bacterio-phage particles without having been knowingly infected with any bacterial viruses. The studies of such 'lysogenic' bacteria since their discovery in the early 1920's have revealed the fascinating situation of a more or less permanent association of phage with the host bacterium. This association consists in the addition of the phage genetic material to the bacterial chromosome. The complex of bacterial and phage genetic material usually multiplies at the rate of the bacterial dupli-cation and while such cells do not show any phage infectivity, they possess and transmit to their progeny the power to produce, under certain conditions, infective phage particles.

Our present picture of the phenomena dealing with lysogeny is based on a great amount of experimental data mainly accumulated in the last decade from the work initiated by Lwoff and his collaborators (Lwoff & Gutmann, 1950; Lwoff, Siminovitch & Kjeldgaard, 1950). As we do not pretend to discuss here all the aspects of lysogeny, we should like to refer the reader for more detailed information to the excellent review articles on this subject published by Lwoff (1953), Jacob (1954, 1960), Bertani (1958), Jacob & Wollmann (1957, 1959, 1961).

THE ESTABLISHMENT OF LYSOGENY

Lytic and lysogenic cycles

If bacteria sensitive to a given strain of phage are infected with particles of this strain, two types of response can be observed. In the lytic response the phage DNA enters the 'vegetative' state subsequent to its injection, i.e. it replicates much faster than does bacterial DNA. New phage proteins required for the formation of the phage envelopes are produced and in some 20–60 min., depending on the phage type and growth conditions used, the bacterium bursts and a progeny of some hundred new infective phage particles is liberated. In the non-lytic response the production of an infective phage progeny is not observed and massive multiplication of the injected phage DNA cannot be demonstrated. The bacteria survive the infection and produce progeny lysogenic for the phage with which their ancestors had been infected. In this lysogenic cycle, the phage genome is propagated in a non-infective form, called *prophage* and the production of infective phage particles is only occasionally induced. At such time, the phage genome enters the vegetative state and its multiplication and the production of phage-specific proteins occur as in the lytic cycle.

Temperate phage

As a general rule, phages able to establish lysogeny are called 'temperate' to distinguish them from the 'virulent' phages which only provoke the lytic response upon the infection of sensitive bacteria. This classification is useful but not truly fundamental as shown by the fact that temperate phage may easily mutate to virulent and vice versa. However, some virulent strains have never been found to give temperate mutants, as for example the well-known T phages (Adams, 1959). The most actively studied temperate phages and their most frequently used bacterial hosts are: phage λ with *Escherichia coli* K12 (Lederberg & Lederberg, 1953); phage P2 with *Shigella dysenteriae* or *E. coli* C, K12 or B (Bertani, 1951); phage P1 with the same hosts as phage P2 (Bertani, 1951) and phage P22 with *Salmonella typhimurium* (Zinder & Lederberg, 1952). If we discuss here, in particular, the results obtained by investigations with these selected temperate phages, it should be borne in mind that lysogeny is a very common phenomenon and that we have to be prepared to find among the vast number of currently less well-known lysogenic systems, situations that are not identical to the ones described here.

The frequency of lysogenization

This is defined by the ratio of the number of cells establishing lysogeny to the total number of cells infected with temperate phage, and it depends on the experimental conditions as well as on the genetic properties of phage and bacterial genomes. Factors such as growth medium, incubation temperature, age of the bacteria, multiplicity of infection and presence of specific inhibitors (e.g. chloramphenicol inhibiting protein synthesis) have been shown to influence the response to infection by temperate phage. For phage λ lysogeny is established with almost 100 % probability if old (post log-phase) bacteria, starved in $0 \cdot 01$ M-MgSO$_4$, are infected at 37° at a multiplicity higher than 5 phages per cell, diluted in tryptone broth and incubated at 37° (Fry, 1959; Séchaud, 1960). Lowering the multiplicity of infection to 1 phage per cell shifts the response of about half of the infected complexes from lysogenic to lytic. The use of young bacteria further favours the lytic response, as does a temperature shock at 42–44° just after DNA injection (Lieb, 1953). For phage P1, incubation of the infected complexes at 20–25° (Bertani & Nice, 1954) or in presence of chloramphenicol (Christensen, 1957; Bertani, 1957) favours the lysogenic response. These few examples demonstrate the lability of the decision between lytic and lysogenic responses which is made after injection of phage DNA. In general, influencing this decision by variation of external conditions is easiest shortly after injection and more difficult as growth of the infected complexes proceeds. This indicates a certain degree of irreversibility of the decision, corresponding to the expectation that vegetative phage multiplication could not easily be stopped in its later stages, or if it could, the cell would probably not survive and yield a lysogenic progeny. On the other hand, a shift to the lytic response of a complex having already decided to establish lysogeny is easier to imagine and has in fact been demonstrated. Such a shift is analogous to the induction of phage production in fully established lysogenic cells.

The genetic determinants for lysogeny

More important than any external influence on the establishment of lysogeny are the genetic characters of both phage and bacterial chromosomes. Here, experiments have been carried furthest with phage λ since on the one hand its host *Escherichia coli* K 12 is genetically well known and on the other hand phage mutants with various lysogenizing capabilities have been isolated.

Clear plaque-type mutants

Temperate phages usually form more or less turbid plaques upon plating on solid media with sensitive indicator bacteria. The turbidity is due to the fact that the bacteria which became lysogenic are immune to a superinfecting, homologous phage. These lysogenized bacteria thus grow in the phage plaque formed by the lysis of the other indicator bacteria. Phage mutants forming clear plaques (*c* mutants) on sensitive indicator occur at relatively high frequencies. Kaiser (1957) showed that the *c* mutants of phage λ can be mapped in a short segment of the phage genome in at least three functional units (cistrons). The λ*c* mutants, also called weak virulent, either give no lysogenization at all or lysogenize at very low rates which depend on the mutant considered. The clear appearance of the plaque is the consequence of this absence of lysogenic cells. However, upon mixed infection of sensitive bacteria with two c mutants belonging to different cistrons, lysogenization occurs again with high efficiency. Similar situations are found with other phages (Levine, 1957).

Jacob (1960) interpreted these findings in the following way. The injected DNA of the wild-type (c^+) phage produces a repressor which acts on a specific structure of the phage genome and thereby prevents the protein synthesis necessary for the initiation of the vegetative phage reproduction. Mutants of the c-type would be defective in the production of such repressor, but its synthesis could again be accomplished by the co-operation of two different c mutants, as long as the mutation sites on the two genomes lay in different cistrons. The 'decision' between lytic and lysogenic response would be the result of a competition between two different actions, both initiated by the injected phage genome: (1) production of the specific repressor and (2) initiation of the early protein synthesis leading to the vegetative growth. From this point of view the observed ease with which the decision is influenced by external factors would reflect the preferred synthesis of either repressor or early protein under given conditions.

DNA content mutants

The density gradient centrifugation technique allowed the isolation of phage mutants showing a different buoyant density. Kellenberger, Zichichi & Weigle (1960, 1961) described such a mutant of phage λ, called λb_2 which contains about 15 % less DNA than the wild-type λ. The genetic determinant for this mutant lies on the map of the phage genome close to the m_5 plaque-type marker and is thus only slightly linked to the c region. It remains unclear whether in the wild-type genome the supplementary DNA is located at the site of the b_2^+ marker; neither is the kind of physical linkage between phage genome and the b_2 piece known. No other genetic markers are yet known to map on this additional piece of DNA in λb_2^+ phages and thus to be missing in λb_2 mutants. However, for this discussion the interesting feature is that λb_2 mutants are unable to establish stable lysogeny in sensitive K12 bacteria. Nevertheless, they give turbid plaques and this fact has been explained by the formation of abortive lysogenics. Following infection with λc^+b_2 phages, the mechanism of the production of repressor by the c^+ region seems to function as in the case of $\lambda c_2^+b_2^+$ wild type. But, while the decision is in favour of the establishment of lysogeny, some other step in the formation of a stable lysogenic complex is never accomplished. Instead the phage genome stays in the cell, is not replicated as bacterial DNA replicates and is thus only linearly transmitted into the progeny of the infected bacterium. As we shall see later, there is good evidence that a $\lambda\, b_2$ genome cannot attach to the bacterial chromosome and one might wonder if in the case of $\lambda\, b_2^+$ the supplementary piece of DNA is responsible for the link between phage and bacterial DNA.

The bacterial chromosome

In the establishment of lysogeny the bacterial chromosome undoubtedly plays a role comparable to that of the phage genome. First, it seems that for the attachment of a phage genome particular regions of the bacterial genome are predetermined, as discussed in the next section. Thus, it is also expected that bacteria of different genetic content may respond differently to infection with temperate phage. Indeed, Arber & Lataste-Dorolle (1961) described a bacterial strain able to yield stably lysogenic cells after infection with $\lambda\, b_2$.

It seems also that the expression of the functions governed by certain c mutants depends on the host strain. Bertani (1960) showed such a dependance for phage P2 and Thomas & Lambert (1962) for phage λ. Presumably bacterial mutations can affect the production of the repressor of vegetative phage growth.

Kinetics of the establishment of lysogeny

The establishment of the lysogenic condition is a rather complex series of events. It may be oversimplified if we assume that two steps are essential, namely (1) the decision, or production of a sufficient amount of repressor, and (2) the fixation of the phage genome as prophage on the bacterial chromosome. The first step must be accomplished relatively quickly, or vegetative growth and lysis of the cell are irreversibly initiated. The second phase, the formation of the prophage-carrying bacterial chromosome has, however, in certain investigations been shown to take as long as several bacterial generations. The clearest demonstration of the complexity of these events comes from the pedigree analysis of single cells surviving phage infection (Luria, Fraser, Adams & Burrous, 1958; Luria, 1959). In these investigations the progeny sister cells were isolated after each duplication for five or six generations and only after that time were the subclones analysed for their content of phage genomes. Such experiments show that sensitive, non-lysogenic cells are usually segregated from the infected complex. This could reflect simply the multinuclear nature of the bacterial cell, but in that case segregation should not last longer than three generations. Many observations indicate, however, that the so-called early segregation may last much longer. This suggests that in such cases the phage genome, called preprophage, has not yet attained its final prophage state, where it replicates synchronously with and at the rate of the bacterial genetic material. It is not known to what degree the fixation of the phage genome to the bacterial chromosome and the accomplishment of the transition from preprophage to prophage state as defined above are identical or related events.

Following multiple infection of a sensitive cell with genetically marked homologous phages, the formation of *polylysogeny* seems to be the rule for certain phages (e.g. phage λ, Appleyard, 1954; Arber, 1960a) but rather the exception for others yielding predominantly singly lysogenic cells (e.g. phage P22, Luria *et al.* 1958). Experiments of this type, however, have frequently been done with inadequate genetic marking of the phages. For the interpretation of the experimental results, attention should be paid to the phage genotype used. The final lysogenic state may be completely different if only phage genotypes are used which are able to give rise to lysogenization upon single infection than if also either c mutants or b_2 type mutants are involved.

Zichichi & Kellenberger (1962) presented experiments in which they infected sensitive K 12 simultaneously with $\lambda\ cb_2^+$ and $\lambda\ c^+b_2$ phages. The expression of c^+ is dominant over that of c, so that repressor is produced in all doubly infected cells and the lytic cycle does not occur. As the infected bacteria grow and divide, one is able to follow the distribution of phage genomes and to measure if they are replicated. These authors found that in their conditions a rather low proportion of doubly lysogenic clones K 12 ($\lambda\ cb_2^+$, $\lambda\ c^+b_2$) were formed. More frequently the b_2^+ genome attached to the bacterial chromosome, but the b_2 genome did not, as indicated by the results of bacterial crosses (zygotic induction). The attached cb_2^+ genome seems to be replicated along with the replication of the bacterial DNA, unlike the c^+b_2 genome which is diluted out as the bacteria divide. This dilution gives rise to bacterial cells that no longer carry a c^+ gene. Repressor is thus no longer made and, although previously attached and established as prophage, the cb_2^+ genome can no longer be maintained in its prophage state and enters the lytic cycle.

THE LYSOGENIC CONDITION

Immunity

Immunity to superinfection with homologous phage is perhaps the best known property of lysogenic bacteria. The genetic markers of phage λ determining the immunity, map in the c region, as shown by Kaiser & Jacob (1957). By repeated backcrossing these authors produced a hybrid phage genome consisting essentially of the λ genetic material but containing the c region of 434, a phage related to λ but showing no cross-immunity with it. This hybrid strain grows on K 12 lysogenic for phage λ, but is sensitive to the immunity substance produced by K 12 lysogenic for the 434-hybrid.

Only certain so-called strong virulent mutants can multiply in a lysogenic strain. Strong virulent mutants of phage λ ($= \lambda$ vir) occur very infrequently and they have been shown by Jacob & Wollman (1954) to be produced in several successive mutational steps. Again the c region is involved in the formation of such mutants.

Jacob & Wollmann (1961) propose that immunity may be explained by the repressor model already described. Repressor, or immunity substance, would be continuously produced in a growing culture of lysogenic bacteria. This repressor, which might be the same as the one involved in the decision phase following the infection, not only prevents the prophage genome from entering into the state of vegetative replication, but by its presence in the cytoplasm also prevents any super-infecting, homologous phage from multiplying. The strong virulent mutants would be phages which are no longer sensitive to this repressor or have a reduced sensitivity. This last case could explain the observations made by Bertani & Six (1958) on certain virulent mutants of phage P2, which grow on hosts singly lysogenic for P2 but not on double lysogenics, as if the double lysogenics produce a greater amount of repressor. This interpretation is supported by the fact that upon multiple infection more repressor is produced than after single infection (Boyd, 1951; Prell & Prell, 1959). The level of immunity depends not only on the apparent number of prophages per cell but also on the genetic content of the prophage genome. Bertani (1961) has shown that immunity due to certain mutant prophages is more easily overcome than that due to the wild type of phage P2.

Prophage interference

While 'immunity' is used as a very specific term and accounts for the inhibition of vegetative growth of homologous superinfecting phage, all other manifestations of complete or partial inhibition of growth of unrelated, superinfecting phage by the presence of a prophage have been called prophage interference. This phenomenological classification probably includes many distinct effects. It is not yet known, however, to what extent the immunity substance repressing the growth of homologous phage also affects the growth of unrelated phages.

Recently two cases of prophage interference have been investigated in detail. Garen (1961) studied the rII mutants of phage T4 which do not produce phage in λ-lysogenic K12 (λ) in medium with low Mg^{2+} concentration. However, under these conditions rII mutants do initiate vegetative growth and the production of phage-related enzymes as in non-lysogenic K12, and only after about 10 min. of growth is phage activity abruptly halted. These activities can be restored by the addition of Mg^{2+} to the medium. The reason for this effect is not yet understood.

Another case of prophage interference has been described by Lederberg (1957). A number of different phages grown previously on non-lysogenic bacterial hosts show a very low efficiency of plating on P1 lysogenic derivatives of the same host strains. Dussoix & Arber (1962) showed that in K12 (P1) bacteria this restriction of phage λ.K ($= \lambda$ grown on K12) is a consequence of degradation of the infecting phage DNA soon after its injection into the new host cell. They concluded that K12 receives, by the presence of the P1 genome, the ability to recognize and degrade infecting λ.K DNA. From the rare λ plaques which do form on K12 (P1) indicator, modified phages ($= \lambda$.K (P1)) can be isolated which are accepted by K12 (P1) without restriction. This 'host-controlled modification' (Bertani & Weigle, 1953; Luria, 1953) is another consequence of the presence of the P1 genome. Arber & Dussoix (1962) found that the phage element which is changed by the host in host-controlled modification, and which is probably also the element recognized by restricting hosts, is physically linked to the phage DNA molecule, and they called this principle, of still unknown chemical nature, 'host specificity'.

Lysogenic conversion

Lysogenic conversion is an arbitrary grouping of all cases in which the phage changes the bacterial phenotype in ways not obviously related to either maintainance of the lysogenic state or accomplishment of the lytic cycle. The previously discussed effects of prophage interference may be considered as examples of lysogenic conversion. In other cases lysogeny has been shown to change cell-surface structures and somatic antigens (Iseki & Sakai, 1953) or to enable the cell to produce toxin (Freeman, 1951; Groman, 1955). In certain cases, it has been shown that the conversion occurs not only after establishment of lysogeny but soon after phage infection and in cases giving exclusively lytic response (Uetake, Luria & Burrous, 1958). Such early 'phage' conversion is also found for P1 controlled modification of DNA host specificity (Arber & Dussoix, 1962).

It is not known whether the genetic determinants for lysogenic conversion exist originally on phage genomes or whether they result from genetic recombination of the phage genome with the bacterial chromosome. However, it should be remembered that examples of such recombination are observed in phage-mediated transduction.

Transduction

Transduction occurs if bacterial genetic material is carried by a phage particle from a phage-producing donor cell into a subsequently infected recipient cell (Zinder & Lederberg, 1952). The transduction of bacterial markers determining the fermentation of galactose (*gal* markers) by phage λ (Morse, Lederberg & Lederberg, 1956) is of particular interest in this context. Such transducing phages, called λ *dg* (Arber, 1958) can be obtained only from phage lysates made from λ-lysogenic bacteria; they cannot be obtained upon lytic infection of a sensitive donor strain. This suggests a closer relation between prophage genome and bacterial chromosome than occurs in lytic phage development. λ *dg* particles have been found to carry defective phage genomes, i.e. certain gene regions necessary for phage development are missing (Arber, Kellenberger & Weigle, 1957; Arber, 1958). On the other hand, this loss is compensated by the presence of bacterial *gal* markers so that the entire genome is really a hybrid formed between phage and bacterial genetic material. These λ *dg* are still able to lysogenize with a low efficiency and thereby form defective lysogenic 'heterogenotes', carrying the intact *gal* region of the recipient cell in the bacterial chromosome ('endogenote') and the *gal* markers acquired from the donor cell as 'exogenote' on the prophage genome. Upon growth or induction of such heterogenotes, recombination between the homologous *gal* regions of endo- and exogenotes can be observed.

The structures of DNA molecules involved in *stable* and *abortive* phage-mediated *transduction* of bacterial markers are less well known than in the case of the transduction by λ *dg*. These two types of transduction are common in phages P22 and P1 where all markers tested have proved to be transducible. Clones resulting from stable transduction carry the transduced marker integrated into the chromosome in place of the original marker, instead of attached as part of an exogenote. In abortive transduction the exogenote is neither integrated nor does it multiply together with the bacterial chromosome. It is simply unilinearly inherited and functions continuously. It is not known whether in every transducing DNA molecule some region of the genome of the active phage is needed or not. If it is, one could assume that in abortive transduction the transducing DNA piece is so defective that it cannot lysogenize but that, on the other hand, something hinders its recombination with the bacterial genome and thus prevents stable transduction. This hindrance seems to be partly overcome by ultra-violet irradiation of the transducing phage prior to infection (Hartman & Goodgal, 1959). A similar shift from transduction by lysogenization to stable transduction occurs if λ *dg* is irradiated with ultra-violet (Arber, 1958).

Prophage location on the bacterial chromosome

In most lysogenic strains, the prophage replicates at the same rate as the bacterial chromosome. For λ-lysogenic K12 it has been shown by bacterial crosses that λ prophage is stably fixed to the bacterial chromosome and that λ occupies in different strains the same position near the *gal* region (Lederberg & Lederberg, 1953; Wollman, 1953). The same results were obtained by Jacob (1955) and Lennox (1955), using the method of transduction with non-related phage. Arber (1960*b*) showed by the same method that in doubly lysogenic cells both prophages are fixed at the same place or that they are at least very closely linked. In P2-lysogenic *Escherichia coli* the situation is different. Bertani & Six (1958) found three different prophage locations of which one is preferred. In P2 doubly lysogenic cells, the two prophages do not occupy the same location. In an analysis of different independently isolated temperate phages, Jacob & Wollman (1961) found definite locations for most of them. However, these authors describe another case, where no definite location could be attributed to the prophage genome, and they relate this fact to the capacity of this particular phage to give general, stable transduction. They suggest that this prophage may either occupy a variable and transient position on the bacterial chromosome or that it is perpetuated as a non-chromosomal element in the autonomous state.

The phage genome as an episomic element

The dynamic relationship between bacterial chromosome and phage genome provided Jacob & Wollman (1958) with a model for their definition of the 'episome' as a genetic element which may be either present or absent in a cell. It may be acquired by a non-carrier bacterium only from an external source. If present in a bacterium, the episome may be either in an autonomous or in an integrated state. If one applies the notion of the episome to the genome of phage λ, it is obvious that it may be either present or absent and that a sensitive cell can acquire it only by external infection, which may occur through phage infection, transfer upon bacterial conjugation (Jacob & Wollman, 1956b), upon transduction of λ prophage with an unrelated phage (Arber, 1960b) or transformation with free λ DNA (Kaiser & Hogness, 1960). The lytic cycle with vegetative phage multiplication corresponds to an autonomous state of the episome which then multiplies at a very fast rate and assumes all functions needed for phage reproduction. The lysogenic cycle of wild-type λ corresponds to an integrated state of the episome: all functions leading to vegetative multiplication are repressed, but the functions assuring this repression and the peaceful replication at the rate of bacterial division are assumed. The probability of finding an episome in one or the other of these states depends on the phage genome, on the host genome and on the physiological state of the complex. Some combination of these three factors may lead to complete absence of integration. In such cases the extrachromosomal element is perhaps better called plasmid (Lederberg, 1952) than episome, although the distinction is slight. Phage mutants of the b_2 type, e.g. carrying less DNA than the wild-type λ, do not attach to the chromosome of non-lysogenic K 12 cells (Zichichi & Kellenberger, 1962), although they can attach to the chromosome of *Escherichia coli* BB (Arber & Lataste-Dorolle, 1961) and they can form doubly lysogenic cells if the second phage is of b_2^+ genotype. In the first case, injected b_2 genomes do not replicate measurably as their host cells divide, but they assume the functions of the c^+ region, producing repressor which prevents the initiation of vegetative growth, and abortive lysogenization results. It is not known if the non-replication and the non-fixation of the phage genome to the bacterial chromosome are strictly related. Comparison with other episomic elements or plasmids, particularly with the already cited case of the non-localizable prophage, which nevertheless multiplies about synchronously with the bacterium, suggest that the non-fixation is perhaps not alone responsible for non-replication of the prophage in abortive lysogenics. Similar reasons may apply to the case of gene pieces which give abortive transduction and the integration of which into the bacterial chromosome seems to be inhibited, although in these cases absence of stable integration can certainly not be attributed to non-homology. It is thus not certain whether in the case of $\lambda\, b_2$ the most attractive explanation for the non-fixation is right, namely that the missing DNA piece is the part of the λ DNA which in wild-type $\lambda\, b_2^+$ assures the link between prophage genome and bacterial chromosome.

The molecular link of the prophage to the bacterial chromosome

The nature of this is not yet known despite many experimental approaches (see Bertani, 1958; Jacob & Wollman, 1961; Calef & Licciardello, 1960). Different models of attachment and insertion have been examined, but none of them has been proved to correspond to reality. It has frequently been believed, but never demonstrated, that attachment due to homology in the phage and bacterial gene regions is the best explanation. A new hypothesis of insertion has been proposed recently by Campbell (cited by Stahl, 1961), who assumes a single crossing over between the bacterial chromosome and a supposedly circular phage genome.

RUPTURE OF THE LYSOGENIC ASSOCIATION

In many cases the established lysogenic condition is very stable and can be maintained in cultures over many years. Other, not so stable, lysogenic systems are known and when permitted to grow, non-lysogenic segregants may accumulate. It seems that here the prophage genome is not transmitted to every progeny cell upon division of the descendants of an originally lysogenic clone. A more dramatic rupture of the lysogenic association occurs upon induction of phage production, where the prophage genome goes over into the vegetatively multiplying state, and the host cell is lost by lysis.

Segregation of non-lysogenic cells

The segregation of non-lysogenic cells from a lysogenic strain cannot be easily measured because of the steady re-infection of such segregants with phage particles produced by spontaneous induction in the course of bacterial growth. Meaningful measurements can only be obtained in special conditions, for instance with low bacterial concentration and presence of anti-serum or with strains resistant to the adsorption of phage homologous to the carried prophage, or with defective lysogenics producing no infective phage particles. From the experimental data actually available one can conclude that between the extreme cases of perfectly stable lysogeny and abortive lysogeny a full series of lysogenic systems may be found showing various levels of intermediate stability. The stability is certainly determined by phage genetic characters and probably also by bacterial gene regions; the stability can probably also be influenced by external conditions.

To explain the segregation of non-lysogenic cells one is tempted to draw the following picture. As episomes, phage genomes can exist in different states in the bacterial host. Transition from one state to another is regulated by a rather labile mechanism. So, a lysogenic bacterial culture may, at any given moment, be composed of a certain proportion of cells having the prophage either (1) stably attached (or integrated) to the bacterial chromosome, or (2) released from its association with the chromosome or (3) perhaps in some intermediate, less firmly attached state. From our knowledge of abortive lysogeny one may further assume that in none of these cases is the vegetative growth cycle initiated as long as repressor-producing functions of the c^+ region are expressed. Finally, it is possible that the rate of replication depends on the episomal state of the phage genome, e.g. it could be that a non-attached λ genome does not replicate or only much more slowly than the bacterial chromosome and that a well-attached λ genome replicates exactly at the rate of the bacterial cell division. If the prophage genome replicates more slowly than the host DNA, it is inevitably diluted out of certain progeny cells produced upon cell division.

In polylysogenic strains, segregation of single phages may occur although the strains would appear fully stable if tested only for the release of non-lysogenic segregants (Arber, 1960a). Some of these segregant prophage genomes are of recombinant types. One might speculate that segregation, in general, is the product of a recombinational event with elimination of the outcrossed complementary genome fragments. But other factors may also cause the phage genome to change from one episomal state to another. It may be relevant to this discussion to mention that upon superinfection of lysogenic cells with genetically marked homologous phage, doubly lysogenic cells are formed with a rather low probability and that these frequencies correspond to about the rate of segregation in polylysogenic strains.

Induction of phage production

The induction of phage production occurs spontaneously with a certain low probability during growth of lysogenic strains. It has been shown by Lwoff & Gutman (1950) that infective phage particles are not secreted at slow rates one by one from every cell of a lysogenic culture, but that phages are produced only when the phage genome enters the vegetative growth cycle which then resembles exactly the lytic cycle and brings the bacterial host to an abrupt end in lysis which liberates the infective phage progeny. The frequency of induction can in certain lysogenic strains be increased by external intervention (Lwoff *et al.* 1950), for example by irradiation with ultra-violet or ionizing radiation, by certain chemicals or by growth conditions. This controlled induction provided favourable conditions for a systematic quantitative study of induction phenomena. Indeed, induction frequencies of practically 100 % are easily obtained. The latent period in induced phage production is in general longer than the latent period of phage production as measured after infection in the same growth conditions (Weigle & Delbrück, 1951). There seems to be a transition phase between the lysogenic condition and the establishment of the vegetative pool. Its length may depend on the nature of the inducing agent and varies between approximately 10 and 30 min. for phage λ. Superinfecting phages can start to multiply more or less at the same time as does the induced prophage genome, i.e. only after the end of the transition phase.

The inducing agents are known to disturb the growth of the bacterial cell, particularly by temporarily inhibiting nucleic acid replication. Again using the repressor model, Jacob & Wollman (1961) suggested that transition from the prophage to the vegetative state is a consequence of a transient disturbance of the regulation system. Inducing agents would then act by stopping or slowing down the synthesis of a labile repressor. The transition period would last until the level of the repressor inside the cell fell below a certain threshold at which time vegetative replication would start.

Abortive lysogenic bacteria are inducible under the same conditions as stably lysogenic ones (Kellenberger *et al.* 1961). This result is to be expected according to the explanation given above, but it could not be easily explained by other theories such as the one equalizing induction with prophage detachment. As a matter of fact, it is not even known whether an attached or inserted prophage needs to separate from the bacterial genome in order to initiate vegetative phage development. On the other hand, autonomous phage genomes are repressed, as we

ly seen, as long as a fully functioning c^+ region is in the same

cible λ mutants have been described by Jacob & Campbell
,. Such mutations are dominant over the wild type, they map in
the c region and seem to affect the formation of repressor either in
quality (greater stability) or in quantity (larger amount). In addition,
non-inducibility can be caused by mutations occurring on the bacterial
genome, as shown for Salmonella phage ϵ 15 by Uetake (1959).

If λ prophage is transferred together with a piece of the bacterial
chromosome, into a non-lysogenic recipient cell, phage production by
the initiation of the lytic cycle occurs as upon infection. Examples of
such a process are given by zygotic induction in bacterial crosses (Jacob
& Wollman, 1956*b*) and by transfer induction in general transduction
(Jacob, 1955; Arber, 1960*b*). Evidently phage development is allowed
by the absence of repressor in the cytoplasm of the recipient cells, at
least for that part of phage genomes 'deciding' to enter the lytic cycle.

Defective lysogenic strains

Defective lysogenic strains have repeatedly been described for phage λ,
e.g. by Appleyard (1956), Jacob & Wollman (1956*a*), Jacob, Fuerst &
Wollman (1957), Arber & Kellenberger (1958), Campbell (1961). They
usually arise from mutations occurring in the prophage genome affecting
some important step in the vegetative growth, but not affecting its
maintenance of the prophage state. Upon induction, phage development
is blocked at the mutant-specific step. This function can be provided by
the presence in the same cell of another phage genome not defective in
the same cistron (complementation).

More complex defective genomes can be formed by deletion mutations
or by some interaction with non-related genetic material. Defective λ *dg*
and perhaps every transducing phage represent a complex of some
functional units from a phage genome and some from a bacterial genome.

CONCLUSIONS

Bacteriophage lysogeny is a relatively stable association of two organisms,
founded on the viral nature of one, the bacteriophage, which cannot
propagate itself without the hospitality of the bacterial cell. Upon
phage infection of a sensitive host two responses are possible: (1) reproduction of phage particles in a lytic cycle which is lethal to the host, and
(2) lysogenization, in which the cell survives and propagates the phage
genome as latent virus or prophage in its progeny. Only phage genetic

material is contained in the prophage and its DNA behaves as a piece of the bacterial chromosome rather than as an infectious particle. Under certain conditions, vegetative replication can be induced and infective phages produced, but this rupture of the lysogenic association is lethal to the host bacterium.

Several factors govern the decision between lytic and lysogenic response and determine the stability of the lysogenic state. Apparently after infection genetic characters of the phage produce a repressor, which if formed in sufficient quantity prevents initiation of vegetative growth. The same or similar repressors may be responsible for the immunity of lysogenic cells against superinfection with homologous phage and—if its production or stability is insufficient—for the induction of vegetative phage growth. Other genetic characters, both from the phage and the bacterial genome allow physical attachment or insertion of the prophage into the bacterial chromosome. It is not known if it is these same characters, or still others that control the replication of the prophage synchronously with the bacterial DNA in stably lysogenic cells and its slower growth rate or lack of growth in unstable and abortive lysogenic strains.

Phage genomes frequently carry genetic information enabling them to assume functions which have no obvious relation either to phage reproduction or to lysogenization. These functions are recognized by the conversion or transduction of the bacterial host and demonstrate the intimate relations between the bacterial chromosome and the genome of temperate phage. Exchange of genetic material between them and formation of episomal units devoid of obvious phage characteristics may be the consequence of—as well as the reason for—associations of the lysogenic type.

REFERENCES

ADAMS, M. H. (1959). *Bacteriophages*. New York: Interscience Publications Inc.

APPLEYARD, R. K. (1954). Segregation of new lysogenic types during growth of a doubly lysogenic strain derived from *Escherichia coli* K12. *Genetics*, **39**, 440.

APPLEYARD, R. K. (1956). The transfer of defective lambda lysogeny between strains of *Escherichia coli*. *J. gen. Microbiol.* **14**, 573.

ARBER, W. (1958). Transduction des caractères gal par le bactériophage Lambda. *Arch. Sci., Genève*, **11**, 259.

ARBER, W. (1960a). Polylysogeny for bacteriophage Lambda. *Virology*, **11**, 250.

ARBER, W. (1960b). Transduction of chromosomal genes and episomes in *Escherichia coli*. *Virology*, **11**, 273.

ARBER, W. & DUSSOIX, D. (1962). Host specificity of DNA produced by *Escherichia coli*. I. Host controlled modification of bacteriophage λ. *J. mol. Biol.* **5**, 18.

ARBER, W. & KELLENBERGER, G. (1958). Study of the properties of seven defective-lysogenic strains derived from *Escherichia coli* K 12 (λ). *Virology,* **5**, 458.

ARBER, W., KELLENBERGER, G. & WEIGLE, J. (1957). La défectuosité du phage lambda transducteur. *Schweiz. Z. Path.* **20**, 659.

ARBER, W. & LATASTE-DOROLLE, C. (1961). Erweiterung des Wirtsbereiches des Bakteriophagen λ auf *Escherichia coli* B. *Path. Microbiol.* **24**, 1012.

BERTANI, G. (1951). Studies on lysogenesis. I. The mode of phage liberation by lysogenic *Escherichia coli*. *J. Bact.* **62**, 293.

BERTANI, G. (1958). Lysogeny. *Advanc. Virus Res.* **5**, 151.

BERTANI, G. & NICE, S. J. (1954). Studies on lysogenesis. II. The effect of temperature on the lysogenization of *Shigella dysenteriae* with phage P 1. *J. Bact.* **67**, 202.

BERTANI, G. & SIX, E. (1958). Inheritance of prophage P 2 in bacterial crosses. *Virology,* **6**, 357.

BERTANI, G. & WEIGLE, J. J. (1953). Host controlled variation in bacterial viruses. *J. Bact.* **65**, 113.

BERTANI, L. E. (1957). The effect of the inhibition of protein synthesis on the establishment of lysogeny. *Virology,* **4**, 53.

BERTANI, L. E. (1960). Host-dependent induction of phage mutants and lysogenization. *Virology,* **12**, 553.

BERTANI, L. E. (1961). Levels of immunity to superinfection in lysogenic bacteria as affected by prophage genotype. *Virology,* **13**, 378.

BOYD, J. S. K. (1951). Observations on the relationship of symbiotic and lytic bacteriophage. *J. Path. Bact.* **63**, 445.

CALEF, E. & LICCIARDELLO, G. (1960). Recombination experiments on prophage host relationships. *Virology,* **12**, 81.

CAMPBELL, A. (1961). Sensitive mutants of bacteriophage λ. *Virology,* **14**, 22.

CHRISTENSEN, J. R. (1957). Effect of chloramphenicol on lysogenization by temperate phage P 1. *Virology,* **4**, 184.

DUSSOIX, D. & ARBER, W. (1962). Host specificity of DNA produced by *Escherichia coli*. II. Control over acceptance of DNA from infecting phage λ. *J. mol. Biol.* **5**, 37.

FREEMAN, V. J. (1951). Studies on the virulence of bacteriophage-infected strains of *Corynebacterium diphtheriae*. *J. Bact.* **61**, 675.

FRY, B. A. (1959). Conditions for the infection of *Escherichia coli* with lambda phage and for the establishment of lysogeny. *J. gen. Microbiol.* **21**, 676.

GAREN, A. (1961). Physiological effects of *r*II mutations in bacteriophage T 4. *Virology,* **14**, 151.

GROMAN, N. B. (1955). Evidence for the active role of bacteriophage in the conversion of non-toxigenic *Corynebacterium diphtheriae* to toxin production. *J. Bact.* **69**, 9.

HARTMAN, P. E. & GOODGAL, S. H. (1959). Bacterial genetics (with particular reference to genetic transfer). *Annu. Rev. Microbiol.* **13**, 465.

ISEKI, S. & SAKAI, T. (1953). Artificial transformation of O antigens in Salmonella E group. II. Antigen transforming factor in bacilli of subgroup E 2. *Proc. imp. Acad. Japan,* **29**, 127.

JACOB, F. (1954). *Les bactéries lysogènes et la notion de provirus* (Monogr. Inst. Past.). Paris: Masson et Cie Editeurs.

JACOB, F. (1955). Transduction of lysogeny in *Escherichia coli*. *Virology,* **1**, 207.

JACOB, F. (1960). Genetic control of viral functions. *Harvey Lect.* **54**, 1.

JACOB, F. & CAMPBELL, A. (1959). Sur le système de répression assurant l'immunité chez les bactéries lysogènes. *C.R. Acad. Sci., Paris,* **248**, 3219.

JACOB, F., FUERST, C. R. & WOLLMAN, E. L. (1957). Recherches sur les bactéries lysogènes défectives. II. Les types physiologiques liés aux mutations du prophage. *Ann. Inst. Pasteur*, **93**, 724.

JACOB, F. & WOLLMAN, E. L. (1954). Étude génétique d'un bactériophage tempéré d'*Escherichia coli*. I. Le système génétique du bactériophage λ. *Ann. Inst. Pasteur*, **87**, 653.

JACOB, F. & WOLLMAN, E. L. (1956a). Recherches sur les bactéries lysogènes défectives. I. Déterminisme génétique de la morphogenèse chez un bactériophage tempéré. *Ann. Inst. Pasteur*, **90**, 282.

JACOB, F. & WOLLMAN, E. L. (1956b). Sur le processus de conjugaison et de recombinaison chez *Escherichia coli*. I. L'induction par conjugaison ou induction zygotique. *Ann. Inst. Pasteur*, **91**, 486.

JACOB, F. & WOLLMAN, E. L. (1957). Genetic aspects of lysogeny. In *The Chemical Basis of Heredity*, p. 468. Ed. W. D. McElroy and B. Glass. Baltimore: The Johns Hopkins Press.

JACOB, F. & WOLLMAN, E. L. (1958). Les épisomes, éléments génétiques ajoutés. *C.R. Acad. Sci., Paris*, **247**, 154.

JACOB, F. & WOLLMAN, E. L. (1959). Lysogeny. In *The Viruses*, 2, 319. Ed. F. M. Burnet and W. M. Stanley. New York and London: Academic Press.

JACOB, F. & WOLLMAN, E. L. (1961). *Sexuality and the Genetics of Bacteria*. New York and London: Academic Press.

KAISER, A. D. (1957). Mutations in a temperate bacteriophage affecting its ability to lysogenize *Escherichia coli*. *Virology*, **3**, 42.

KAISER, A. D. & HOGNESS, D. S. (1960). The transformation of *Escherichia coli* with deoxyribonucleic acid isolated from bacteriophage λ *dg*. *J. mol. Biol.* **2**, 392.

KAISER, A. D. & JACOB, F. (1957). Recombination between related temperate bacteriophages and the genetic control of immunity and prophage localization. *Virology*, **4**, 509.

KELLENBERGER, G., ZICHICHI, M. L. & WEIGLE, J. (1960). Mutations affecting the density of bacteriophage λ. *Nature, Lond.* **187**, 161.

KELLENBERGER, G., ZICHICHI, M. L. & WEIGLE, J. (1961). A mutation affecting the DNA content of bacteriophage Lambda and its lysogenizing properties. *J. mol. Biol.* **3**, 399.

LEDERBERG, E. M. & LEDERBERG, J. (1953). Genetic studies of lysogenicity in *Escherichia coli*. *Genetics*, **38**, 51.

LEDERBERG, J. (1952). Cell genetics and hereditary symbiosis. *Physiol. Rev.* **32**, 403.

LEDERBERG, S. (1957). Suppression of the multiplication of heterologous bacteriophages in lysogenic bacteria. *Virology*, **3**, 496.

LENNOX, E. S. (1955). Transduction of linked genetic characters of the host by bacteriophage P1. *Virology*, **1**, 190.

LEVINE, M. (1957). Mutations in the temperate phage P22 and lysogeny in Salmonella. *Virology*, **3**, 22.

LIEB, M. (1953). The establishment of lysogenicity in *Escherichia coli*. *J. Bact.* **65**, 642.

LURIA, S. E. (1953). Host-induced modifications of viruses. *Cold Spr. Harb. Symp. quant. Biol.* **18**, 237.

LURIA, S. E. (1959). Lysogeny and lysogenization—studies in infectious heredity. In *A Symposium on Molecular Biology*, p. 152. Ed. R. E. Zirkle. University of Chicago Press.

LURIA, S. E., FRASER, D. K., ADAMS, J. N. & BURROUS, J. W. (1958). Lysogenization, transduction, and genetic recombination in bacteria. *Cold Spr. Harb. Symp. quant. Biol.* **23**, 71.

LWOFF, A. (1953). Lysogeny. *Bact. Rev.* **17**, 269.

Lwoff, A. & Gutmann, A. (1950). Recherches sur un bacillus megatherium lysogène. *Ann. Inst. Pasteur*, **78**, 711.

Lwoff, A., Siminovitch, L. & Kjeldgaard, N. (1950). Induction de la production de bactériophages chez une bactérie lysogène. *Ann. Inst. Pasteur*, **79**, 815.

Morse, M. L., Lederberg, E. M. & Lederberg, J. (1956). Transduction in *Escherichia coli* K12. *Genetics*, **41**, 142.

Prell, H. H. & Prell, H. H. M. (1959). Der Einfluss der Infektionsmultiplizität auf die Lysogenisierung im System *Salmonella typhimurium*—Phage P22. *Arch. Mikrobiol.* **34**, 211.

Séchaud, J. (1960). Développement intracellulaire du coliphage Lambda. *Arch. Sci., Genève*, **13**, 427.

Stahl, F. (1961). A chain model for chromosomes. *J. Chim. phys.* **58**, 1072.

Thomas, R. & Lambert, L. (1962). On the occurrence of bacterial mutations permitting lysogenization by clear variants of the temperate bacteriophages. *J. mol. Biol.* **5**, 373.

Uetake, H. (1959). The genetic control of inducibility in lysogenic bacteria. *Virology*, **7**, 253.

Uetake, H., Luria, S. E. & Burrous, J. W. (1958). Conversion of somatic antigens in Salmonella by phage infection leading to lysis or lysogeny. *Virology*, **5**, 68.

Weigle, J. J. & Delbrück, M. (1951). Mutual exclusion between an infecting phage and a carried phage. *J. Bact.* **62**, 301.

Wollman, E. L. (1953). Sur le déterminisme génétique de la lysogénie. *Ann. Inst. Pasteur*, **84**, 281.

Zichichi, M. L. & Kellenberger, G. (1962). Two distinct functions in the lysogenization process: the repression of phage multiplication and the incorporation of the prophage in the bacterial genome. *Virology* (in the Press).

Zinder, N. D. & Lederberg, J. (1952). Genetic exchange in Salmonella. *J. Bact.* **64**, 679.

EXPERIMENTAL STUDIES OF
LICHEN PHYSIOLOGY

D. C. SMITH

University Department of Agriculture, Oxford

INTRODUCTION

Lichens are generally considered to be one of the outstanding examples of symbiosis in the plant kingdom, and there has been much speculation about the nature of the association between their constituent algae and fungi. Unfortunately, this speculation has not been accompanied by much experimental investigation. A belief that lichens are 'difficult' material to use in laboratory experiments has discouraged research into their physiology, and difficulties in their taxonomy have discouraged botanists from becoming familiar with the group. The recent revival of interest and research into lichens has begun to demonstrate that the obstacles to their study are not as great as originally envisaged.

The object of this article is to review the experimental investigations that have been made into lichen physiology, particularly those concerning the nature of the association between the symbionts. Since such investigations have been scanty, one cannot expect a clear and comprehensive picture to emerge. It is not the object of this article to add to the wealth of theories that already exist about lichen symbiosis, except to formulate working hypotheses to guide future research. Quispel (1943, 1959) has already summarized and discussed the more important theories.

Many different kinds of fungi and algae may be components of lichens. It is therefore unwise to assume too readily that results of experiments with one lichen species are necessarily applicable to all others. There is a similar danger in thinking that there is the same type of relationship between the algae and fungi of all lichen species, or indeed that the type of relationship is necessarily the same as in other associations which are termed 'symbiotic'.

It is assumed in this article that a lichen is composed only of an alga and a fungus. Theories that bacteria are also a component of the lichen thallus have been reviewed by des Abbayes (1951) and Scott (1956), and there is as yet no good evidence that any lichen species constantly contains a particular bacterium. It is very probable that lichens, like other macroscopic plants, bear a surface microflora under natural conditions, but there is no evidence to show that it is an essential feature of the lichen symbiosis.

Most investigations into the nature of the association between the algae and fungi of lichens have begun with isolation of the symbionts, continued with observations of their behaviour in pure culture, and ended with attempts at synthesis of the original lichen. It will be shown later in this article that apart from certain of the experiments on lichen synthesis, such investigations have not been particularly fruitful. Two reasons for this are: (a) the behaviour of the isolated symbionts in pure culture probably differs from their behaviour in the thallus; and (b) a full understanding of lichen symbiosis presumably involves some appreciation of the ecological advantages inherent in the association—and this can only come from studies of the intact thallus as well as of the isolated symbionts.

This article is written from the standpoint that knowledge of the physiological properties and behaviour of lichen thalli is a necessary prerequisite to any adequate consideration of the nature of the association between the symbionts.

THE LICHEN THALLUS AS A FUNCTIONAL UNIT IN NATURE

The thallus formed by the association of a lichen fungus with its alga behaves in nature as a self-reproducing, functional unit. This is illustrated by the fact that in most aspects of their taxonomy and ecology, lichen thalli can be treated as if they were single plants.

I have recently reviewed what is known of the biology of lichen thalli (Smith, 1962), and some of the major points established in that article may be summarized as follows:

(1) *Water relations.* The water relations of a lichen thallus closely resemble those of a hydrophilic gel such as gelatin. Water can be absorbed over all parts of the thallus surface, either very rapidly from liquids or more slowly from vapour. The rates of absorption are similar to those of gels. Water is lost by evaporation from the thallus surface, and both absorption and loss of water appear to be entirely physical processes.

The saturated water content of a lichen thallus is generally less than that of other kinds of cryptogamic plants, and for most species it lies between 100 and 300 % of the dry weight. Much of the water in a saturated thallus is held externally to the cytoplasm in the interhyphal spaces and in the thick cell walls of the fungal hyphae. In some species under conditions of drought in nature, the water content may fall to below 5 % of the dry weight.

Variations in the water content of the thallus in nature are closely correlated with variations in moisture conditions of the habitat, so that frequent alternations between desiccation and saturation of the thallus normally occur throughout the long lives of the majority of species.

(2) *Rates of respiration and photosynthesis.* Although the respiration rates of lichens under optimum conditions are comparable per unit surface area to those of mature angiosperm leaves, the rates of photosynthesis are much lower, probably because lichens contain proportionately much less photosynthetic pigment. Consequently, the net rate of carbon dioxide assimilation per unit surface area is much lower in lichens than in most leaves. Maximum values so far recorded for net assimilation in the light lie in the range 0·2–3·2 mg. carbon dioxide/50 cm.^{2}/hr.

The rates of both photosynthesis and respiration are closely dependent upon the water content of the thallus. The optimum water content for respiration is generally higher than for photosynthesis, and in completely saturated thalli photosynthesis (measured as carbon dioxide uptake) may become strongly depressed although respiration similarly measured is not reduced. Hence, maximum rates of net assimilation are achieved at water contents somewhat below saturation. For example, Ried (1960) found that the optimum water contents for net assimilation in eleven species of lichens lay between 65 and 90 % of complete saturation; the denser the thallus, the lower the optimum water content.

(3) *Nutrition and metabolism.* Little is known of the nutrition and metabolism of lichens. Existing experimental evidence suggests that the following features may be common to lichens as a group:

(*a*) They appear to have very efficient mechanisms for the active accumulation of both organic and inorganic nutrients from solution. The extent to which lichens utilize organic nutrients in nature is not known, and it probably varies with species and habitat.

(*b*) It is probable that most of those lichens which contain blue-green algae can fix atmospheric nitrogen.

(*c*) Rates of protein turnover and of utilization of absorbed nutrients may be relatively very slow.

(*d*) Sugar alcohols, particularly mannitol, may occupy an important role in carbohydrate metabolism. Mannitol is very widespread and may even be universally present in lichens: it is found more abundantly than simple sugars. In *Peltigera polydactyla*, externally supplied glucose is rapidly converted to mannitol after absorption, and much of the carbon fixed in photosynthesis initially appears as mannitol (Smith, 1961).

(4) *'Lichen substances'.* These are insoluble organic compounds typically found encrusting external surfaces of hyphae, and they may comprise 1–8 % or even more of the dry weight of the thallus. More than 80 different 'lichen substances' are known and much information is available on their chemical structure. Only very few have been isolated from free-living fungi. Little is known of their biosynthesis, and most authors believe that 'lichen substances' are excretory products. Many possess antibiotic properties, especially one of the commonest and most abundant of them, usnic acid. They may therefore assist in protecting lichens from pathogenic attack.

(5) *Growth rates and longevity.* The great majority of lichens are characteristically very slow growing plants. In temperate climates, mature

crustaceous lichens have a radial expansion growth which rarely exceeds 1–2 mm. per year and it is very often much less. Some non-crustaceous lichens may have faster growth rates, up to and occasionally exceeding 5 mm. per year. Since lichens grow so slowly, large thalli—especially of crustaceous saxicolous forms—are presumably of great age. Approximate estimates of the maximum life-span of lichens made by various authors have ranged from some hundreds to a thousand or more years.

(6) *Drought resistance.* Lichens can usually survive much longer periods of drought than occur in their natural habitats. For example, many lichens of temperate climates can probably survive several months drought without permanent damage. Immediately after a drought, there is a temporary stimulation of respiration accompanied by a depression in photosynthesis, so that carbon dioxide assimilation is temporarily reduced to a low or negative value. The marked correlation between the distribution of particular species in nature and the moisture conditions of the habitat is probably related more to the frequency than to the length of droughts.

(7) *Resistance to other environmental conditions.* When dry, lichen thalli have a marked resistance to high temperatures; for example, Lange (1953) recorded maximum thallus temperatures under natural conditions for twelve lichen species in Europe ranging between 53° and 69°. When saturated, thalli are no more resistant to high temperatures than other kinds of plant tissue. Nearly all lichens are very sensitive to atmospheric pollution, a phenomenon which may well be related to their very efficient mechanisms for the accumulation of substances from solution. Lichens are relatively free from attack by pathogenic organisms.

(8) *Seasonal variation in physiology.* In Western Europe, the optimum season for growth and assimilation is probably the winter, and the unfavourable season the summer.

One of the most interesting problems in the biology of lichen thalli is their long persistence in habitats where environmental conditions are unfavourable for the growth of most other kinds of plants. Their pattern of water relations and drought resistance enables them to tolerate erratic and marked fluctuations in the moisture conditions. The fact that uptake of liquid water is by imbibition means that it is rapid and that it allows for a rapid rise in metabolic rates when environmental conditions are favourable. Since water vapour can also be absorbed, lichens are particularly suited to habitats with high atmospheric humidity such as tropical forests and misty places.

Many lichens live in habitats where the level of nutrient supply is presumably very low, and the hypothesis has been advanced (Smith, 1960a) that their successful existence in such habitats is due partly to their efficient mechanisms for nutrient accumulation, and partly to slow rates of growth and nutrient utilization.

THE PHYSIOLOGY OF DIFFERENT REGIONS OF
THE THALLUS

In most lichens the thallus is differentiated internally into a number of distinct anatomical regions. The bulk of the thallus is composed of fungal hyphae and the algal cells are mainly confined to a thin layer beneath the surface of the thallus (Pl. 1, figs. 1, 2). The hyphae form a cortex above the algal layer, and beneath it they form a medulla; additional hyphal tissues are often formed, such as a lower cortex, hypothallus, rhizinae, etc. The hyphae may vary markedly in form and arrangement between the different regions: for example, in the medulla they are characteristically thick-walled and somewhat loosely arranged; in the algal layer they are thin-walled; and in the cortex they are arranged very compactly to give a dense tissue which may be pseudo-parenchymatous, fibrous, etc., depending upon the species. Lichens having a thallus with distinct internal differentiation of tissues are termed 'heteromerous'. There are only a few species in which the hyphae and algal cells are distributed at random throughout the thallus, and these are termed 'homoiomerous'.

There is hardly any tendency for a well-differentiated vegetative thallus to be formed by any of the kinds of free-living ascomycetes which are related to lichen fungi. The evolution of a well-differentiated thallus in lichens is presumably associated with some of the factors which distinguish lichens from free-living ascomycetes, such as the presence of algal cells, the great longevity of lichens, the kinds of habitats in which they live, etc. At all events, it is reasonable to assume that the different anatomical regions have different functions.

The physiology of different regions of the thallus of *Peltigera polydactyla* has been investigated in experiments involving dissection techniques (Harley & Smith, 1956; Smith, 1960*b*, 1961). By making an incision immediately beneath the algal layer, the thallus could be dissected into two separate tissue zones: (*a*) the medulla, entirely fungal in composition; and (*b*) the 'algal zone', comprising the algal layer together with the upper cortex (Table 1; Pl. 1, figs. 1, 2). The 'algal zone' had higher rates of respiration, stronger powers of nutrient absorption and a higher nitrogen content than the medulla. Since it is difficult to believe that the upper cortex makes an appreciable contribution to the metabolic activity of the 'algal zone', it is probable that the algal layer itself is the region of greatest metabolic activity in the thallus.

The medulla occupies approximately 80 % of the volume of the thallus in *Peltigera polydactyla*, and it is likely that it has a skeletal

Table 1. *Physiological properties of 'algal zones' and medullae of* Peltigera polydactyla *after separation by dissection (Harley & Smith, 1956; Smith, 1960b, 1961)*

(Data calculated as mean values for normal thalli.)

	Dry wt. per cm.² thallus mg.	Composition per 100 mg. dry weight				Absorption from 5 mM solutions (per 100 mg. dry wt./24 hr. at 20°.)[4]			Oxygen uptake per 100 mg. dry wt. at 20°[7] μl./hr.
		N[1] mg.	Carbohydrate[2] mg.		Saturated water content mg.	Glucose[5] mg.	Asparagine mg. N	Phosphate[6] mg. P	
			Sol.[3]	Insol.					
Algal zones	3·6	4·5	7·6	18·8	305	23·4	1·7	1·11	121
Medullae	2·3	2·5	4·0	24·8	375	11·8	1·4	0·86	70

Transverse section of the thallus

Upper cortex · Algal layer · Medulla · Dissection · 25μ · 50μ · 400μ · Algal zones · Medullae

Notes.   [1] Measured in November, when the nitrogen content of thallus is at its highest.
[2] Measured in May–June, just after season of maximum carbohydrate content.
[3] Does not include sugar alcohols.
[4] Aqueous solutions, and pH maintained at 5·6–5·8 by 10 mM phthalate buffer.
[5] Total absorption by tissue zones is 23% less than undissected controls.
[6] Total absorption by tissue zones is 17% less than undissected controls.
[7] Total oxygen uptake by tissue zones is 56% greater than undissected controls.

function. The medulla may also have a role in the water relations of the thallus. It has a higher saturated water content than other regions of the thallus, presumably because the hyphae have thicker walls and are more loosely arranged than elsewhere, and it has been suggested (Smith, 1962) that the medulla provides water for the metabolically more active algal layer when the thallus is drying out. Relatively high carbohydrate contents have been found in the medulla of *P. polydactyla*, and since it has been demonstrated in this lichen that the products of photosynthesis move from the algal layer to the medulla (Smith, 1961) it is possible that the medulla may function as a region of carbohydrate storage, although considerably more experimental evidence is required for conclusive proof of this function.

As regards other lichen species, it is of interest to note that Sosa-Bourdouil (1944) found that the algal region of *Usnea barbata* contained more nitrogen and had a higher enzymic activity than the central axis, and she believed that this latter tissue is skeletal in function. Ertl (1951) found that the transparency of the cortices of a number of species decreased markedly upon drying out, particularly if the cortex was pigmented. It is therefore possible that the cortex may protect the algal layer in dry thalli from the effects of strong insolation. Zukal (1895) believed that the cortex retarded water loss, but later authors (e.g. Smyth, 1934) disagree with this.

A relatively small amount of research has given strong indications that the different regions of a heteromerous lichen thallus possess different physiological properties. In considering the nature of the lichen symbiosis, it must therefore be remembered that not only does the product of this symbiosis—the thallus—behave as a functional unit in nature, but also that it is a complex structure whose component tissues may each have distinct functions. It is probable that the lichen fungus exists in different physiological states in different parts of the thallus, but it is not known whether any of these states are the same as that shown by the fungus when it is grown in isolated culture.

INTERRELATIONSHIPS BETWEEN LICHEN ALGAE AND FUNGI

Although this section is primarily concerned with the physiology of lichens, so little is known of the subject that it may be helpful to consider also certain aspects of lichen biology which are not purely physiological. Some consideration is therefore given below to questions of specificity in the association and to the types of morphological contact between the symbionts.

Specificity

It was at one time believed that a lichen fungus was highly specific towards its algal partner (cf. Chodat, 1930), but many modern lichenologists now doubt the validity of this belief (cf. Ahmadjian, 1960b). The experimental evidence is sometimes difficult to interpret because of the complexities and confusion inherent in the taxonomy of unicellular algae—especially in the concept of what constitutes a species. The situation may be summarized as follows. (a) In most lichen species the alga belongs to only one genus of either green or blue-green algae. However, when the algae from different thalli of the same lichen species are isolated into pure culture, they sometimes show different cultural characteristics (Chodat, 1913; Waren, 1920; Jaag, 1929; Thomas, 1939), although no important morphological differences have been reported. Thomas also showed that different strains of algae could be isolated from the same thallus. (b) There are a few lichen species in which the fungus enters into an intimate relationship with both a green and a blue-green alga, e.g. *Solorina crocea*, certain Stictaceae (such as the *Lobaria amplissima–Dendriscocaulon bolani* complex) and lichens with cephalodia. Whether this phenomenon is related to the ability of blue-green algae to fix atmospheric nitrogen is not known. In the above examples, each lichen fungus is associated in the thallus with only one kind of green alga and one kind of blue-green alga. (c) The commonest algal genus in lichens is *Trebouxia*. Raths (1938), Zeitler (1954) and Ahmadjian (1960a) showed that the same morphological and physiological type of *Trebouxia* may be the symbiont of widely different lichen species. (d) When attempts are made to synthesize lichens in culture, it is observed that some lichen fungi do not have a high specificity for their algal partners during the early stages of synthesis. This topic is discussed further below.

While examples are therefore known of lichen algae and fungi not showing a high degree of specificity towards each other, it is not known whether the same degree of specificity is shown in all lichen species. Probably, specificity varies from species to species, and those lichens which are the least specifically constituted may well be those which rely on synthesis in nature for their mode of reproduction rather than upon dispersal by fragmentation or soredia. At present, one can only conclude that the great majority of lichen species consist each of one species of fungus associated with a particular morphological taxon of a genus of algae; it is possible that in some lichen species the specificity of the fungus may extend to physiological strains within the morphological taxon.

Morphological contact between the symbionts

It can be easily observed that the symbionts of lichens are usually in close contact with each other in the thallus. In most lichen species, modified hyphae or hyphal branches are closely appressed to the algal cells. In the remaining few species, the association is somewhat looser with unmodified hyphae either ramifying over the surface of the algal cells (e.g. *Coenogonium*, Graphidiaceae) or within the gelatinous sheaths of blue-green algae (e.g. *Ephebe*, Collemaceae).

It has long been known that in some lichens the algal cells may be penetrated by fungal haustoria (e.g. Bornet, 1873), but earlier authors considered that this phenomenon occurred in very few species—nearly all of them homoiomerous lichens (cf. Nienburg, 1926). In recent decades, the only extensive and detailed studies of this problem are those of Geitler and his pupils (Geitler, 1933, 1934, 1937, 1955; Tschermak, 1941, 1943; Plessl, 1949; Schiman, 1958), who consider that penetration of the algal cells by fungal hyphae is a widespread phenomenon in lichens as a whole. The type of penetration which they describe varies widely from species to species, and in some cases its frequency varies greatly even within the same thallus at different times of the year.

Tschermak (1941) recognized two main types of fungal penetration:

(*a*) Intracellular haustoria, which actually penetrate the protoplast of the algal cell. This type occurs mainly in homoiomerous lichens and is known from only a few heteromerous species such as *Lecidea parasema* (Geitler, 1934) and *Lecania candicans* (Fry, 1928). Infected algal cells typically become enlarged and they may also develop thicker walls; in most cases they eventually lose their cell contents and die, but in a few lichens—such as *Lecidea parasema*—the algal cells normally survive infection by forming a wall which severs the haustorium from the parent hypha, and the haustorium then disintegrates. The frequency of intracellular haustoria is very variable: in *Lecidea parasema* and *Lecania candicans* almost every cell is regularly infected, while in some *Lempholemma* species (Geitler, 1955) only a very small proportion of the cells is infected.

(*b*) Intramembranous haustoria, which penetrate the external wall of the algal cell but do not penetrate the protoplast. This type is characteristic of many heteromerous lichens as well as certain homoiomerous forms such as *Collema*. Infected algal cells generally recover, and are able to reconstitute their external walls in the places where they have been punctured. The frequency of intramembranous haustoria may show large variations, even within the same thallus at different times of the year (Tschermak, 1941, 1943; Schiman, 1958) although the variation does not appear to be entirely seasonal.

Moore & McAlear (1960) published electron micrographs of what are probably intramembranous haustoria in *Cladonia cristatella* and a *Lecidea* species. On the basis of admittedly very limited investigation, they suggest that this type of haustorium may not be 'as uncommon as previously supposed'. Others who have recently reported observing haustoria in lichens are Degelius (1954), Steiner (1959) and Ahmadjian (1959). Ahmadjian (1962) has also observed formation of intramembranous haustoria during the early stages of the artificial synthesis of *Acarospora fuscata*. Ahmadjian & Henriksson (1959) found that the isolated fungus of *Collema tenax*, which may penetrate its *Nostoc* partner with intramembranous haustoria (Degelius, 1954), formed intracellular haustoria when mixed with the alga *Trebouxia impressa* isolated from the lichen *Physcia stellaris*.

Apart from the investigators cited, modern lichenologists have not commented on the occurrence of haustoria, but they are probably widespread.

There is no evidence of the function or significance of haustoria in lichens, and it is not known to what extent they are an important or essential feature of the symbiosis. It is also not known in the majority of species what proportion of algal cells in the thallus are infected. The fact that structures termed 'haustoria' can be observed in lichens is not necessarily an indication that lichen fungi were originally (or are now) parasitic. In some cases at least, formation of these haustoria might well represent a stage in the evolution of a lichen from a condition where the ancestral fungi and algae lived commensally, and they may represent a device for increasing the area of intimate contact between the algae and fungi rather than an indication of parasitism.

In most healthy lichen thalli, dead algal cells are rare. Excluding the very few species with abundant intracellular haustoria, it appears that the lichen symbiosis does not normally involve digestion of the algal cells by the fungus.

Physiological interrelationships within the thallus

Most theories of lichen symbiosis postulate an exchange or movement of substances between the symbionts within the thallus, although direct evidence that this occurs is almost completely lacking. It is certainly reasonable to assume that carbon compounds produced in photosynthesis by the alga may move to the fungus, but the only direct experimental investigation of this so far has been in *Peltigera poly-dactyla* (Smith, 1961), where it was demonstrated that when $H_2{}^{14}CO_3$ was supplied to the lichen in the light, organic compounds containing ^{14}C first accumulated in the 'algal zone', and later appeared in the medulla. Scott (1956) has produced good indirect evidence that nitrogen fixed by the *Nostoc* symbiont of *Peltigera praetextata* is passed to the fungus. Ahmadjian's observations (1962) led him to suppose that during the early stages of synthesis of *Acarospora fuscata*, organic compounds moved from the alga to the fungus. The only other type of experimental evidence adduced in support of the various theories of exchanges of substances between the symbionts is that derived from studies of the isolated components in pure culture. The nature and value of this type of evidence is discussed later.

Under certain environmental conditions, the algae and fungi of lichens begin to grow independently of each other, with the result that they lose intimate contact and the thallus disintegrates. Tobler (1925) observed that this occurred when lichens come into contact with substrates that are very rich in organic substances. Scott (1960), in studies of the culture of tissue discs of *Peltigera praetextata*, concluded that the breakdown of the symbiotic relationship could result not only from excessive nutrient supply, but also from the conjoint effects of excessive moisture and light.

A number of authors have investigated the possibility that the lichen fungus protects the alga from unfavourable environmental conditions such as drought or extremes of temperature. The results of their experiments have been either inconclusive or have not shown that protection by the fungus is essential; this topic has been recently reviewed by Quispel (1959). The results of some of Ahmadjian's (1962) attempts to synthesize *Acarospora fuscata* in culture under a range of light intensities and nutrient conditions suggest that the fungus may protect the alga from excessive light intensities in certain circumstances.

The symbionts in isolated culture

Pure culture of the symbionts of most lichen species presents no great technical problem; Quispel (1943, 1959) has reviewed the various methods that have been used. The isolated symbionts—like many free-living fungi and terrestrial algae—are generally able to utilize a wide range of organic carbon and nitrogen sources.

Quispel (1943) observed that the addition to the medium of certain growth factors such as aneurin and β-alanine were essential for the growth of fungi isolated from lichens or lichenized algal covers; other substances had a marked stimulatory effect on the autotrophic development of the algae. These and other observations led Quispel to advance the hypothesis that there were exchanges of growth factors between the symbionts within the lichen thallus itself.

Subsequent investigations have given conflicting results: Zehnder (1949) found that fungi from ten out of eleven different lichen species did not require growth factors; the eleventh, from *Placodium saxicola*, required thiamine for good growth. None of the lichen algae showed any such requirements. Quispel (1959) pointed out that Zehnder used rather heavy inocula in his experiments, and that this may have obscured partial requirements for growth factors. However, Quispel agreed that absolute dependence on certain growth factors may not be a general characteristic of the lichen symbionts in pure culture. Hale (1958), in a careful study of the requirements of three rapidly growing lichen fungi, found that those from *Acarospora fuscata* and *Buellia stillingiana* were almost totally dependent on added biotin and thiamine, while the fungus of *Sarcogyne similis* had a partial requirement for thiamine only. Ahmadjian (1961) also found that the fungus of *Acarospora fuscata* required biotin and thiamine, as did the fungi of *A. smaragdula*, *Baeomyces roseus*, *Sarcogyne simplex*, *Cladonia cristatella* and *C. pleurota*. Tobler (1944) was unable to germinate ascospores of *Xanthoria parietina* without thiamine, although a number of other authors have succeeded in doing so (e.g. Quispel, 1943). Apart from the work of Quispel and of Zehnder, few investigators have studied the growth factor requirements of lichen algae in culture. Ahmadjian (1962) reported that unlike the fungus, the alga of *Acarospora fuscata* has no requirement for biotin. Hale (1961) believes that a partial requirement for thiamine is probably very common in lichen fungi—but it must be remembered that this is also the commonest requirement amongst free-living fungi.

Although no clear and comprehensive picture emerges of the growth-factor requirements of the lichen symbionts, it seems probable that at least partial deficiencies are widespread amongst lichen fungi in pure culture. This does not necessarily prove that growth factors move between the symbionts in the thallus. Requirements for growth factors are commonly observed in free-living fungi when grown in culture, and it is well known that some of these only appear under certain cultural conditions; under natural conditions, the fungi may be able to synthesize the growth factors for themselves (cf. Hawker, 1950; Cochrane, 1958).

Henriksson (1951, 1957, 1958, 1960, 1961) has studied the excretion of substances into the medium by pure cultures of the symbionts of the lichen *Collema tenax*. The alga, a *Nostoc* species which can fix atmospheric nitrogen, excretes a variety of substances into the medium including vitamins, polysaccharides, nitrogenous compounds and substances which inhibit the growth of the lichen fungus. However, as Henriksson herself points out, all these substances may also be excreted by free-living blue-green algae in culture. She also showed that the lichen fungus could exist on the excretion products of the alga. When pure cultures of the symbionts were mixed, the fungus had a lethal effect upon the lichen alga (as it did on free-living blue-green algae). Although Henriksson has not yet established any clear differences between lichen symbionts and their free-living counterparts, her studies touch upon an aspect of lichen physiology which has been relatively neglected in the many theories of the symbiosis. If, as seems probable, organic substances move from the alga to the fungus, the question would arise as to whether this movement was due to excretion by the alga, absorption by the fungus (such as by haustoria or appressoria), or by a combination of both processes. Henriksson's work draws attention to the possible importance of algal excretion.

Compared to free-living fungi, a general characteristic of lichen fungi in culture is that most have very slow growth rates. For example, Thomas (1939) found that the maximum diameters of fungal colonies from fifteen different lichen species grown on four types of agar media at optimum temperatures ranged from 4·9 to 22·6 mm. after 180 days. Some lichen fungi grow appreciably faster than this, and three species studied by Hale (1958) had growth rates and yields approaching those of free-living saprophytic discomycetes on the same medium (the fastest lichen fungus, that from *Buellia stillingiana*, gave a yield in a liquid medium of 110 mg. dry weight of mycelium in 2–4 weeks). Ahmadjian (1961) also found relatively rapid growth in some lichen fungi (yields of 50–60 mg. dry weight of mycelium in 60 days) though others were extremely slow (e.g. a monospore culture of the fungus of *Physcia orbicularis* var. *rubropulchra* reached a diameter of 1–2 mm. in 9 months). Although it is probable that improvements in cultural techniques may give faster growth rates for lichen fungi, most of them must still be considered as very slow growing. No explanation has been offered for this, and it is of interest that Hess (1959) found no effect of gibberellic acid on the growth of his cultures of lichen fungi. Much less is known about the growth rates of lichen algae in culture. It appears that although they may also grow relatively slowly, they generally have faster growth rates than lichen fungi.

In pure culture on a solid medium, lichen fungi generally form dense, compact colonies which are often elevated. There is usually no resemblance to the characteristic shape of the lichen thallus, although Ahmadjian (1961) reported a gross external similarity between cultures of the fungi of two *Acarospora* species and the areolae of the parent thalli. The internal differentiation of a lichen thallus into distinct anatomical regions such as a cortex, medulla, rhizoids, etc., has never been observed in cultures of lichen fungi. Ahmadjian (1961) commented that no pseudoparenchymatous tissue could be observed in any of his cultures of fungi from eighteen different lichen species although it is very commonly found in lichen thalli. As regards the reproductive behaviour of lichen fungi, it is of interest that those fungi which produce a conidial stage in culture never show it in the thallus. Furthermore, although most lichen fungi produce perithecia or apothecia in the thallus, these have never been observed in culture.

Lichen algae do not show such marked differences between their morphology in culture and in the thallus. There may be differences in the size and colour of the cells (Ahmadjian, 1962), and the *Nostoc* species in the thalli of heteromerous lichens usually lack a gelatinous sheath and the cells are arranged in packets instead of chains as in free-living forms. However, lichen algae may show marked differences in behaviour: for instance reproduction within the thallus is nearly always by simple cell division, but in culture other forms of reproduction such as zoospore formation may occur abundantly. Lange (1953) showed that the heat resistance of algae may be lowered after they are brought into isolated culture.

The differences between the morphology and behaviour of the symbionts in culture and in the thallus probably imply differences in physiology. It is essential to bear this in mind when making assumptions about the physiology of the symbionts in the lichen thallus based upon their physiology in isolated culture.

Artificial synthesis of lichens

Almost every attempt to synthesize a lichen in pure culture from the isolated algal and fungal components has failed. Bonnier as early as 1889 claimed to have effected a number of completely successful syntheses, but his experiments have been so heavily criticized by subsequent authors that serious doubts now exist as to the validity of his results (Chodat, 1913; Thomas, 1939; Quispel, 1943, 1959; Ahmadjian, 1959, 1962). Stahl (1877) claimed to have achieved a successful synthesis of *Endocarpon pusillum*, but as Ahmadjian (1962) points out, Stahl's numerous illustrations do not contain adequate and convincing figures of the various developmental stages of the synthesis, and the drawing of a 5-month-old culture shows only a rhizoidal system of the fungus covered with a layer of algae. Although Stahl reported formation of perithecia in his culture, there are no drawings to illustrate them. Ahmadjian also points out that Stahl's cultures were susceptible to airborne contamination.

A more or less successful artificial synthesis was reported by Thomas (1939). After placing the symbionts of *Cladonia* species together in culture flasks, he often observed subsequent formation of soredia, isidia and structures resembling thallus lobes. In one flask containing the symbionts of *Cladonia pyxidata* he even observed the formation of small podetia with scyphae; however, his attempts to repeat the formation of podetia in a further 800 culture flasks all failed. Unfortunately, Thomas did not investigate the internal morphology of these various synthetic structures, and it is also not known if the artificially produced *C. pyxidata* podetia contained apothecial initials.

Nearly all other attempts at the artificial synthesis of lichens have either completely failed or not progressed beyond the initial stages of synthesis in which the fungal hyphae have encircled algal cells to form soredia-like masses (Tobler, 1909; Werner, 1931; Bartusch, 1932; Lange de la Campe, 1933; Hérisset, 1946; Quispel, 1943; Ahmadjian, 1959). In none of these experiments was anything resembling a lichen thallus produced, but although these attempts did not succeed, they have given some interesting results which may be summarized as follows. On agar media, the algae and fungi usually grow quite independently of each other when organic nutrients are present. When organic nutrients are absent, and the level of mineral nutrients is low, the initial stages of synthesis are frequently observed. Thomas (1939) and Quispel (1959) therefore maintained that sub-optimum nutrient levels may be essential for successful synthesis. On certain substrates other than agar—such as wood and elder pith—the initial stages of synthesis are often better developed and there is a greater tendency towards the formation of soredia-like masses. Such observations led Thomas and Quispel to suggest that another essential condition for successful synthesis may be the maintenance of correct humidity—perhaps even involving variations in humidity during synthesis. It is also possible that the physical nature of the substratum may be important. Certainly, the known sensitivity of lichens to atmospheric pollution makes it probable that purified air may be another essential condition for successful synthesis.

Ahmadjian (1962) has recently published a detailed description of his attempts to synthesize the lichen *Acarospora fuscata*, which support some of the suggestions made by Thomas and Quispel. Ahmadjian inoculated the symbionts together on agar media in culture flasks under a range of light intensities and nutrient conditions. On media containing Bristol's solution, soil infusion water and dextrose, the symbionts grew quite independently of each other. On media containing Bristol's solution and soil infusion water but no dextrose, some hyphal encirclements and haustorial penetrations of algae could be observed at the lowest light intensities (30 foot-candles), although no real synthesis had occurred at the end of $3\frac{1}{2}$ months. On media containing Bristol's solution only, many contacts between algal cells and fungal hyphae were observed although the synthesis did not progress beyond these initial stages. However, on media of pure agar with no added nutrients, the initial stages of encirclement and haustorial penetration developed rapidly (particularly at the lower light intensities), and within 30 days the synthesis had progressed to a stage where hyphae enclosing groups of algal cells had formed pseudoparenchymatous tissue. In regions where there were no algal cells, no pseudoparenchyma developed. After 9 months, Ahmadjian reported that the cultures on pure agar were still progressing. He also noted that the rate of lichenization increased if the cultures were subjected to a simple system of alternate wetting and drying. Addition of biotin also increased the rate of lichenization, and this was ascribed to its effect on the fungus.

Studies of the early stages of synthesis in culture suggest that the initial contact between the symbionts is accidental and not the result of any factor causing directional growth such as a hormone (Ahmadjian, 1960b). Furthermore, the fungus does not appear to be very specific towards the alga at this stage. For example, Lange de la Campe (1933) observed that the fungus of *Xanthoria parietina* reacted towards *Coccomyxa* as towards its normal *Trebouxia* partner—although it showed no tendency to associate with *Stichococcus*. Other evidence of this type has been reviewed by Ahmadjian (1960b).

DISCUSSION

Ahmadjian (1962) concluded from his attempts to synthesize *Acarospora fuscata* that a lichen association will not form under conditions which support the independent growth of either or both components. The observations of Tobler (1925) and the experiments of Scott (1960) certainly suggest that such conditions—particularly excessive nutrient supply—cause the breakdown of the symbiosis. It therefore appears that both the initiation and the maintenance of the lichen symbiosis require environmental conditions which do not favour the independent growth of the symbionts.

It is not known why adverse conditions should stimulate the algae and fungi of lichens to enter into symbiosis. It may well be a simple question of the exchange of nutrients between the symbionts, but the morphological and behavioural changes shown by the symbionts when they form lichens indicate that more complex factors are involved.

The thallus in most lichen species is a complex structure differentiated into anatomical regions which have different physiological properties. Since lichen fungi are of such diverse kinds—including both pyreno-mycetes and discomycetes—and since none of the related free-living ascomycetes possess a well-differentiated thallus, it seems likely that in the course of evolution, differentiation of the thallus occurred after the phenomenon of lichenization. Parallel to the evolution of such morphological changes, there was presumably also an evolution in physiological characteristics—perhaps those characteristics enabling lichens to exist in habitats with more extreme conditions of low nutrient supply and drought than those in which their free-living ancestral algae and fungi existed. Although such speculations about the origin and evolution of lichens can be easily carried too far, they help to stress the need to regard the lichen thallus as an integrated structure and not simply as an association of a particular alga with a particular fungus.

The amount of information that can be gained about the physiology of lichens from a study of the symbionts in pure culture is necessarily limited because the nature and extent of the differences between the condition of the symbionts in pure culture and their condition in the thallus is not fully understood. Although some valuable information has been gained from the attempts to synthesize lichens in culture, it is clear that future research into the nature of the association should lay more emphasis upon the behaviour of the individual symbionts as they occur in the thallus. There are obvious difficulties to such studies, but the complete range of relevant experimental techniques—such as dis-section, radioactive tracers, autoradiography—has yet to be fully applied to the problem. As regards the study of the isolated symbionts in pure culture, it now appears that a profitable avenue of future research may be the investigation of the effects of unfavourable conditions upon them.

Many of those who have recently investigated the physiology of lichens have considered their very slow growth rates to be significant or important. Although no one has yet offered a clear and precise explanation, a number of suggestions have been advanced. Scott (1960) and Ahmadjian (1962) consider that the slow growth rate of lichens results from the need for a balanced growth between the algal and fungal components. Scott considers that the nutrient, moisture and light conditions act jointly as a growth regulating system. Another view— not necessarily incompatible with that of Scott and Ahmadjian—has been offered by those who have regarded the lichen more as a functional unit than as an association between two organisms. For example, Stålfelt (1939) suggested that the slow growth rate might result from low rates of net carbon assimilation. Smith (1961, 1962) doubts that this is the prime cause of slow growth, and suggests that factors such as the slow rates of protein turnover or the low availability of metabolic energy for synthetic processes might be more important.

Perhaps the most important conclusion to be drawn from recent research into the physiology of lichens is that many different kinds of experimental techniques can be satisfactorily applied to them. The discovery of much valuable information about the nature of the association between the symbionts awaits the application of these techniques.

REFERENCES

ABBAYES, H. DES (1951). *Traité de Lichénologie. Encycl. biol.*, vol. 41. Paris: Lechevalier.

AHMADJIAN, V. (1959). A contribution towards lichen synthesis. *Mycologia*, **57**, 56.

AHMADJIAN, V. (1960*a*). Some new and interesting species of *Trebouxia*, a genus of lichenized algae. *Amer. J. Bot.* **47**, 677.

AHMADJIAN, V. (1960*b*). The lichen association. *Bryologist*, **63**, 250.

AHMADJIAN, V. (1961). Studies on lichenized fungi. *Bryologist*, **64**, 168.

AHMADJIAN, V. (1962). Investigations on lichen synthesis. *Amer. J. Bot.* **49**, 277.

AHMADJIAN, V. & HENRIKSSON, E. (1959). Parasitic relationship between two culturally isolated and unrelated lichen components. *Science*, **130**, 1251.

BARTUSCH, H. (1932). Beiträge zur Kenntnis der Lebensgeschichte des *Xanthoria*-pilzes. *Arch. Mikrobiol.* **3**, 122.

BONNIER, G. (1889). Recherches sur la synthèse des lichens. *Ann. Sci. nat.*, Sér. 7, **9**, 1.

BORNET, E. (1873). Recherches sur les gonidies des lichens. *Ann. Sci. nat.*, Sér. 5, **17**, 45.

CHODAT, R. (1913). *Monographie d'algues en culture pure.* Matériaux pour la flore cryptogamique Suisse, vol. 4, fasc. 2. Bern: K.-J. Wyss.

CHODAT, R. (1930). La symbiose des lichens et la théorie de la specificité en géneral. *Verh. schweiz. naturf. Ges.* p. 221.

COCHRANE, V. W. (1958). *Physiology of Fungi.* New York: John Wiley and Sons.

DEGELIUS, G. (1954). The lichen genus *Collema* in Europe. Morphology, taxonomy, ecology. *Symb. bot. upsaliens*, **13**, 1.

ERTL, L. (1951). Über die Lichtverhältnisse in Laubflechten. *Planta*, **39**, 245.

FRY, E. J. (1928). The penetration of lichen gonidia by the fungal constituent. *Ann. Bot., Lond.* **42**, 141.

GEITLER, L. (1933). Beiträge zur Kenntnis der Flechtensymbiose. I–III. *Arch. Protistenk.* **80**, 378.

GEITLER, L. (1934). Beiträge zur Kenntnis der Flechtensymbiose. IV–V. *Arch. Protistenk.* **82**, 51.

GEITLER, L. (1937). Beiträge zur Kenntnis der Flechtensymbiose. VI. *Arch. Protistenk.* **88**, 161.

GEITLER, L. (1955). Gehäufte Haustorien bei einer Collematacee. *Öst. bot. Z.* **102**, 317.

HALE, M. E. (1958). Vitamin requirements of three lichen fungi. *Bull. Torrey bot. Cl.* **85**, 182.

HALE, M. E. (1961). *Lichen Handbook*. Washington, D.C.: Smithsonian Institute.

HARLEY, J. L. & SMITH, D. C. (1956). Sugar absorption and surface carbohydrase activity of *Peltigera polydactyla* (Neck.) Hoffm. *Ann. Bot., Lond.*, N.S. **20**, 513.

HAWKER, L. E. (1950). *Physiology of Fungi*. London: University of London Press.

HENRIKSSON, E. (1951). Nitrogen fixation by a bacteria-free symbiotic *Nostoc* strain isolated from *Collema*. *Physiol. Plant.* **4**, 542.

HENRIKSSON, E. (1957). Studies in the physiology of the lichen *Collema*. I. The production of extracellular nitrogenous substances by the algal partner under various conditions. *Physiol. Plant.* **10**, 943.

HENRIKSSON, E. (1958). Studies in the physiology of the lichen *Collema*. II. A preliminary report on the isolated fungal partner with special regard to its behaviour when growing together with the symbiotic alga. *Svensk. bot. Tidskr.* **52**, 391.

HENRIKSSON, E. (1960). Studies in the physiology of the lichen *Collema*. III. The occurrence of an inhibitory action of the phycobiont on the growth of the mycobiont. *Physiol. Plant.* **13**, 751.

HENRIKSSON, E. (1961). Studies in the physiology of the lichen *Collema*. IV. The occurrence of polysaccharides and some vitamins outside the cells of the phycobiont. *Physiol. Plant.* **14**, 813.

HÉRISSET, A. (1946). Démonstration expérimental du rôle du *Trentepohlia umbrina* (Kg.) Born. dans la synthèse des Graphidées corticoles. *C.R. Acad. Sci., Paris*, **222**, 100.

HESS, D. (1959). Untersuchungen über die Bildung von Phenolkörpern durch isolierte Flechtenpilze. *Z. Naturf.* **14***b*, 345.

JAAG, O. (1929). Recherches expérimentales sur les gonidies des lichens appartenant aux genres *Parmelia* et *Cladonia*. *Bull. Soc. bot., Genève*, **21**, 1.

LANGE, O. L. (1953). Hitze- und Trockenresistenz der Flechten in Beziehung zu ihrer Verbreitung. *Flora, Jena*, **140**, 39.

LANGE DE LA CAMPE, M. (1933). Kulturversuche mit Flechtenpilzen (*Xanthoria parietina*). *Arch. Mikrobiol.* **4**, 379.

MOORE, R. T. & MCALEAR, J. H. (1960). Fine structure of mycota. 2. Demonstration of the haustoria of lichens. *Mycologia*, **52**, 805.

NIENBURG, W. (1926). *Anatomie der Flechten*. Berlin: Borntraeger.

PLESSL, A. (1949). Beziehungen von Organisationshöhe und Haustorientypus bei *Lecanora* und anderen Krustenflechten. *Öst. bot. Z.* **96**, 145.

QUISPEL, A. (1943). The mutual relations between algae and fungi in lichens. *Rec. Trav. bot. néerl.* **40**, 413.

QUISPEL, A. (1959). Lichens. *Encycl. Plant Physiol.* **11**, 577.

RATHS, H. (1938). Experimentelle Untersuchungen mit Flechtengonidien aus der Familie der Caliciaceen. *Ber. schweiz. bot. Ges.* **48**, 329.

RIED, A. (1960). Thallusbau und Assimilationshaushalt von Laub- und Krustenflechten. *Biol. Zbl.* **79**, 129.

SCHIMAN, H. (1958). Beiträge zur Lebensgeschichte homoeomerer und heteromerer Cyanophyceen-Flechten. *Öst. bot. Z.* **104**, 409.

50   D. C. SMITH

SCOTT, G. D. (1956). Further investigations of some lichens for fixation of nitrogen. *New Phytol.* **55**, 111.

SCOTT, G. D. (1960). Studies of the lichen symbiosis. I. The relationship between nutrition and moisture content in the maintenance of the symbiotic state. *New Phytol.* **59**, 374.

SMITH, D. C. (1960*a*). Studies in the physiology of lichens. 2. Absorption and utilization of some simple organic nitrogen compounds by *Peltigera polydactyla. Ann. Bot., Lond.,* N.S. **24**, 172.

SMITH, D. C. (1960*b*). Studies in the physiology of lichens. 3. Experiments with dissected discs of *Peltigera polydactyla. Ann. Bot., Lond.,* N.S. **24**, 186.

SMITH, D. C. (1961). The physiology of *Peltigera polydactyla* (Neck.) Hoffm. *Lichenologist,* **1**, 209.

SMITH, D. C. (1962). The biology of lichen thalli. *Biol. Rev.* **37**, 537.

SMYTH, E. S. (1934). A contribution to the physiology and ecology of *Peltigera canina* and *P. polydactyla. Ann. Bot., Lond.* **48**, 781.

SOSA-BOURDOUIL, C. (1944). Sur la biologie et le chimisme d'un lichen (*Usnea barbata* Web.). *C.R. Acad. Sci., Paris,* **218**, 475.

STAHL, E. (1877). *Beiträge zur Entwicklungsgeschichte der Flechten. II. Ueber die Bedeutung der Hymenialgonidien.* Leipzig: Arthur Felix.

STÅLFELT, M. G. (1939). Der Gasaustausch der Flechten. *Planta,* **29**, 11.

STEINER, M. (1959). *Maronella laricina* n.gen., n.spec. (*Acarosporaceae*) eine neue Flechte aus Tirol. *Öst. bot. Z.* **106**, 440.

THOMAS, E. A. (1939). Über die Biologie von Flechtenbildnern. *Beitr. Kryptogamenfl., Schweiz.* **9**, 1.

TOBLER, F. (1909). Das physiologische Gleichgewicht von Pilz und Alge in den Flechten. *Ber. dtsch. bot. Ges.* **27**, 421.

TOBLER, F. (1925). *Biologie der Flechten.* Berlin: Borntraeger.

TOBLER, F. (1944). Die Flechtensymbiose als Wirkstofffrage. I. Die Keimung von Flechtensporen und ihre Anregung durch Wirkstoffe. *Planta,* **34**, 34.

TSCHERMAK, E. (1941). Untersuchungen über die Beziehungen von Pilz und Alge im Flechtenthallus. *Öst. bot. Z.* **90**, 234.

TSCHERMAK, E. (1943). Weitere Untersuchungen zur Frage des Zusammenlebens von Pilz und Alge in den Flechten. *Öst. bot. Z.* **92**, 15.

WAREN, H. (1920). Reinkulturen von Flechten gonidien. *Öfvers. finska VetenskSoc. Forh.* **61**, 1.

WERNER, R. G. (1931). Histoire de la synthèse lichénique. *Mem. Soc. Sci. nat. Maroc.* **27**, 1.

ZEHNDER, A. (1949). Über den Einfluss von Wuchstoffen auf Flechtenbilder. *Ber. schweiz. bot. Ges.* **59**, 201.

ZEITLER, I. (1954). Untersuchungen über die Morphologie, Entwicklungsgeschichte und Systematik von Flechtengonidien. *Öst. bot. Z.* **101**, 453.

ZUKAL, H. (1895). Morphologische und biologische Untersuchungen über die Flechten. *S.B. Akad. Wiss. Wien,* **104**, 529, 1303.

EXPLANATION OF PLATE

PLATE 1

Fig. 1. Transverse section through the thallus of *Peltigera polydactyla.* The algal layer appears as the dark band beneath the upper cortex. Under the algal layer is the relatively thick medulla of loosely arranged thick-walled hyphae.

Fig. 2. Transverse section through the algal region of the thallus of *Peltigera polydactyla.* Groups of algal cells (*Nostoc* sp.) can be observed beneath the upper cortex of pseudoparenchyma. Thick-walled medullary hyphae occur beneath the algal layer.

PLATE 1

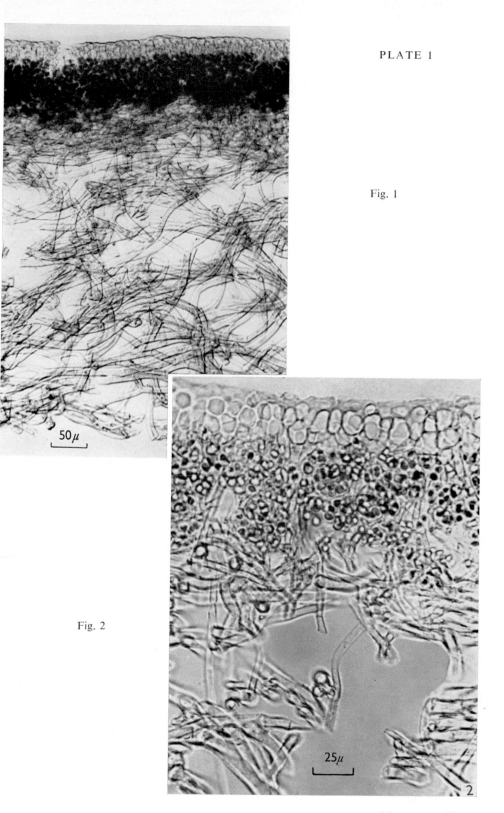

Fig. 1

Fig. 2

50 μ

25 μ

FACTORS INFLUENCING THE BAL
OF MUTUAL ADVANTAGE IN LEGUME
SYMBIOSIS

P. S. NUTMAN
Soil Microbiology Department, Rothamsted Experimental Station

In many symbioses involving micro-organisms, the mutual benefits and the means by which they are secured are obscure. This seems not to apply to legume root-nodule symbiosis; the host's advantage clearly comes from the nitrogen fixed in the nodule, and the bacteria's from the provision within the host of a favourable environment and nutrition. The soil population of nodule bacteria is also increased wherever legumes are grown. The benefit to the host from nitrogen fixation has been the more studied because of its agricultural importance and significance in the nitrogen economy of nature. Donald (1960) reviewed the evidence and assigned a value of the order of 10^8 tons N_2 for the world's annual income of biologically-fixed nitrogen (mainly from symbiotic sources). Estimates for individual crops vary widely, indicating that fixation is often much less than it could be. For the effectively nodulated legume, growing vigorously under optimal conditions, the amount of nitrogen fixed seems to be regulated only by the quantity of combined nitrogen the plant can obtain from the soil. When the soil provides none the nodules can supply the full requirement of the host for nitrogen. The nodulated legume's freedom from the nutritional deficiency most suffered by other plants may have contributed to the success of the Leguminosae as one of the largest and most cosmopolitan of plant families.

Under natural conditions, even more than in agriculture, the full nitrogen-fixing potential of the nodulated legume is rarely realized because of limiting factors, either intrinsic to the symbiosis or in the environment. The legumes' independence of soil nitrogen tends to unmask deficiencies not only in other major elements that would otherwise be adequate, particularly phosphorus and calcium, but also in the trace elements Mo, B, Zn, Cu and Co, some of which are needed in larger amounts by legumes than by the non-legumes (see Allen & Allen, 1958; Hallsworth, 1958; Anderson, 1956; Shaukat-Ahmed & Evans, 1961; Riceman & Jones, 1960).

The benefits to both host and bacteria are derived from nitrogen fixation. It will therefore be the purpose of this chapter first to discuss some recent work on the origin and function of the host and bacterial structures associated with nitrogen fixation. Secondly, an attempt will be made to bring together, with some new data, the rather scattered evidence of ecological benefit to the bacteria in the soil. Thirdly, as instructed by the convenors of the symposium, the evolution of the nodulating habit of legumes will be discussed. To discuss the first and last of these topics it will be necessary to outline the normal processes of infection and nodule formation (for earlier literature see Fred, Baldwin & McCoy, 1932; Nutman, 1956; Allen & Allen, 1958; Nutman, 1958). The following description is inclusive; all stages are not found in all symbioses.

THE THREE STAGES OF NODULE DEVELOPMENT

(1) *The infection of the root*

The early pre-infection interactions between plant and bacteria take place in the immediate vicinity of the plant root and are initiated in some way by substances exuded or secreted therefrom.

Among the exudation products is tryptophan which is released in very small amounts (Rovira, 1959). This is converted by nodule bacteria to indoleacetic acid—IAA—(Kefford, Brockwell & Zwar, 1960). A characteristic deformation of the host's root-hairs, generally believed to be caused by IAA, is often a prelude to actual infection. Exudation of tryptophane, IAA formation and hair curling occur whether or not the bacteria involved in IAA production can form nodules on the host in question.

The first specific reaction between host and bacteria seems to be caused by the extracellular polysaccharide slime produced by the bacteria (Fåhraeus & Ljunggren, 1959). This induces the plant to secrete polygalacturonase (PG). Whenever infection occurs PG can be detected in the rhizosphere, but polysaccharide produced by strains of nodule bacteria unable to infect a given host does not induce this host to secrete PG. Most PG is produced by those combinations of bacteria and host that give rise to most infections. Nitrate in the root milieu stops infection and also inhibits PG production (Ljunggren & Fåhraeus, 1961). Bacterial polysaccharide type, infective virulence and PG-inducing activity all appear to be genetically transformable characters (Balassa, 1957, 1960; Ljunggren, 1961; Lange & Alexander, 1961).

The function of the PG is not known, but it may act with IAA to affect the plasticity of the primary wall of the young root hair and thus assist penetration.

An infected hair is usually recognized by the presence within it of a hypha-like infection thread containing the bacteria and consisting largely of cellulose laid down by the host cell. The infection thread grows at its tip, which is free from cellulose; tip-growth occurs only when the host-cell nucleus is nearby (Fåhraeus, 1957; Nutman, 1959a).

(2) *The initiation and organization of the nodule*

The nodule itself is initiated when the penetrating infection thread approaches a pre-formed tetraploid cell in the cortex. This cell and neighbouring diploid cells are stimulated to repeated division, and the mass of cells so formed rapidly differentiates into the young nodule. The infection threads branch and spread throughout the tetraploid cells as they are formed, but do not penetrate diploid cells of the apical meristem. An endodermis separates the central infected zone from the rest of the nodule. The uninfected diploid tissue differentiates into nodule cortex, vascular traces, etc. Further nodule growth is provided for by the organization of a nodule meristem of both diploid and tetraploid cells.

(3) *The intracellular phase*

Intracellular invasion of the tetraploid cells of the central zone of the nodule by the bacteria occurs by the rupture of vesicles formed on the infection threads. The bacteria so released multiply rapidly in the host cell and are then transformed into swollen and sometimes branched forms called bacteroids. Nitrogen is fixed only in tissue containing bacteroids.

The bacteroids were thought to develop freely within the host cytoplasm, but recent work suggests that this description may not accord strictly with the facts. Electron-microscopy of thin sections of soya bean nodule tissue indicates that the bacteria from the thread-vesicles become enclosed, singly or in small groups, within folds of the outermost protoplasmic membrane of the host cytoplasm (the plasmalemma) where they multiply and change to bacteroids. In this way there arises in the mature infected nodule cell a very complex system of host membrane envelopes, each one surrounding a small group of bacteroids (Bergersen & Briggs, 1958).

Bacteroid formation consists of enlargement and change in cell shape, a large increase in the perinuclear region, fragmentation of chromatin, and prominent development of respiratory mitochondria. The bacteroid does not divide within the nodule, and has not yet been cultivated on artificial media. Concomitantly with bacteroid formation, haemoglobin appears in the nodule and nitrogen is fixed. After some weeks or months the haemoglobin is replaced by bile pigments, the bacteroids lyse and the degenerative processes so initiated usually spread to the whole nodule. The rod-shaped bacteria which have lain dormant in the infection thread invade and multiply in the old nodule. As the old nodule decays the bacteria escape into the soil. In some legumes, usually tropical species, the bacteroid-containing tissue degenerates at the end of the growing season, but this degeneration does not spread throughout the nodule, and nodule growth is resumed periodically, sometimes over several years.

Symbiotic development in legumes thus involves a complex series of interactions between bacteria and plant. Eight linked stages can be distinguished in the pre-infection phase (secretion of growth factors etc. by the plant, rhizosphere stimulation, secretion of tryptophane, IAA production, root-hair curling, production of specific bacterial polysaccharide, PG induction and secretion); a further five stages in the initiation and elaboration of the nodule (hair infection, thread initiation, co-ordinated thread growth and host nucleus migration, activation of tetraploid cells, co-ordinated thread growth and differentiation of nodule); and six more in the intracellular phase (bacterial release, membrane folding, bacterial multiplication, bacteroid formation— associated with haemoglobin formation and nitrogen fixation— bacteroid lysis, senescence or renewal of nodule growth).

Each of the above stages seems independently variable either in morphology or function, and in many hosts certain of them are abbreviated or omitted altogether without impairing symbiotic efficiency. For example *Arachis* becomes infected not by way of the root hair, but at points where lateral roots emerge. In certain legumes the bacteria spread passively from cell to cell during host-cell division without the intervention of the infection-thread mechanism. In some species nodules are initiated deep in the cortex or in the pericycle, in others the tetraploid initial may arise in the hypodermis.

The breakdown of the symbiosis caused by incompatibility between host and bacteria constitutes a further large category of variation. This may altogether prevent nitrogen fixation, as with the so-called 'ineffective' symbioses, or may reduce it. Incompatibility may show itself at any stage; in the pre-infection phase thus preventing infection (the so-called cross-inoculation group incompatibility), or at any point in nodule development, or in the 'intracellular' phase. In the late stages several distinct kinds of defect are known which prevent bacteroid development, and some of these are caused by single gene changes in plant or bacteria (Nutman, 1959 b).

THE NITROGEN-FIXING PROCESS

In spite of so much variability in nodule development and structure, all fully effective nodules nevertheless possess a characteristic and very similar bacteroid-containing tissue. Nitrogen fixation is strictly correlated with the formation of this specialized infected tissue; the amount fixed in any particular symbiosis is regulated by the total amount of bacteroid containing tissue and the duration of its active life (Chen & Thornton, 1940; Nutman, 1958; Bergersen, 1961 c). These correlations suggest very strongly that the bacteroid is directly involved in the actual fixing process, and that the nitrogen fixed in the bacteroid is later transferred to the host plant. This view, which recent work has challenged, led to much experimentation on nitrogen fixation by bacteroids separated from the host, to attempts to induce fixation by nodule bacteria in culture, and to extensive work on bacteroid metabolism (see Wilson & Burris, 1953; Bergersen, 1958). Although several other soil microorganisms can fix nitrogen in culture, nodule bacteria are unable to do so; none of the claims to the contrary that has been made from time to time has stood up to critical examination.

All efforts to obtain *in vitro* fixation by bacteroids separated from the nodule have also so far failed. The destruction of the joint structures in the bacteroid-containing cell stops fixation and, even in the excised nodule, fixation continues only briefly (Aprison & Burris, 1952; Magee & Burris, 1954). In this respect the non-legume root nodules may be less sensitive and continue to fix nitrogen for some hours after removal from the plant, see Dr Bond's chapter in this volume, p. 72. Work on bacteroid metabolism also failed initially to elucidate the fixing process. The bacteroid taken from the nodule has many of the properties of the bacteria in culture and does not significantly differ in metabolism (Bergersen, 1958). However, Bergersen's (1955, 1958, 1960, 1961 a, 1962) cytochemical and respiration studies have enlarged our understanding of the host's and bacteroid's carbohydrate metabolism by pointing out their co-ordination in the nodule, and the relation of bacteroid respiration to fixation.

The onset of fixation in the young nodule coincides with a rise in respiration of the isolated bacteroids. Decreasing partial pressures of oxygen depress both fixation by whole nodules and respiration by isolated bacteroids. When fixation does not occur, the nearby uninfected host cell accumulates starch and the bacteroid stores glycogen. Bergersen (1957) also showed that infected cells containing bacteroids, but not fixing nitrogen, accumulate a hitherto undescribed storage carbohydrate the function of which is not known.

The elucidation of the site of fixation followed work on the fine structure of the bacteroid tissue. By differentially centrifuging crushed nodules and using ^{15}N analysis in short-time experiments, Bergersen & Wilson (1959 a) separated the plant membranes and bacteroids and showed that enrichment with ^{15}N first occurs in the plant membranes. The plant membrane and not the bacteroid is thus the likely site of fixation. A host site for fixation was also suggested by Turchin's (1956) studies on ^{15}N enrichment of different nodule fractions. A function for haemoglobin is suggested by the work of Hamilton, Shug & Wilson (1957), Bergersen & Wilson (1959 b) and Bergersen (1962), who showed that nitrogen (in the absence of oxygen) can oxidize haemoglobin to haemiglobin, which in its turn is reduced by isolated bacteroids. In summarizing these results Bergersen (1960), elaborating a suggestion by Parker (1954), proposed that bacteroid respiration makes two contributions to nodule function: (1) the provision of reducing power which the haemoglobin–haemiglobin system then carries to the site of nitrogen fixation on the plant membranes, and (2) the provision of ammonia acceptors for synthesizing amino acids.

The nature of the links in the electron transport chain between bacteroid respiration and haemiglobin is still unknown. Falk, Appleby & Porra (1959) and Appleby & Bergersen (1958) suggest that cytochrome-like pigments in the plasma membrane of bacteroids may be involved. Rhizobium bacteria and ineffective bacteroids produce cytochromes a, b and c, whereas effective bacteroids have b and c only.

The location of the haemoglobin of nodules in the space between bacteroids and plant membranes (Bergersen, 1960) is further evidence for the function proposed above. The site of synthesis of haemoglobin is unknown; enzymes for porphyrin synthesis are active both in crushed nodules and in isolated bacteroids (Falk $et al.$ 1959). Mature nodules resemble mature erythrocytes in their loss of the enzymes catalysing the later step in haem synthesis.

The first compound containing fixed nitrogen remains to be discovered, and no convincing explanations have been proposed for the specific functions of Mo and Co, but recent work has revolutionized earlier ideas on the functions of haemoglobin and of the bacteroids. It co-ordinates many of the previously disconnected facts of metabolism and structure and suggests new lines of investigation.

THE BIOLOGICAL SIGNIFICANCE OF
THE BACTEROID-ENCLOSING MEMBRANE

It is clear from Bergersen & Brigg's (1958) work that the term 'intra-cellular', used to describe the final stages of symbiotic development is not strictly correct. If the membranes enclosing the bacteroids become 'pinched off' as they arise, the bacteroids are vacuolar. If the membranes remain as folds the bacteroids should be thought of as occupying an extended system of the plant's free space, which connects directly with the rest of the plant's cell-wall reticulum and thus with the external environment.

Before the invaginated membranes of the bacteroid tissue were identified, Nutman (1956) suggested that the infection thread itself may arise by invagination of the primary wall of the root hair. There is no direct evidence that infection-thread formation actually involves the penetration of the boundary layer of cytoplasm in the root hair. All the structures seen in the light microscope are consistent with an invagination hypothesis. Whenever actual, and probably accidental, penetration by the bacteria into the lumen of the root hair has been observed, the cytoplasm of the hair collapses and no infection threads are formed (Nutman, 1959). None of the facts of infection and nodule development conflicts with the view that the legume symbiosis depends on a special kind of intimate association between plant and bacteria that stops short of cytoplasmic miscegenation, but direct ultramicroscope evidence is lacking for the infection phase. The technical difficulties are formidable because of the smallness of the region to be observed, its uncertain location among thousands of uninfected root hairs and the probably rapid sequence of events it would be necessary to record.

Such invagination processes as are proposed for the legume symbiosis are well known in animal physiology, as in phagocytosis and pinocytosis—the uptake of liquid droplets by macrophages in tissue culture (Lewis, 1931; Holter, 1959). Buvat (1958) was the first to suggest that plant cells may take up material by pinocytosis, although Bernard in 1909, with characteristic prescience, described the digestion of haustoria in orchid mycorrhiza as phagocytosis. Using autoradiography and other techniques McLaren, Jensen & Jacobsen (1960) showed that the plant cell can take up large molecules such as haemoglobin and ribonuclease which first diffuse through the plant free-space and then enter rapidly into the cytoplasm and nucleus. Jensen & McLaren (1960) proposed pinocytosis as the probable mechanism of uptake. Pinocytosis in the plant cell has been confirmed by Weiling in a series of thin-section electron-microscope studies on a range of plant material. Of particular interest was his demonstration that material secreted by the tapetum of the tomato anther is taken up by the pollen mother-cell by pinocytosis (Weiling, 1961). Suganuma (1961) has described pinocytosis in *Staphylococcus aureus*.

The demonstration of pinocytosis in plants and the recent work on processes of invagination in legume symbiosis may have far-reaching implications for symbiosis, and also for plant pathology. Some animal parasites are already known to be actively taken up by the host cell, either by phagocytosis or pinocytosis (Trager,1960). Chick embryo fibroblasts in tissue culture can be seen to ingest *Brucella* which then multiply within the cells (Holland & Pickett, 1956), but phagocytosis is usually inferred and the development of the limiting membranes has been little studied. In fine-structure studies on leprosy, Imaeda & Convit (1962) showed the bacilli in vesicular leprous lesions to be surrounded by enclosing membranes, thought to arise by phagocytosis, and in deer *Anaplasma*, Ristic (1960) described marginal inclusions surrounded by double membranes. 'La symbiose est manifestement à la frontière de la maladie' (Bernard, 1909).

Corresponding studies on intracellular plant parasites remain to be done, but if membrane invagination is a general phenomenon this mode of intracellular 'infection' offers a way out of a dilemma often not explicitly recognized in pathology; viz. to account for the lack of gross cytoplasmic disturbances including antibody reactions in symbioses involving warm-blooded animals, where the micro-symbiont is apparently intracellular. Until immunological reactions are better understood, this problem cannot be more clearly defined, but current hypotheses (Burnet, 1960; Boyden, 1960) suggest that a requirement for true intracellular existence may be that all exposed antigens on the micro-organism should have their counterparts in the host. So stringent a requirement may indeed only be met in an ovum at fertilization and even here the 'penetration' of the egg membrane takes place indirectly. Comparison can usefully be made between fertilization in *Hydroides*, recently described fully by Colwin & Colwin (1961 *a, b, c*), and the bacterial symbiont of the cockroach egg (Bush & Chapman, 1961). The spermatozoon penetrates the outer vitelline layers of the egg of *Hydroides* mechanically, by the folding of part of the acrosomal structure; at the same time a vitelline-dissolving lysin is produced—which may be compared with polygalacturonase produced in legumes, or with permeases in phage infection of *Escherichia coli* (Puck & Lee, 1958), and the infection of red blood cells by influenza virus (Howe, 1951). As the sperm head approaches the boundary membrane of the egg, pre-formed villi of the acrosome membrane (continuous with the boundary membrane of the rest of the spermatozoon) rapidly elongate (within a few minutes) and become intermingled with similar ultramicroscopic tubules which simultaneously arise from the egg membrane. It is only after intermingling that membranes fuse, the outer membrane of the zygote then becoming a mosaic of egg and sperm origin.

In the bacterial symbiosis in the cockroach, which is transmitted through the egg, the bacteria closely invest the egg and remain external to its limiting membrane, lying between it and the outer layers which are cellular. The symbiont is thus intercellular and, like sperm in the penultimate stage of fertilization, the bacterium is also invested in a mass of microvilli originating from the egg membrane, but unlike the sperm the membranes of the bacteroid do not fuse with those of the sperm. It is of interest that this insect symbiont was originally called a 'bacteroid' because its bacterial nature was in doubt (see Gier, 1947).

An electron microscope study of the rickettsia-like micro-organisms in the Malpighian tubule cells of the fowl tick shows very similar structural relationships (Roshdy, 1961). Both within the host cell, and in the lumen of the tubule (where the symbiont was before thought to be free), the rickettsia-like bodies are enclosed, usually in small groups, within typical double membranes of the endoplasmic reticulum.

Not enough is known to generalize upon the fine structure of symbiosis but the tentative hypothesis might perhaps be advanced that symbiosis may be essentially a mutualistic relationship involving profound structural adaptations of intimately mixed rather than joint origin, and that in the legume root nodule symbiosis, some of the crucial reactions of nitrogen fixation occur on, and involve transport across, the external plant cytoplasmic membrane.

Moreover, common features can be recognized between the initiation of symbiosis, intracellular parasitism and fertilization. In all a property of the host's outer wall is affected enzymically and, except possibly for viruses, penetration of the host cell involves invagination of the host membrane, which, except only in the zygote, remains intact during functional symbiosis.

THE SOIL POPULATION OF *RHIZOBIUM*

The advantages of root nodule symbiosis seem to be mostly on the side of the host. At the cost of providing special structures and an energy source, the legume is supplied with a major nutrient, whereas the independent existence of the bacteria as bacteroids is sacrificed. Only when nodules break down do those bacteria that have not become bacteroids multiply greatly in the host, and the progeny of these cells soon escape from the nodule and become a part of the soil population of nodule bacteria.

Except for work on the numbers of nodules found on plants in the field, the ecology of host and bacteria in natural and agricultural habitats has been little studied (Petrosyan, 1959; Masefield, 1957); in particular there is little experimental work about the populations of nodule bacteria in the soil.

Wherever a nodulating species of legume occurs naturally, there its nodule bacteria are found. Conversely, nodule bacteria do not occur in soil where their host plants are not naturalized or have only recently been introduced. It is for this reason that crop legumes grown for the first time in regions free from their near botanical relations often fail because they do not nodulate. The few bacteria carried as seed contaminants (Wilson, 1926) are usually insufficient for rapid colonization and pioneer legumes require artificial inoculation to ensure successful establishment; for example, inoculation of lucerne is necessary in the U.K. (Thornton, 1931), clovers in Australia (Vincent, 1954) and soyabean in U.S.A. (Allen & Allen, 1958). The effect of successful pasture improvement helped by inoculation can be large. In southern Australia the value of the increment in soil nitrogen contributed by the legumes of top-dressed pastures between 1935 and 1960 is estimated at £$1 \cdot 6 \times 10^9$ (Donald, 1960). In spite of this success in practical field inoculation, very little has been done to study the establishment of the bacteria in the soil. This may be because no selective medium for counting nodule bacteria is known. Estimates have to be made indirectly using aseptically grown host plants which are then inoculated with dilutions made from the soil being examined. Using this technique, Wilson (1930, 1931) counted the two cross-inoculation group species, *Rhizobium trifolii* and *R. legumino-sarum*, in plots of long-term manurial experiments (excluding nitrogenous fertilizers) in three soils. All plots had previously been free from leguminous crops for at least ten years, and were sown in the test year with either peas, clover, cowpea, potatoes or oats. The numbers of nodule bacteria per g. dry wt. soil varied from < 1 to > 10^6. Fewest were found in plots without legumes or with low pH or low exchangeable Ca. Both kinds of nodule bacteria were stimulated least by cowpea and most by clover. Pea plants stimulated their own bacteria (*R. leguminosarum*) more than did clover, and vice versa; *R. trifolii* populations were generally larger than those of *R. leguminosarum*. Numbers were unaffected by a previous year's fallow and little influenced by mineral fertilizers. Numbers increased while the crop was growing and tended to decline after harvest. This study was not continued into the following year.

Similar results are reported by Krasilnikov (1958) who showed that *Rhizobium trifolii* in a podsol increased in 1 year under clover from a negligible number to 10^7/g. dry soil, whereas under lucerne, peas and wheat the maximum counts were 10^5/g. dry soil, and under corn 10/g. dry soil. Cotton, however, increased numbers of *R. trifolii* as much as did clover. Tsuzimura & Watanabe (1959) found that there were 10^5 nodule bacteria of *Astragalus sinicus* per g. dry soil under rotation, and only 10 per g. dry soil in neighbouring rice-wheat plots.

Hely, Bergersen & Brockwell (1957) also showed that nodule bacteria multiply rapidly in the rhizosphere of the inoculated legume, even when other soil conditions are initially unfavourable for establishment. They used large inocula to overcome the early rapid decline in the inoculum size, and populations in excess of 10^6 bacteria per cm. length of root were then established.

The influence of host is also shown by work (unpublished) on trefoil (*Medicago lupulina*) and beans (*Vicia faba*). Uninoculated trefoil seed was sown on light acid soil at Woburn that initially contained very few *Rhizobium meliloti* (< 1/g. dry soil). Six months after sowing, the rhizosphere population exceeded 10^5/g. dry soil, although the increase had come too late for the host to benefit from early effective nodulation. In neighbouring plots sown to rye-grass, the rhizosphere population remained at about 1/g. soil. On similar soil, trefoil in an established rotation produced rhizosphere counts often in excess of 10^7/g. dry soil.

At Rothamsted (heavy clay loam), *Rhizobium leguminosarum* was counted in early spring from plots: (i) fallow after cereal, not before sown to beans, (ii) 2nd year fallow after mixed arable, (iii) in cereal sown to spring beans the previous year, (iv) in winter beans after potatoes, and (v) in winter beans sown to spring beans the previous year. The numbers of bacteria per g. dry soil (or rhizosphere soil) were respectively: 20, $1\cdot0 \times 10^3$, $3\cdot7 \times 10^4$, $7\cdot5 \times 10^4$, and $1\cdot0 \times 10^5$, showing that the rhizobial population increased mostly in the first crop.

Studies in sand, water and agar culture (Krasilnikov, 1958) have with some exceptions, confirmed the general selectively stimulating effect of the host towards its own particular strain of nodule bacteria. Under conditions of restricted competition this effect may be much increased, as in Purchase & Nutman's (1957) experiments with clover and lucerne bacteria in mixed culture in the rhizosphere of clover plants grown in agar. In 3 weeks clover bacteria reached 10^5–10^6/ml.

rhizosphere independently of the size of the inoculum, which varied from < 1 to about 100 cells per ml. The lucerne bacteria inoculated at a constant 10^6/ml. did not increase significantly over the same period.

All these results suggest that under very different conditions the legume rhizosphere can support a mean maximum population of nodule bacteria of the order of 10^5–10^7 per g. of adhering soil. Rovira (1961) showed that nodule bacteria are stimulated 10–20 mm. from the root surface, so that where a legume is grown from season to season, as in a pasture, the whole soil in the root zone becomes rhizosphere. An acre under pasture to a depth of 6 in. may therefore contain as many as 10^{14} free-living nodule bacteria. Although these are large numbers they represent only a small proportion of the total bacterial population. In grass–clover associations giving a total bacterial count of between 10^7 and 10^{10}/g. dry rhizosphere soil, Rovira & Stern (1961) picked and characterized 273 isolates; none proved to be *Rhizobium*. Jagnow (1961) reports maximum rhizosphere populations for *Holcus lanatus* of 8×10^{12}/g. dry soil and suggests that many of the bacteria may occur as L-forms.

Nodules can arise from single initial infections (Hughes & Vincent, 1942) so that the host's need for such high densities of nodule bacteria in the rhizosphere is not clear, unless there is severe restriction on the number and siting of the susceptible root hairs. Purchase & Nutman (1957) showed that inoculum size up to about 10^4 per ml. of rhizosphere greatly affects the number of nodules formed on red clover seedlings grown in agar; above this the effect of inoculum size declined sharply. Bhaduri (1951) also found that the number of nodules formed on the roots of *Phaseolus radiatus* grown in water culture was only moderately increased by a large increase in inoculum size.

The level at which an increase in number of bacteria in the rhizosphere fails to increase nodule number may also be affected by the proportion of original infections that develop into nodules. This proportion differs greatly with species of host and strain of bacteria (Nutman, 1962; Lim, 1963). The full nodulating capacity of the host is satisfied by relatively few bacteria; low bacterial densities simply produce fewer surplus infections.

Even when sparse infection restricts the number of nodules formed, these fewer nodules grow to a large size and so maintain the same volume of nitrogen-fixing tissue. This self-compensatory mechanism appears to be host-controlled (Nutman, 1958).

Unless, therefore, the enormous rhizosphere populations of nodule bacteria perform some other useful function than as an inoculum for nodule formation, their value to the symbiosis may lie only in providing a reserve to survive unfavourable soil conditions, such as drought, which might cause a decline to a point where numbers would again severely limit infection.

A possible additional benefit of large numbers would be to increase in some way the availability to plants of the nutrients in the soil. Gerretsen (1948), Sperber (1958), Louw & Webley (1959 a, b) and Myskow (1961) have shown that the rhizosphere flora as a whole has an important effect on the solution and uptake by plants of mineral phosphate. Webley (personal communication) showed that nodule bacteria may take part in this process. Strains from three cross-inoculation groups produced clear haloes on plates containing dicalcium phosphate and released soluble phosphate from Gafsa rock phosphates. The amount of phosphate brought into solution was small compared to that released by organisms able to produce 2-ketogluconic acid, but may nevertheless be important because of the very large rhizosphere populations involved.

In established plant associations containing legumes, the bacteria released from decaying nodules also contribute to the soil population. The importance of this contribution compared to multiplication in the soil is not known, but for reasons already discussed, it is probably small.

Without legumes the soil population of nodule bacteria declines. *Rhizobium trifolii, R. leguminosarum* and *R. meliloti* have been counted in permanent clean fallows in light sandy soil (Woburn) and heavy clay (Rothamsted) over a period of several years. In both soils the relative rate of fall in numbers of each species during fallow was very similar and fairly constant, except for some fluctuations mainly in the first year, and the half-life of a population was about 35 days. The clay loam supported the larger populations. The clover bacteria were most abundant and the lucerne bacteria least abundant. It is not known to what extent the presence of non-legumes may arrest this decline. Survival in soil or peat culture is of practical interest because some commercial inoculants employ soil or peat carriers. The relationship between log number and age of such cultures also tends towards linearity, the death rate depending on temperature and rate of drying during storage (Vincent, 1958), and on soil type (Bonnier, 1955). Nodule bacteria are killed in culture at temperatures only a little above 40°, strains from tropical and temperate regions being about equally sensitive (Bowen & Kennedy, 1959). Rhizobium produces no spores but nevertheless some cells can survive in neutral soil under favourable conditions for many years. Sen & Sen (1956) record the survival of *Rhizobium japonicum* in air-dry soils for 19 years, and Jensen (1961) the survival of *R. meliloti* in sterilized soil culture for 35–40 years.

THE NATURE OF THE RHIZOSPHERE STIMULATION

Lochhead (1952), Starkey (1958) and Katznelson, Lochhead & Timonin (1948) have reviewed work on the effect of host and environment on the microbial population of rhizosphere and root surface. Legumes generally stimulate all groups of rhizosphere organisms more than plants of other families, but as we have seen, a given legume tends to promote the multiplication of bacteria able to infect it more than others. This specific stimulation is in the rhizosphere proper and not on the root surface which seems to harbour rather few nodule bacteria (Sperber & Rovira, 1959). The immediate neighbourhood of the young seedling may indeed be somewhat unfavourable for rhizobia; seeds of both *Trifolium subterraneum* and *Centrosema pubescens* produce diffusable antibiotics against bacteria, including rhizobia (Thompson, 1960; Bowen, 1961).

Formidable difficulties face the analysis of rhizosphere stimulation in terms of the chemistry of root exudations. The mechanism of exudation itself is obscure. Kurosumi (1961), reviewing secretion in general, proposes that substances may be extruded from tissues either by whole cell loss, by pinching-off of macro- or micro-villi, by vacuolar opening or by diffusion through intact membranes. Tesar & Kutacek (1955) conclude, from comparing the amino acid contents of roots and exudates, that the mechanism is exosmosis; Rovira (1959) presents data against this view. Direct observation of roots in soil show that root hairs may actively guttate under certain conditions (Rogers, 1939). The amount of material that comes into the rhizosphere from cells ruptured by lateral roots emerging, from the short-lived separated root-cap cells, or from other causes, is not known; nor can the contribution from plasmotypsed (burst) hairs be estimated. Hairs burst spontaneously (Nutman, 1959), but the number that do so is increased by change in pH, presence of narcotics, etc. (Strugger, 1926; Ekdahl, 1953). There may also be an effect from antibiotics produced by soil organisms, for Norman (1955) found that these cause exosmosis from plant roots. All such mechanisms and probably many others may be involved in the complex and changing environment of the soil and rhizosphere.

Because root exudates are rapidly degraded in soil by microbial activity, critical chemical work on them is not possible. Instead root exudates from plants grown in aseptic culture have been studied, although this has its own drawbacks. Many different substances have been shown to be lost from roots in small amounts; these include carbohydrates, organic acids, amino acids, proteins, enzymes, and accessory growth factors (Lundegårdh & Stenlid, 1944; Ratner & Samoilova, 1955; Tesar & Kutacek, 1955; Rovira, 1959; Rovira & Harris, 1961). The quantities reported differ between excreting plants and with the experimental conditions.

None of this work has provided an explanation for the fact that some microbes are preferentially stimulated in the legume rhizosphere and *Rhizobium* more than most. The larger stimulatory effect of legumes compared with that of other plants is not correlated with a more copious exudation (Rovira, 1959). Some of the rhizosphere bacteria are characterized by their ability to use simple carbohydrates (Brisbane & Rovira, 1961) and their need for amino acids, but these are not criteria of *Rhizobium* (Bergersen, 1961 *b*). Even the known growth-factor requirements of *Rhizobium* (biotin by most strains and thiamine by some) are not better provided by legumes. Biotin seems to be excreted by all plants and more by some non-legumes than by legumes (Rovira & Harris, 1961).

Extrapolation from aseptic culture to soil is such a large step that it is perhaps understandable that work on root exudation has not explained the great increase of *Rhizobium* in the rhizosphere, especially when it is known that even a population of *Rhizobium* of 10^7/g. soil may be only a small proportion of the total microbial population. Whatever may be the exact relationship between stimulatory exudates, the environment (including its biological interactions) and the response of *Rhizobium*, the free-living nodule bacteria undoubtedly derive large ecological advantage from the presence of the legume host. So far as is known the advantage to the host of this increased population is only indirect in providing a reservoir for infection.

EVOLUTION OF LEGUME SYMBIOSIS

There is no fossil record of nodules to indicate the way in which the legume symbiosis may have arisen. The oldest fossil legumes would be classified today in the Eucaesalpiniodeae, which contains both nodulating and non-nodulating species, suggesting that the nodulating habit arose in the early development of the family. Except for specially bred lines of resistant red clover (Nutman, 1949) and soya bean (Williams & Lynch, 1954), the non-nodulating habit is restricted to the subfamilies Caesalpinioidea and Mimosoidea which, for floristic reasons, are thought to be phylogenetically more primitive than the uniformly nodulating Papilionateae. An early origin is also indicated by the fact that some of the families of nodulating non-legumes have some affinity with the Leguminosae in Hutchinson's (1959) phylogenetic arrangement of the flowering plants. Most root-nodulating plants may therefore have arisen from some common stock possibly represented today by the Dilleniales. Norris (1959) has pointed out that the less specialized symbioses are to be found among little-studied tropical calcifuge hosts, which are botanically more primitive than the more studied calcicole hosts of temperate zones; this may have biased our ideas on nodule evolution. Certainly more work should be undertaken on the primitive legumes, particularly those which still retain the regular flower structure of their near relatives and are woody in habit. Bond, MacConnell & McCallum (1956) suggested that the original nodulated plants (legumes and others) were woody rather than herbaceous. Manil (1958) put forward the novel idea that the individuality of the Leguminosae may be determined by the symbiosis.

The features of the primordial legume symbiosis can only be guessed. Parker (1957) suggests that symbiosis began as a simple association between a free-living nitrogen-fixing bacterium and the plant root, entry into the plant and the differentiation of the nodule following at a later stage. That the bacteroid has now been shown to be intercellular rather than intracellular may support this view, but it is difficult to see why a closer association with the host should have led always to the bacteria losing its independent ability to fix nitrogen, unless the appropriate enzyme systems were transferred in some way *in toto* to the plant membrane. None of the free-living nitrogen-fixing bacteria much resembles *Rhizobium*.

Gäumann's (1950) theory of origin through parasitism begs the question and raises the difficulty of intermediate stages. Many kinds of so-called parasitic (ineffective) nodule symbioses are known but these are generally structurally complete and are often defective in quite simple ways; they fit in better as degenerate forms.

A third possible mode of origin would be to suppose that the bacterium's capacity to induce nitrogen fixation in the host was acquired at a very early stage, possibly even as a first step. Such an origin would explain the restriction of bacterial root nodules to a single host family, the non-legume root-nodule symbioses with other micro-organisms possibly having arisen similarly, at about the same time. A further point in its favour is that it avoids the serious difficulty raised by the non-viability of the bacteroid cell—upon which all the adaptive advantages of symbiosis depends. The bacteroid cell cannot itself transmit any property to a succeeding generation; all inheritance of symbiotic factors is collateral.

Leaving aside the question of origin, the great variation in symbiosis, summarized in a preceding section, lends itself to evolutionary speculation. But the uncertainties here are so great that the detailed elaboration and discussion of possible evolutionary sequences cannot be undertaken without departing from the frugal, matter-of-fact style allowed by the Society's editors. Our story of symbiosis should begin 'Once upon a time' and end 'and the plant and bacteria lived happily ever after'; even these cardinal points of a good fairy story are denied by the facts.

REFERENCES

ALLEN, E. K. & ALLEN, O. N. (1958). Biological aspects of symbiotic nitrogen fixation. *Encycl. Pl. Physiol.* **8**, 43. Berlin: Springer Verlag.

ANDERSON, A. J. (1956). The role of molybdenum in plant nutrition. *Symp. Inorganic Nitrogen Metabolism*, no. 3. Ed. W. D. McElroy and Bentley Glass. Baltimore: Johns Hopkins Press.

APPLEBY, C. A. & BERGERSEN, F. J. (1958). Cytochromes of Rhizobium. *Nature, Lond.* **182**, 1174.

APRISON, M. H. & BURRIS, R. H. (1952). Time course of fixation of N_2 by excised soybean nodules. *Science*, **115**, 264.

BALASSA, R. (1957). Durch Deoxyribonukleinsäuren induzierte Veränderungen bei Rhizobien. *Acta microbiol. Acad. Sci. Hungaricae*, **4**, 77.

BALASSA, R. (1960). Transformation of strain of *Rhizobium lupini*. *Nature, Lond.* **188**, 246.

BERGERSEN, F. J. (1955). The cytology of bacteroids from root nodules of subterranean clover (*Trifolium subterraneum* L.). *J. gen. Microbiol.* **13**, 411.

BERGERSEN, F. J. (1957). The occurrence of a previously unobserved polysaccharide in immature infected cells of root nodules of *Trifolium ambiguum* M. Bieb. and other members of the Trifolieae. *Aust. J. biol. Sci.* **10**, 7.

BERGERSEN, F. J. (1958). The bacterial component of soybean root nodules; changes in respiratory activity, cell dry weight and nucleic acid content with increasing nodule age. *J. gen. Microbiol.* **19**, 312.

BERGERSEN, F. J. (1960). Biochemical pathways in legume root nodule nitrogen fixation. *Bact. Rev.* **24**, 246.

BERGERSEN, F. J. (1961a). Nitrate reductase in soybean root nodules. *Biochim. biophys. Acta*, **52**, 206.

BERGERSEN, F. J. (1961b). The growth of Rhizobium in synthetic media. *Aust. J. biol. Sci.* **14**, 349.

BERGERSEN, F. J. (1961c). Haemoglobin content of legume root nodules. *Biochim. biophys. Acta*, **50**, 576.

BERGERSEN, F. J. (1962). Oxygenation of leghaemoglobin in soybean root nodules in relation to external oxygen tension. *Nature, Lond.* **194**, 1059.

BERGERSEN, F. J. & BRIGGS, M. J. (1958). Studies on the bacterial component of soyabean root nodules: cytology and organization of the host tissue. *J. gen. Microbiol.* **19**, 482.

BERGERSEN, F. J. & WILSON, P. W. (1959a). The location of newly fixed nitrogen in soy nodules. *Bact. Proc. Amer. Soc. Bact.* **59**, 25.

BERGERSEN, F. J. & WILSON, P. W. (1959b). Spectrophotometric studies of the effects of nitrogen on soybean nodule extracts. *Proc. nat. Acad. Sci., Wash.* **45**, 1641.

BERNARD, N. (1909). L'évolution dans la symbiose. *Ann. sci. nat. Bot.* **9**, 1.

BHADURI, S. N. (1951). The influence of the number of Rhizobium supplied on the subsequent nodulation of the legume host plant. *Ann. Bot., Lond.* **15**, 209.

BOND, G., MacCONNELL, J. T. & McCALLUM, A. H. (1956). The nitrogen-nutrition of *Hippophae rhamnoides* L. *Ann. Bot., Lond.*, N.S. **20**, 501.

BONNIER, C. (1955). La conservation du *Rhizobium* en sols stériles. *Bull. Inst. agron. Gembl.* **23**, 359.

BOWEN, G. D. (1961). The toxicity of legume seed diffusates towards rhizobia and other bacteria. *Plant & Soil,* **15**, 155.

BOWEN, G. D. & KENNEDY, M. M. (1959). Effect of high soil temperature on *Rhizobium* spp. *Qd J. agric. Sci.* **16**, 177.

BOYDEN, S. V. (1960). Antibody production. *Nature, Lond.* **185**, 724.

BRISBANE, P. G. & ROVIRA, A. D. (1961). A comparison of methods for classifying rhizosphere bacteria. *J. gen. Microbiol.* **26**, 379.

BURNET, F. M. (1960). Immunity as an aspect of general biology. *Proc. Symp. Mechanisms of Antibody Formation.* Czechoslovak Acad. Sci. Prague.

BUSH, G. L. & CHAPMAN, G. B. (1961). Electron microscopy of symbiotic bacteria in developing oocytes of the American cockroach *Periplaneta americana. J. Bact.* **81**, 267.

BUVAT, R. (1958). Recherches sur les infrastructures du cytoplasme, dans les cellules du meristème apical, des ébauches foliaires et des feuilles developpées d'*Elodea canadensis. Ann. sci. Nat. Bot.* (IIe Ser.), **19**, 121.

CHEN, H. K. & THORNTON, H. G. (1940). The structure of ineffective nodules and its effect on nitrogen fixation. *Proc. Roy. Soc.* B, **127**, 208.

COLWIN, A. L. & COLWIN, L. H. (1961a). Fine structure of the spermatozoon of *Hydroides hexagonus* (Annilida), with special reference to the acrosomal region. *J. biophys. biochem. Cytol.* **10**, 211.

COLWIN, L. H. & COLWIN, A. L. (1961b). Changes in the spermatozoon during fertilization in *Hydroides hexagonus* (Annilida). I. Passage of the acrosomal region through the vitilline membrane. *J. biophys. biochem. Cytol.* **10**, 231.

COLWIN, A. L. & COLWIN, L. H. (1961c). Changes in the spermatozoon during fertilization in *Hydroides hexagonus* (Annilida). II. Incorporation with the egg. *J. biophys. biochem. Cytol.* **10**, 255.

DONALD, C. M. (1960). The impact of cheap nitrogen. *J. Aust. Inst. Agric. Sci.* **26**, 319.

EKDAHL, I. (1953). Studies on the growth and osmotic condition of root hairs. *Symb. bot. upsaliens,* **11**, 5.

FÅHRAEUS, G. (1957). The infection of clover root hairs by nodule bacteria, studied by a simple glass slide technique. *J. gen. Microbiol.* **16**, 374.

FÅHRAEUS, G. & LJUNGGREN, H. (1959). The possible significance of pectic enzymes in root-hair infection by nodule bacteria. *Physiol. Plant.* **12**, 145.

FALK, J. E., APPLEBY, C. A. & PORRA, R. J. (1959). The nature, function and biosynthesis of the haem compounds and porphyrins of legume root nodules. *Symp. Soc. exp. Biol.* **12**, 73.

FRED, E. B., BALDWIN, I. L. & McCOY, E. (1932). *Root Nodule Bacteria and Leguminous Plants.* University of Wisconsin, Madison.

GÄUMANN, E. (1950). *Principles of Plant Infection.* English ed. by W. B. Brierley. London: Crosby Lockwood.

GERRETSEN, F. C. (1948). The influence of microorganisms on the phosphate intake by the plant. *Plant & Soil,* **1**, 51.

GIER, H. T. (1947). Intracellular bacteroids in the cockroach *Periplaneta americana* (Linn.). *J. Bact.* **53**, 173.

HALLSWORTH, E. G. (1958). *Nutrition of the Legumes.* London: Butterworths Scientific Publications.

HAMILTON, P. B., SHUG, A. L. & WILSON, P. W. (1957). Spectrophotometric examination of hydrogenase and nitrogenase in soybean nodules and Azotobacter. *Proc. nat. Acad. Sci., Wash.* **43**, 297.

HELY, F. W., BERGERSEN, F. J. & BROCKWELL, J. (1957). Microbial antagonism in the rhizosphere as a factor in the failure of inoculation of subterranean clover. *Aust. J. agric. Res.* **8**, 24.

HOLLAND, J. J. & PICKETT, M. J. (1956). Intracellular behaviour of Brucella variants in chick embryo cells in tissue culture. *Proc. Soc. exp. Biol., N.Y.* **93**, 476.

HOLTER, H. (1959). Pinocytosis. *Int. Rev. Cytol.* **8**, 481.

HOWE, C. (1951). The influenza virus receptor and blood group antigens of human erythrocyte stroma. *J. Immunol.* **66**, 9.

HUGHES, D. Q. & VINCENT, J. M. (1942). Serological studies of the root nodule bacteria. III. Tests of neighbouring strains of the same species. *Proc. Linn. Soc. N.S.W.* **67**, 142.

HUTCHINSON, J. (1959). *The Families of Flowering Plants*, Vol. I. *Dicotyledons*, 2nd ed. Oxford: Clarendon Press.

IMAEDA, T. & CONVIT, J. (1962). Electron microscope study of *Mycobacterium leprae* and its environment in a vesicular leprous lesion. *J. Bact.* **83**, 43.

JAGNOW, G. (1961). Numbers and types of bacteria from the rhizosphere of pasture plants: possible occurrence of L-forms. *Nature, Lond.* **191**, 1220.

JENSEN, H. L. (1961). Survival of *Rhizobium meliloti* in soil culture. *Nature, Lond.* **192**, 682.

JENSEN, W. A. & McLAREN, A. D. (1960). Uptake of proteins by plant cells—the possible occurrence of pinocytosis in plants. *Exp. Cell Res.* **19**, 414.

KATZNELSON, H., LOCHHEAD, A. G. & TIMONIN, M. I. (1948). Soil micro-organisms and the rhizosphere. *Bot. Rev.* **14**, 543.

KEFFORD, N. P., BROCKWELL, J. & ZWAR, J. A. (1960). The symbiotic synthesis of auxin by legumes and nodule bacteria and its role in nodule development. *Aust. J. biol. Sci.* **13**, 456.

KRASILNIKOV, N. A. (1958). *Soil Microorganisms and Higher Plants.* (Acad. Sci. U.S.S.R. Moscow. English ed. National Science Foundation.)

KUROSUMI, K. (1961). Electron microscopic analysis of the secretion mechanism. *Int. Rev. Cytol.* **11**, 1.

LANGE, R. T. & ALEXANDER, M. (1961). Anomalous infections by Rhizobium. *Canad. J. Microbiol.* **7**, 959.

LEWIS, W. H. (1931). Pinocytosis. *Johns Hopk. Hosp. Bull.* **49**, 17.

LIM, G. (1963). Studies on the physiology of nodule formation. VIII. The influence of the size of the rhizosphere population of nodule bacteria on root hair infection in clover. *Ann. Bot.* (in the Press).

LJUNGGREN, H. (1961). Transfer of virulence in *Rhizobium trifolii. Nature, Lond.* **191**, 623.

LJUNGGREN, H. & FÅHRAEUS, G. (1961). The role of polygalacturonase in root-hair invasion by nodule bacteria. *J. gen. Microbiol.* **26**, 521.

LOCHHEAD, A. G. (1952). The nutritional classification of soil bacteria. *Proc. Soc. appl. Bact.* **15**, 15.

LOUW, H. A. & WEBLEY, D. M. (1959a). The bacteriology of the root region of the oat plant grown under controlled pot culture conditions. *J. appl. Bact.* **22**, 216.

LOUW, H. A. & WEBLEY, D. M. (1959b). A study of soil bacteria dissolving certain mineral phosphate fertilizers and related compounds. *J. appl. Bact.* **22**, 227.

LUNDEGÅRDH, H. & STENLID, G. (1944). On the exudation of nucleotides and flavanone from living roots. *Ark. Bot.* **31**A, 1.

McLAREN, A. D., JENSEN, W. A. & JACOBSON, L. (1960). Absorption of enzymes and other proteins by barley roots. *Plant Physiol.* **35**, 549.

MAGEE, W. E. & BURRIS, R. H. (1954). Fixation of N_2^{15} by excised nodules. *Plant Physiol.* **29**, 199.

MANIL, P. (1958). The legume–rhizobia symbiosis. In *Nutrition of the Legumes*, p. 124. Ed. E. G. Hallsworth. London: Butterworths Scientific Publications.

MASEFIELD, G. B. (1957). The nodulation of annual leguminous crops in Malaya. *Emp. J. exp. Agric.* **25**, 139.

MYSKOW, W. (1961). The occurrence of microorganisms solubilizing phosphorus in the rhizosphere of some crop plants. *Acta Microbiol. Polonica*, **10**, 93.

NORMAN, A. G. (1955). The effect of polymixin on plant roots. *Arch. Biochem.* **58**, 461.

NORRIS, D. O. (1959). Legume bacteriology in the tropics. *J. Aust. Inst. agric. Sci.* **25**, 202.

NUTMAN, P. S. (1949). Nuclear and cytoplasmic inheritance of resistance to infection by nodule bacteria in red clover. *Heredity*, **3**, 263.

NUTMAN, P. S. (1956). The influence of the legume in root-nodule symbiosis. A comparative study of host determinants and functions. *Biol. Rev.* **31**, 109.

NUTMAN, P. S. (1958). The physiology of nodule formation. In *Nutrition of the Legumes*, p. 87. Ed. E. G. Hallsworth. London: Butterworths Scientific Publications.

NUTMAN, P. S. (1959a). Some observations on root-hair infection by nodule bacteria. *J. exp. Bot.* **10**, 250.

NUTMAN, P. S. (1959b). Sources of incompatibility affecting nitrogen fixation in legume symbiosis. *Symp. Soc. exp. Biol.* **13**, 42.

NUTMAN, P. S. (1962). The relation between root hair infection by *Rhizobium* and nodulation in *Trifolium* and *Vicia*. *Proc. Roy. Soc.* B, **156**, 122.

PARKER, C. A. (1954). The effect of oxygen on the fixation of nitrogen by *Azotobacter*. *Nature, Lond.* **173**, 780.

PARKER, C. A. (1957). Evolution of nitrogen-fixing symbiosis in higher plants. *Nature, Lond.* **179**, 593.

PETROSYAN, A. P. (1959). *Ecological Characteristics of the Nodule Bacteria in the Armenian S.S.R.* [in Russian]. Erevan Inst. Microbiol. Armenia S.S.R.

PUCK, T. T. & LEE, H. H. (1955). Mechanisms of cell wall penetration of viruses. II. Demonstration of cyclic permeability change accompanying virus infection of *Escherichia coli* B cells. *J. exp. Med.* **101**, 151.

PURCHASE, H. F. & NUTMAN, P. S. (1957). Studies on the physiology of nodule formation. VI. The influence of bacterial numbers in the rhizosphere on nodule initiation. *Ann. Bot., Lond.*, N.S. **21**, 439.

RATNER, E. I. & SAMOILOVA, S. A. (1955). Extracellular phosphatase activity of roots. (In Russian.) *Fiz. Rast.* **2**, 30.

RICEMAN, D. S. & JONES, J. B. (1960). Distribution of zinc in subterranean clover (*Trifolium subterraneum* L.) during the onset of zinc deficiency as determined by the use of the radioactive isotope ^{65}Zn. *Aust. J. agric. Res.* **11**, 162.

RISTIC, M. (1960). Structural characterization of *Anaplasma marginale* in acute and carrier infections. *J. Amer. Vet. Med. Ass.* **136**, 417.

ROGERS, W. S. (1939). Root studies. 8. Apple root growth in relation to root stock, soil, seasonal and climatic factors. *J. Pomol.* **17**, 99.

ROSHDY, M. A. (1961). Observations by electron microscopy and other methods on the intracellular rickettsia-like microorganisms in *Argas persicus* Oken (Ixodoidea, Argasidae). *J. Insect Path.* **3**, 148.

ROVIRA, A. D. (1959). Root excretions in relation to the rhizosphere effect. IV. Influence of plant species, age of plant, light, temperature and calcium nutrition on exudation. *Plant & Soil*, **11**, 53.

ROVIRA, A. D. (1961). Rhizobium numbers in the rhizospheres of red clover and paspalum in relation to soil treatment and numbers of bacteria and fungi. *Aust. J. agric. Res.* **12**, 77.

ROVIRA, A. D. & HARRIS, J. R. (1961). Plant root excretions in relation to the rhizosphere effect. V. The exudation of B-group vitamins. *Plant & Soil*, **14**, 199.

ROVIRA, A. D. & STERN, W. R. (1961). The rhizosphere bacteria in grass–clover associations. *Aust. J. agric. Res.* **12**, 1108.

SEN, A. & SEN, A. N. (1956). Survival of *Rhizobium japonicum* in stored air-dry soils. *J. Indian Soc. Soil Sci.* **4**, 215.

SHAUKAT-AHMED & EVANS, H. J. (1961). The essentiality of cobalt for soybean plants grown under symbiotic conditions. *Proc. nat. Acad. Sci., Wash.* **47**, 24.

SPERBER, J. I. (1958). The incidence of apatite-solubilizing organisms in the rhizosphere and soil. *Aust. J. agric. Res.* **9**, 778.

SPERBER, J. I. & ROVIRA, A. D. (1959). A study of the bacteria associated with the roots of subterranean clover and Wimmera rye-grass. *J. appl. Bact.* **22**, 85.

STARKEY, R. L. (1958). Interrelations between microorganisms and plant roots in the rhizosphere. *Bact. Rev.* **22**, 154.

STRUGGER, S. (1926). Untersuchungen über den Einfluss der Wasserstoffionen auf das Protoplasma der Wurzelhaare von *Hordeum vulgare* L. *S.B. Akad. Wiss. Wien* (Abt. 1), **135**, 453.

SUGANUMA, A. (1961). The plasma-membrane of *Staphylococcus aureus*. *J. biophys. biochem. Cytol.* **10**, 292.

TESAR, S. & KUTACEK, M. (1955). Root excretions of higher plants. 1. Excretion of amino-acids by the roots of wheat in water culture. *Ann. Czechoslov. Acad. agric. Sci.* **28**, 927.

THOMPSON, J. A. (1960). Inhibition of nodule bacteria by an antibiotic from legume seed coats. *Nature, Lond.* **187**, 619.

THORNTON, H. G. (1931). Lucerne 'inoculation' and the factors affecting its success. *Imp. Bur. Soil Sci. Tech. Comm.* no. 20. London: H.M. Stationery Office.

TRAGER, W. (1960). Intracellular parasitism and symbiosis. In *The Cell*, **4**, 152. Ed. J. Brachet and A. E. Mirsky. New York and London: Academic Press.

TSUZIMURA, K. & WATANABE, I. (1959). Estimation of numbers of root-nodule bacteria by the nodulation-dilution frequency method and some applications. *J. Sci. Soil Tokyo*, **30**, 292.

TURCHIN, F. V. (1956). The role of mineral and biological nitrogen fixation in the agriculture of the U.S.S.R. *Pochvovedenie*, **6**, 15.

VINCENT, J. M. (1954). The root-nodule bacteria of pasture legumes. *Proc. Linn. Soc. N.S.W.* **79**, 1.

VINCENT, J. M. (1958). Survival of root nodule bacteria. In *Nutrition of the Legumes*, p. 108. Ed. E. G. Hallsworth. London: Butterworths Scientific Publications.

WEILING, F. (1961). Pinocytose bei Pflanzen. *Naturwissenschaften*, **48**, 411.

WILLIAMS L. F. & LYNCH D. L. (1954). Inheritance of a non-nodulating character in the soybean. *Agron. J.* **46**, 28.

WILSON J. K. (1926). Legume bacteria population of the soil. *J. Amer. Soc. Agron.* **21**, 810.

WILSON, J. K. (1930). Seasonal variation in the numbers of two species of Rhizobium in soil. *Soil Sci.* **30**, 289.

WILSON, J. K. (1931). Relative numbers of two species of Rhizobium in soils. *J. agric. Res.* **43**, 261.

WILSON, P. W. & BURRIS, R. H. (1953). Biological nitrogen fixation—a reappraisal. *Ann. Rev. Microbiol.* **7**, 415.

THE ROOT NODULES OF
NON-LEGUMINOUS ANGIOSPERMS

G. BOND

Department of Botany, University of Glasgow

According to present knowledge there are nine genera of non-leguminous angiosperms which bear structures accepted as root nodules.* Recent work indicates that these nodules are very like those of legumes in their physiological properties; see reviews by Bond (1958 *a*, 1959), Allen & Allen (1958) and Schwartz (1959). In the present account particular attention will be paid to those aspects which seemed most relevant to the general subject of the symposium.

TAXONOMY AND OCCURRENCE OF THE NODULE-BEARING PLANTS

Table 1 lists the genera of these nodule-bearing plants (all dicotyledons), and shows the number of species and their present distribution.

Table 1. *The size and present distribution of the non-legume nodule-bearing genera*

Genus	Number of species	Present distribution
Coriaria	10	Mediterranean to Japan, New Zealand, Chile to Mexico
Myrica	45	In many tropical, subtropical and temperate regions, extending in latter nearly to Arctic Circle
Alnus	25	Europe, Siberia, North America, nearly to Arctic Circle; Japan, Andes
Casuarina	35	Australia, tropical Asia, Pacific Islands. Widely introduced elsewhere
Elaeagnus	30	Asia, Europe, North America
Hippophaë	2	Asia, Europe, from Himalayas to Arctic Circle
Shepherdia	3	North America
Ceanothus	40	North America
Discaria	15	Andes, Brazil, New Zealand, Australia

Whilst the occurrence of nodules in some of the genera has been on record in botanical literature for about 100 years, the record for *Discaria* is recent (Morrison & Harris, 1958). There is an obvious possibility that additional non-legume nodulating genera will be found.

* The nodular growths of uncertain affinities first noted on roots of *Dryas Drummondii* by Lawrence (1953), and those reported on the roots of some Zygophyllaceae (Allen & Allen, 1958) will not be considered in this review.

The nodule-bearing habit has by no means been confirmed for all the species included in these genera (Allen & Allen, 1958). Information is particularly lacking on the tropical and subtropical species of *Myrica*. The only species of these genera reported not to bear nodules in its natural habitat is *Coriaria sarmentosa* (Morrison & Harris, 1959).

It is clear from Table 1 that several of the genera are very widely distributed. This has been the subject of comment, particularly in the cases of *Hippophaë*, *Myrica* and *Coriaria*, by writers who were unaware that there were any unusual features in the nutrition of these plants. The group is represented in Britain by *Alnus glutinosa*, *Myrica gale* and *Hippophaë rhamnoides*, the latter now mainly confined to coastal situations, though the pollen record shows that formerly the species occurred inland also.

Table 2. *Classification of non-legume nodule-bearing genera according to Hutchinson (1959)*

Genus	Family and number in classification	Order
Coriaria	Coriariaceae (23)	Coriariales
Myrica	Myricaceae (59)	Myricales
Alnus	Betulaceae (61)	Fagales
Casuarina	Casuarinaceae (67)	Casuarinales
Elaeagnus ⎫ *Hippophaë* ⎬ Elaeagnaceae (200) ⎫ *Shepherdia* ⎭		
Ceanothus ⎱ Rhamnaceae (201) ⎭ *Discaria* ⎰		Rhamnales

While the other great group of root-nodulating plants, the legumes, comprise a single natural order, the Leguminales, the non-legume nodulating genera are not so closely related. In Hutchinson's (1959) classification, however, which claims to show phylogenetic relationships of plants, they are ascribed somewhat greater affinity (Table 2) than in older classifications. It is safe to assume that no attention was paid to nodulating habits in framing the classification. *Myrica* (if taken to include *Comptonia*), *Coriaria* and *Casuarina* are each the only genus in their families. *Betula*, the other genus of the Betulaceae, is not nodule-bearing. The Rhamnaceae includes some forty genera of which only *Ceanothus* and *Discaria* are recorded as bearing nodules. The three genera *Elaeagnus*, *Hippophaë* and *Shepherdia* compose the family Elaeagnaceae, which is thus wholly nodulating, at least as regards genera. Hutchinson remarks of the Coriariaceae that 'it is a very distinct family difficult to place satisfactorily' and that it is 'doubtfully arranged here'. The families containing *Myrica*, *Alnus* and *Casuarina* show fairly close affinity, and the Elaeagnaceae and Rhamnaceae are recognized as closely related.

NODULE STRUCTURE

In early stages of development the nodules of these plants form lateral swellings on the roots (Pl. 1, fig. 1) and show a fairly close external resemblance to those of legumes. Under cultural conditions the young nodules of *Alnus*, *Myrica* and *Casuarina* are often intensely red due to an anthocyanin-type pigment (Bond, 1951); those of the Elaeagnaceae are in the writer's experience always white. A few weeks after their first appearance the original, simple nodules produce new lobes at the apex, as shown in Fig. 1. The frequent repetition of this branching, together with the perennial habit of most of these nodules, results in the eventual formation, possibly from a single original nodule, of a nodule cluster which in *Alnus*, *Casuarina* and *Ceanothus* may be several centimetres in diameter (Pl. 1, fig. 2). In *Myrica* and *Casuarina* an

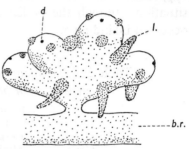

Fig. 1. Diagram of branching nodule of *Alnus glutinosa* grown in water culture. *b.r.* = bearing root; *d.* = small depression over growing point; *l.* = lenticel. (× 15.)

additional feature is that the apex of each nodule lobe gives rise to a normal root, so that the nodule cluster becomes clothed in rootlets which under cultural conditions show upward growth (Bond, 1952, 1956*a*).

The following account of the internal structure of the nodules is based on the work of various authors whose papers are in most cases cited by Hawker & Fraymouth (1951) or by Allen & Allen (1958), and on the writer's own observations. It is often stated that the nodule shows an obviously root-like structure, but it should be noted that there are certain anomalous features including the lack of a root-cap, the absence of root hairs and the development of a superficial cork, while the origin of the nodule may differ somewhat from that of a root (p. 79). The nodule shows a central stele bounded by an endodermis often containing tannin and soon becoming suberized. The cortex is enlarged owing to both hyperplasia and hypertrophy and it is here that the intracellular endophyte is found. Only some of the cortical cells are infected, the smaller uninfected ones often containing starch or tannin. The nodule is bounded by a few layers of cork cells, and there is an apical meristematic region which remains uninfected. Some of these features are illustrated in Pl. 2, fig. 3.

In some species, e.g. *Myrica gale*, the infected cells are scattered through most of the cortex, while in *Myrica cerifera* and *Ceanothus* they have been described as confined to a particular zone in the middle or outer cortex. In *Alnus* the infected cells tend to form concentric zones alternating with uninfected cells, a feature suggested near the apex in Pl. 2, fig. 3. In *Coriaria* the nodule is asymmetric, the infected and hypertrophied cells of the cortex lying to one side so that the stele is no longer central.

From an early stage in nodule development practically the whole organ is encased in the cork layer already mentioned, only the tip being free. In *Alnus* the cork layer is interrupted by frequent lenticels, the complementary tissue of which becomes hypertrophied in water culture, so that prominent conical structures arise (Fig. 1). The nodule roots formed in *Myrica* and *Casuarina* remain free of infection and show a normal root structure (Bond, 1952).

The method of branching of the nodule needs further study. Present information suggests that in certain genera (*Myrica*, *Ceanothus*) the new lobes arise endogenously within the stele of the original lobe, while in others (*Alnus*, *Casuarina*, *Elaeagnus*) a division of the apical meristem is involved.

NODULE CYTOLOGY AND IDENTITY OF ENDOPHYTES

The identity of the endophytes in non-leguminous nodules has been the subject of sporadic investigation for more than 100 years but is still in doubt, because difficulties in the isolation of the endophytes (see p. 76) leave the study of nodule cytology as the main source of evidence on identity. Conclusions drawn from such study have been conflicting, doubtless because the congested condition of the infected cell contents renders elucidation difficult. Most investigators concluded that the endophytes have a hyphal structure, and on account of hyphal dimensions and other features the organisms have frequently been assigned to the actinomycetes, as for example by Shibata (1902), Shibata & Tahara (1917), Schaede (1948), and Fletcher (1955). Hawker & Fraymouth (1951), on the other hand, failed to detect definite hyphae, and held the endophytes to be members of the Plasmodiophorales. The strands observed within and between cells were interpreted as protoplasmic extensions of an endophytic plasmodium. Young infected cells were thought to be filled by an undifferentiated plasmodium which was held to have absorbed the host protoplast. Taubert (1956) noted plasmodium-like structures in recently infected cells of *Alnus*, but thought that they might really consist of coagulated host cytoplasm; in older cells he saw hyphae.

The most prominent endophytic structures in the nodule cells are the vesicles ('Bläschen') which are roughly spherical in *Alnus*, *Hippophaë*, *Elaeagnus* and *Ceanothus*, but club-shaped in *Myrica* and *Coriaria*. They are attached to hyphae or strands of protoplasm and are usually disposed peripherally in the cell (Pl. 2, fig. 4). The significance of these bodies has been left in doubt by most authors, but Hawker & Fraymouth (1951) held them to be sporangia, the contents of which form motile particles that migrate into adjacent cells and spaces and eventually into the soil.

Apart from certain infected cells containing persistent 'bacteroids', the cells eventually undergo a marked loss of content (Pl. 2, fig. 3) due in many authors' view to a digestion of the endophyte, including the vesicles, by the host cells. The latter have been found to be still living with intact nuclei, and in Schaede's work (1948) this was supported by photographic evidence.

ATTEMPTS TO ISOLATE THE ENDOPHYTES

Cultural experience shows that the endophytes of these nodules are not seed-borne, so that plants of each generation must be infected afresh from the soil; this implies that the organisms can at least survive in soils, apart from the host plants, though not necessarily that they actually grow under such conditions.

Considerable effort has gone into attempts to grow the endophytes in pure culture. The nodules, after surface sterilization, followed in some cases by excision of the outer tissues as a further precaution against contaminants, have usually been crushed in sterile water and the resulting suspension plated out. Sections of nodules have also been employed, while other workers have picked out endophytic structures with a micromanipulator. A wide variety of media and cultural conditions have been employed.

Of the various organisms obtained after such procedures some have been obvious contaminants while others have been held to show resemblance to the endophytes. Among the many workers who obtained negative results in re-inoculation tests with isolates of the latter type are Lieske (1921), Krebber (1932), Bouwens (1943), Quispel (1954a) and Pommer (1956), all using *Alnus*, Fletcher (1955) with *Myrica*, and Uemura (1952, 1961) with several genera. The organisms isolated and tested by these workers were mostly actinomycetes. Great persistence has been shown by some of these investigators, particularly Uemura, who has tested some hundreds of isolates of actinomycetes and strepto-mycetes. Quispel (1960) believes that the endophytes may have special nutrient requirements. Using a technique in which crushed-nodule inoculum from surface-sterilized alder nodules was added to various nutrient media, he found that although typically no visible growth of endophyte occurred, in certain media definite increase in the number of infective particles sometimes took place during incubation, as judged by the number of nodules produced by the incubated inoculum on alder test plants.

In contrast, other authors have claimed successful re-inoculation with isolates which grew readily on standard media. Peklo (1910) claimed this for an isolate which he named *Actinomyces alni*, but his tests are open to criticism on the ground that there were too few control plants. Von Plotho (1941) obtained some nodules after inoculation with her isolate to which the above name was again applied, but they developed on only about one-quarter of the plants and not until a year after inocu-lation (cf. Table 3), by which time the control plants had succumbed to nitrogen starvation. Bouwens (1943) and Uemura (1961) failed to secure nodulation with von Plotho's organism. Youngken (1919) claimed successful re-inoculation with an actinomycete isolated from *Myrica cerifera* nodules, but it is clear that typical nodules did not form. Petry's (1925) similar claim in respect of *Ceanothus* is also uncon-vincing. Pommer (1959) incubated sections of surface-sterilized alder nodules on glucose–asparagine–agar, and in many instances obtained growth of an organism which showed affinities with the actinomycetes. In inoculation tests with aseptically grown alder plants, those inoculated developed nodules in a few weeks. The criticisms may be made that there were few test plants, and an undesirable hiatus while the plants lay dormant over winter. Lastly, Niewiarowska (1959, 1961) has claimed isolation of the endophyte (to which the name *Nocardia hippophaë* was assigned) from *Hippophaë* nodules. In inoculation tests in sand culture, done on a large scale though not under aseptic conditions, the majority of the inoculated plants became freely nodulated, while control plants mostly remained nodule-free. It should be noted that the nodules took about a year to appear (cf. Table 3) and could not be secured in water culture; it is possible that the delay arose because test plants were inocu-lated at too young a stage, when the low carbohydrate–nitrogen ratio may prevent quick nodulation (Gardner & Bond, 1957). There is clearly a need for independent confirmation of the nodule-inducing capacity claimed by Pommer and Niewiarowska for their isolates.

Though at the moment there is no fully substantiated isolation from these non-legume nodules, it is possible to regard the future with some optimism. In the absence of pure cultures of the endophytes, a common method of securing nodulation for experimental purposes has been to apply to the roots a water suspension of crushed nodules. This procedure is not unfailingly successful—there is, for example, no record of its success in *Coriaria* (Bond, 1958b). In such cases recourse may be had to sowing seed in naturally infected soil.

ROOT INFECTION AND NODULE INITIATION

The processes culminating in the formation of a macroscopic nodule are completed quite quickly under favourable conditions (Table 3), but only a beginning has been made in the elucidation of their nature.

Table 3. *Minimum times observed for nodulation of plants in water culture*

Host plant	Time of first appearance of macroscopic nodules (days from inoculation)
Alnus glutinosa	12
Myrica gale	18
Hippophaë rhamnoides	11
Shepherdia canadensis	15
Casuarina equisetifolia	26
Ceanothus azureus	26

As noted by Hiltner (1903), Krebber (1932) and Pommer (1956) with *Alnus*, and by Fletcher (1955) with *Myrica* and Bond (1957a) with *Casuarina*, the application of crushed-nodule inoculum to the roots causes distortion and branching in the root hairs (Fig. 2). Comparison with legumes encourages the belief that these effects are due to the action of the endophyte, but this is not proved.

Hiltner (1903) shows a rather convincing drawing of an alder root hair containing a structure interpreted as a slime thread with included bacteria. In a very small proportion of distorted root hairs Pommer (1956) detected invading hyphae, and he provides photographs of these and of hyphae growing from the root hair into the cortex. Taubert (1956) also figures root hairs of *Alnus* with included hyphal structures. The present writer has examined deformed root hairs of several genera without finding undoubted infection structures, and it is certain that nothing as conspicuous as the infection thread of *Rhizobium* is present.

Few studies of the earliest stages in the actual initiation of the nodules are available, doubtless because most workers have relied on limited material from the field or botanic gardens. In a valuable study Pommer (1956) found that the first effect of the presence of the endophyte in the cortex of the parent root is to induce a limited meristematic division in that tissue, followed by the formation of a periderm surrounding the whole of this infected area and linking up with the stele. Meanwhile, a lateral root arises from the stele opposite the infected area and as it pushes through the cortex its structure becomes modified and its own

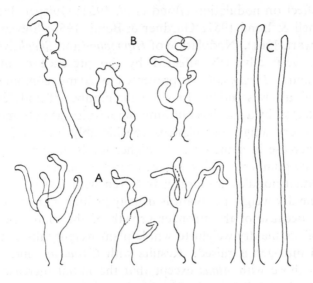

Fig. 2. Drawings of root hairs of: A, *Myrica gale* 3 weeks after inoculation (× 300); B, *Alnus glutinosa* 3 weeks after inoculation (× 600); C, *A. glutinosa* uninoculated (× 600).

cortical region infected. This lateral root develops into the nodule and emerges from the parent root still enclosed in the periderm layer which now becomes the cork layer noted on p. 74. It should be noted that according to Pommer's work the initiation of the nodule commences in the cortex of the parent root, as in many legumes. However, Taubert (1956) gives a somewhat different account according to which all the tissues of the alder nodule are derived from the modified lateral root. Also he notes that it is uncertain whether the presence of the endophyte in a particular region of the parent root cortex induces lateral root development in the vicinity, and that a nodule may only result when the two events happen to coincide. The chances of this are increased by the longitudinal spread of infection which he saw in the parent cortex.

Quispel (1954*b*), on the basis of a study of the effect of inoculum concentration on nodulation in *Alnus*, concluded that nodule initiation is limited to certain sites which are not normally points of lateral root production.

The proper development of nodules depends on the provision of suitable conditions. The pH of the rooting medium has a marked effect (Bond, Fletcher & Ferguson, 1954; Bond, 1957*a*). In *Alnus*, *Myrica*, *Hippophaë* and *Casuarina* nodulation is best near neutrality. In the two last genera nodulation is very sensitive to a lowering of pH, but it is still extensive at pH 4 in *Alnus* and *Myrica*.

The presence of combined nitrogen in the rooting medium also has a marked effect on nodulation (Bond *et al.* 1954; Quispel, 1954*b*, 1958; MacConnell & Bond 1957; Gardner & Bond, 1957; Stewart & Bond, 1961; Stewart, 1961). Nodulation of *Hippophaë* and *Shepherdia* in water culture is very adversely affected by the presence of ammonium-nitrogen, and in *Hippophaë* it is prevented by 20 mg. combined nitrogen per litre of culture solution. In *Alnus* the number of nodules per plant is little affected by a low level of ammonium-nitrogen (10 mg. per litre) but nodule dry weight per plant is considerably increased, coincident with an increase in plant size. At higher levels of combined nitrogen both nodule number and dry weight decrease, particularly the former, i.e. individual nodules are larger; but even with 100 mg. nitrogen per litre nodule dry weight per plant is usually greater than in nitrogen-free solution. Because of the stronger growth of the plant itself the proportion of nodule dry weight to whole plant weight falls as the level of combined nitrogen is raised. Results with *Casuarina* and *Myrica* are similar to those with *Alnus* except that the initial increase in nodule weight per plant is less marked in *Myrica*.

CROSS-INOCULATION BETWEEN SPECIES AND GENERA

Certain information based on the crushed-nodule inoculum technique is available concerning cross-inoculation between species and genera of these plants.

Roberg (1938) demonstrated reciprocal cross-inoculation between four species of *Alnus*, while Mowry (1933) reported that an inoculum from one species of *Casuarina* caused nodulation in eight other species. Gardner (1958*b*) and Bond (unpublished) found that *Myrica gale* inoculum caused nodulation in *M. cerifera* and *M. cordifolia*, though the normal benefit of nodules to plant growth (see later) was not shown. Bond (1962) found no cross-inoculation between a European and a Japanese species of *Coriaria*.

It is almost certain that cross-inoculation is possible between the three genera of the Elaeagnaceae, since Roberg (1934*a*) demonstrated cross-infection between *Elaeagnus* and *Hippophaë*, while Gardner & Bond (1957) showed that a *Hippophaë* inoculum nodulates *Shepherdia*. Except for this family there is no record of cross-inoculation between genera. An *Alnus* inoculum does not cause nodulation in *Casuarina* (Miehe, 1918), *Myrica* (Fletcher, 1955), *Hippophaë* or *Elaeagnus* (Roberg, 1934*a*), and a *Casuarina* inoculum is inoperative on *Myrica* and *Coriaria* (writer's observations). It appears therefore that the endophytes from the various genera differ from each other, at least in infective power. Tests of cross-inoculation between the Elaeagnaceae and the nodule-forming members of the Rhamnaceae would be of interest in view of the close relation between these families.

EVIDENCE THAT THE NODULES FIX NITROGEN

It will be recalled that the classic experiment carried out by Hellriegel and Wilfarth, which led to the recognition of the capacity of the nodulated legume to fix atmospheric nitrogen, was to compare the growth of nodule-bearing and nodule-free plants provided with a rooting medium free of combined nitrogen. Comparable experiments have been done by many investigators with species of the following non-legume genera: *Alnus*, Hiltner (1896, 1898, 1903), Nobbe & Hiltner (1904), Krebber (1932), Roberg (1934*b*), Virtanen & Saastamoinen (1936), Ferguson & Bond (1953), Quispel (1954*a*); *Myrica*, Bond (1951); *Hippophaë*, Servettaz (1909), Bond, MacConnell & McCallum (1956), Bond & Gardner, (1957); *Elaeagnus*, Nobbe & Hiltner (1904), Roberg (1934*a*), Panosjan (1943), Gardner (1958*a*); *Shepherdia*, Nobbe & Hiltner (1904), Gardner & Bond (1957); *Casuarina*, Aldrich-Blake (1932), Mowry (1933), Bond (1957*a*); *Ceanothus*, Petry (1925); *Coriaria*, Kataoka (1930), Bond (1962). All agree that nodule-bearing plants have the capacity which nodule-free plants lack to make satisfactory or even vigorous growth in a rooting medium free of combined nitrogen. Pl. 3, figs. 5 and 6, show typical results for this kind of experiment, and Table 4 shows the actual amounts of nitrogen accumulated in experiments at Glasgow by six different species of nodulated plants. In the writer's experience this capacity for growth in nitrogen-free media is unfailingly shown, for in recent years he and his collaborators have cultured successfully some thousands of nodulated plants, particularly of *Alnus*, *Myrica*, *Casuarina* and *Hippophaë*.

Confirmation of the fixation of atmospheric nitrogen by nodulated plants, implied by the above results, has been provided by ^{15}N tests with whole plants as follows: *Alnus*, Bond *et al.* (1954), Bond (1956*b*); *Myrica* and *Hippophaë*, Bond (1956*b*); *Casuarina*, *Shepherdia* and *Ceanothus*, Bond (1957*b*); *Coriaria*, Stevenson (1958). Harris & Morrison (1958) and Morrison (1961) demonstrated fixation in detached nodules of *Coriaria* and *Discaria* respectively by ^{15}N experiments. Various other ^{15}N studies with detached nodules will be mentioned later.

Table 4. *Accumulation of nitrogen by nodulated plants in rooting media free of combined nitrogen*

Species	Mg. nitrogen accumulated per plant during first season's growth	Subsequent accumulation in mg.
Alnus glutinosa	300	2500 by end of 2nd season
Myrica gale	146	5020 by end of 3rd season
Hippophaë rhamnoides	26	200 by end of 2nd season
Casuarina cunninghamiana	70	1400 after 2½ seasons
Elaeagnus angustifolia	—	186 after 1½ seasons
Coriaria myrtifolia	36	—

Next it must be considered whether the nodules are the site of the fixation. As noted above it is only nodulated plants that grow vigorously and accumulate nitrogen in the absence of combined nitrogen, and such plants have a higher percentage of total nitrogen in the nodule tissues compared with root and shoot. This is consistent with the view that the nodules are the source from which fixed nitrogen migrates to the rest of the plant. More direct evidence is provided by the writer's observation that if the nodules are removed from a plant, the latter soon develops symptoms of nitrogen deficiency which persist until new nodules are formed. A similar observation, in this case on field plants, was made a long time ago by Dinger (1895).

^{15}N tests confirm that the nodules are the site of fixation. Bond (1958*a*) showed that if the root systems in four genera of nodulated plants are exposed for some hours to a gas mixture comprising essentially oxygen and labelled nitrogen, subsequent assay shows greatest enrichment in the nodules. If the nodules are removed prior to the test, and they and the plant exposed separately to the labelled nitrogen, only the nodules show fixation. Further, no evidence of fixation is shown in non-nodulated plants supplied previously with combined nitrogen. In these tests the possibility that the shoot was the site of fixation was discounted. No ^{15}N was applied to the shoot, but there is no doubt that the isotopic nitrogen supplied to the root system in these experiments leaked away through the shoot, so that any faculty for fixation in the latter would have revealed itself.

It remains to be said that with one exception the investigations reviewed in this section relate to plants which had not been grown under aseptic conditions. Crushed-nodule inocula were applied initially, and further opportunities for contamination occurred later. There is thus a possibility that fixation of nitrogen in the nodules might not be due to the symbiotic system of host plant and endophyte, but rather to some foreign organism which, as would have to be assumed, was associated only with the nodules. All available evidence is against this idea—the regularity and magnitude of the fixation, the absence of any appreciable number of contaminants in nodule sections, the rapidity with which the fixed nitrogen becomes available to the host plant (p. 85) and the almost complete loss of fixation in pulverized nodules (Bond, unpublished). Lastly, Quispel (1954a) prepared a crushed-nodule inoculum from surface-sterilized alder nodules which had remained free of contaminant growth after incubation on a favourable medium; the plants inoculated with this were grown aseptically and no contaminants could be detected on the resulting nodulated plants, in which fixation was still shown.

The conclusion seems justified that the fixation occurs in the nodules and is due to the endophyte, or to it and the host-cell cytoplasm jointly.

PHYSIOLOGICAL CHARACTERISTICS OF THE FIXATION

Hewitt & Bond (1961) and Bond & Hewitt (1961) investigated the significance of molybdenum in *Casuarina*, *Alnus* and *Myrica*. With nodulated plants growing in water culture in a solution free of combined nitrogen it was found that fixation per plant was up to ten times greater when molybdenum was supplied. The molybdenum-deprived plants, though well nodulated, were stunted and showed severe nitrogen deficiency, but recovered quickly if supplied with molybdenum. Pl. 4, fig. 7, illustrates the results with *Myrica*; the somewhat better growth of the nodulated plants not supplied with molybdenum, as compared with the nodule-free plants, is doubtless due to molybdenum in the seeds and from uncontrolled sources. The nodule tissues showed a much greater capacity to accumulate molybdenum than the rest of the plant. Non-nodulated plants supplied with nitrate-nitrogen also showed somewhat reduced growth when deprived of molybdenum, due to a failure of the nitrate-reducing mechanism of the plant. These findings suggest that molybdenum has some special role in fixation. Becking (1961) likewise

6-2

found markedly beneficial effects on fixation and growth of nodulated alders rooted in a peat soil deficient in molybdenum, and further found that molybdenum deficiency tended to produce nodules which were unusually numerous and scattered; this, however, seemed to involve a comparison of plants grown in soils differing in other respects besides molybdenum content.

Bond & Hewitt (1962) have demonstrated a cobalt requirement in nodulated *Alnus* and *Casuarina* plants in water culture, and in unpublished observations have extended this to *Myrica*. The lack of cobalt was found to result in symptoms of intense nitrogen deficiency (Pl. 4, fig. 8). In alder plants without cobalt, fixation of nitrogen was only one-quarter of that in plants supplied with cobalt. These effects, though eventually impressive, took longer to appear than in the case of molybdenum deficiency, possibly because the cobalt requirement is smaller. The necessity of cobalt for fixation matches that demonstrated for nodulated legumes by Ahmed & Evans (1959, 1961), Reisenauer (1960), Hallsworth, Wilson & Greenwood (1960), and Delwiche, Johnson & Reisenauer (1961), and may hinge on the role of vitamin B_{12} in the synthesis of the haemoglobin which is present in both legume and non-legume nodules (p. 85).

Stewart & Bond (1961) used a ^{15}N technique to investigate the effect of ammonium-nitrogen in the rooting medium on fixation in the nodules of *Alnus* and *Myrica*. The efficiency of fixation per unit of nodule tissue was reduced, possibly because of a substitution of ammonium-nitrogen for elemental nitrogen at the nitrogen-fixing centres. Despite this, fixation per plant was enhanced in *Alnus* by low levels of ammonium-nitrogen, owing to the greater nodule development (p. 80). At higher levels of ammonium, fixation per plant was still of comparable magnitude to that in nitrogen-free medium, but now only accounted for about one-third of the total nitrogen uptake by the plants.

The study of other aspects of fixation is facilitated by the circumstance that fixation in detached nodules is more easily measured (by a ^{15}N technique) in non-legumes than in legumes (Bond, 1959). The influence on fixation of the proportion of oxygen in the atmosphere has been examined with nodules of four genera (Bond & MacConnell, 1955; Bond, 1959, 1961). Fixation was very small at low levels of oxygen, but rose with increasing oxygen supply and attained a maximum in the range of 12–25 % oxygen in the different genera. Levels of 40–50 % oxygen almost completely inhibited fixation. Detached nodulated roots of pea plants were included for comparison; the optimal oxygen level for fixation was 20 %, with extinction at 70 %. Burris, Magee & Bach

(1955) had previously demonstrated an inhibition of fixation in soya bean nodules at high oxygen levels. Nodular respiration is not inhibited at high oxygen levels (Bond, 1961), so that the inhibiting effect may be on fixation itself. Parker & Scutt (1960) have advanced evidence of a specific and competitive inhibition of fixation in *Azotobacter* at higher oxygen levels, which Bauer (1960) has suggested is due to a displacement of nitrogen by oxygen at nitrogen-fixing sites.

In further studies with detached nodules Bond (1960) demonstrated that fixation in non-legumes, as in legumes, is inhibited by substantial partial pressures of hydrogen, and shows marked sensitivity to carbon monoxide. Subsequently Davenport (1960) undertook a re-examination of non-legume nodules for haemoglobin, previously reported absent, and obtained spectroscopic evidence of its presence in *Casuarina*, *Alnus* and *Myrica*. The detection of the haemoglobin was made more difficult by the presence in the nodules of a very active phenol-oxidase system, and by the close binding of the pigment to cell constituents.

Leaf, Gardner & Bond (1958, 1959) by combined ^{15}N and chromato-graphic techniques showed that in nodules of *Alnus* and *Myrica* the fixed nitrogen accumulates in substances which are usually held to gain their nitrogen directly from ammonia, such as glutamic acid and citrul-line in *Alnus* and amides in *Myrica*.

THE RELATION BETWEEN HOST AND ENDOPHYTE

The non-legume symbiosis is evidently closely comparable to that in legumes. Neither is systemic and plants must be re-infected from the soil in each generation.

The advantage of the association to the higher plant is obvious enough, since it is thereby endowed with the distinctive ability to thrive in media low in combined nitrogen, and hence acquires pioneer status. The rapidity with which nitrogen fixed in the nodules becomes available to the rest of the plant was illustrated in the ^{15}N experiments of Bond (1956c), which indicated, in *Alnus*, that only 6 hr. after the commence-ment of fixation, a substantial part of the products has reached the shoot. The conclusion is supported by observations of Stewart (1962), who sampled a population of alders at 12-day intervals during the first season of growth. From the data it was possible to determine the nitrogen fixed over successive intervals, and also the proportion trans-ferred from the nodules to the rest of the plant. It was found that, as in legumes, there was from the beginning of fixation a steady transfer from the nodules of approximately 90 % of the nitrogen fixed.

These results focus attention on the question of the actual site of the fixation within the nodule cells. If it is assumed that the fixation occurs within the endophyte, then most of the products must be immediately passed on to the nodule cell, because, in spite of the eventual degeneration of the endophyte in older parts of the nodule, Pommer (1956) verified that some months typically elapse before this occurs in a given cell. A more satisfactory explanation is that the fixation occurs outside the endophyte, for which there is direct evidence in the case of the legumes (Bergerson, Wilson & Burris, 1959). This also assists in the explanation of the fact that in a nodulated plant the scale of fixation is attuned to the growth requirements of the host plant rather than to those of the endophyte. Stewart (1962) was able to show that fixation per unit growth of the alder endophyte was several times greater than that shown by *Azotobacter* growing under favourable conditions.

Non-legumes, like legumes, are not obligate symbionts but will grow perfectly well without nodules provided that adequate supplies of combined nitrogen are available to the roots. Under these conditions growth is initially much faster than in inoculated plants in nitrogen-free solution, owing to the time required for the nodules to become functional. Later the nodulated plants grow strongly, and may overtake the combined nitrogen plants even in the first season of growth. This is the writer's experience with *Alnus* and *Myrica*, but not with *Hippophaë*, *Casuarina* and other genera, for which optimal conditions may not have been provided.

Too little is known about the endophytes of these nodules to say much about any benefit they may derive from the association. They are probably heterotrophic and dependent on the host for organic carbon and other nutrients. While they can apparently survive apart from the host, it remains uncertain whether they can grow under these conditions.

EVOLUTIONARY ASPECTS

All the nodule-bearing Angiosperms are woody plants or, as in the case of the herbaceous legumes, are in a family which includes woody ancestral types. This suggests that the nodule-forming habit may have arisen in an era when the angiosperms were represented by woody members only, and when conditions may have been specially favourable to the establishment of this type of symbiotic association (Bond *et al.* 1956). While the sources of fixed nitrogen for earlier floras are conjectural, it is possible that the cycadophytes, which may have borne nitrogen-fixing nodules as do modern cycads, were an important source in the Jurassic and early Cretaceous eras. If so, the period of transition in which the existing flora was gradually replaced by new angiosperms might have been one of increasing deficiency of soil nitrogen, with a resulting evolutionary pressure favouring the establishment of nitrogen-fixing types. This same period may also have been one of intensive evolution and re-adjustment in the lower, heterotrophic organisms, making it conceivable that the nodule endophytes were evolved at this time.

Parker (1957) supposes that the nodule symbiosis began as a casual superficial association between free living nitrogen-fixing soil organisms and plant roots, with a subsequent evolution of a more efficient symbiosis. As already indicated (p. 86) more recent studies throw some doubt on the view that the fixation mechanism in nodules is provided entirely by the endophyte. Rather they suggest that it may have arisen through a pooling of the resources of both symbionts. In this case the original lower symbiont must be thought of as possessing some parasitic tendencies, coupled with a metabolism that can combine with that of a higher plant to produce a nitrogen-fixing unit.

REFERENCES

AHMED, S. & EVANS, H. J. (1959). The effect of cobalt on the growth of soybeans in the absence of supplied nitrogen. *Biochem. Biophys. Res. Comm.* 1, 271.

AHMED, S. & EVANS, H. J. (1961). The essentiality of cobalt for soybean plants grown under symbiotic conditions. *Proc. nat. Acad. Sci., Wash.* 47, 24.

ALDRICH-BLAKE, R. N. (1932). On the fixation of atmospheric nitrogen by bacteria living symbiotically in root nodules of *Casuarina equisetifolia*. *Oxford For. Mem.* 14.

ALLEN, E. K. & ALLEN, O. N. (1958). Biological aspects of symbiotic nitrogen fixation. In *Encyclopedia of Plant Physiology*, 8. Ed. W. Ruhland. Berlin: Springer-Verlag.

BAUER, N. (1960). A probable free-radical mechanism for symbiotic nitrogen fixation. *Nature, Lond.* 188, 471.

BECKING, J. H. (1961). A requirement of molybdenum for the symbiotic nitrogen fixation in alder. *Plant & Soil*, 15, 217.

BERGERSON, F. J., WILSON, P. W. & BURRIS, R. H. (1959). Biochemical studies on soybean nodules. *Abstr. Proc. IX Int. bot. Congr.* 2, 29.

BOND, G. (1951). The fixation of nitrogen associated with the root nodules of *Myrica gale* L. with special reference to its pH relation and ecological significance. *Ann. Bot., Lond.,* N.S. 15, 447.

BOND, G. (1952). Some features of root growth in nodulated plants of *Myrica gale* L. *Ann. Bot., Lond.,* N.S. 16, 467.

BOND, G. (1956a). A feature of the root nodules of *Casuarina*. *Nature, Lond.* 177, 192.

BOND, G. (1956b). An isotopic study of the fixation of nitrogen associated with nodulated plants of *Alnus, Myrica,* and *Hippophaë*. *J. exp. Bot.* 6, 303.

BOND, G. (1956c). Some aspects of translocation in root nodule plants. *J. exp. Bot.* **7**, 387.

BOND, G. (1957a). The development and significance of the root nodules of *Casuarina*. *Ann. Bot., Lond.*, N.S. **21**, 373.

BOND, G. (1957b). Isotopic studies of nitrogen fixation in non-legume root nodules. *Ann. Bot., Lond.*, N.S. **21**, 513.

BOND, G. (1958a). Symbiotic nitrogen fixation by non-legumes. In *Proc. 5th Easter School Agric. Sci., Univ. of Nottingham.* London: Butterworth's Scientific Publications.

BOND, G. (1958b). Root nodules of *Coriaria*. *Nature, Lond.* **182**, 474.

BOND, G. (1959). Fixation of nitrogen in non-legume root-nodule plants. In *Symp. Soc. exp. Biol.* **13**. Cambridge University Press.

BOND, G. (1960). Inhibition of nitrogen fixation in non-legume root nodules by hydrogen and carbon monoxide. *J. exp. Bot.* **11**, 91.

BOND, G. (1961). The oxygen relation of nitrogen fixation in root nodules. *Z. allg. Mikrobiol.* **1**, 93.

BOND, G. (1962). Fixation of nitrogen in *Coriaria myrtifolia*. *Nature, Lond.* **193**, 1103.

BOND, G., FLETCHER, W. W. & FERGUSON, T. P. (1954). The development and function of the root nodules of *Alnus, Myrica* and *Hippophaë*. *Plant & Soil*, **5**, 309.

BOND, G & GARDNER, I. C. (1957) Nitrogen fixation in non-legume root nodule plants. *Nature, Lond.* **179**, 680.

BOND, G. & HEWITT, E. J. (1961). Molybdenum and the fixation of nitrogen in *Myrica* root nodules. *Nature, Lond.* **190**, 1033.

BOND, G. & HEWITT, E. J. (1962). Cobalt and the fixation of nitrogen by root nodules of *Alnus* and *Casuarina*. *Nature, Lond.* **195**, 94.

BOND, G. & MACCONNELL, J. T. (1955). Nitrogen fixation in detached non-legume root nodules. *Nature, Lond.* **176**, 606.

BOND, G., MACCONNELL, J. T. & MCCALLUM, A. H. (1956). The nitrogen-nutrition of *Hippophaë rhamnoides* L. *Ann. Bot., Lond.*, N.S. **20**, 501.

BOUWENS, H. (1943). Investigation of the symbiont of *Alnus glutinosa, Alnus incana* and *Hippophaë rhamnoides*. *Leeuwenhoek ned. Tijdschr.* **9**, 107.

BURRIS, R. H., MAGEE, W. E. & BACH, M. K. (1955). The pN_2 and the pO_2 function for nitrogen fixation by excised soybean nodules. *Ann. Acad. Sci. Fenn.* **60**, 190.

DAVENPORT, H. E. (1960). Haemoglobin in the root nodules of *Casuarina cunninghamiana*. *Nature, Lond.* **186**, 653.

DELWICHE, C. C., JOHNSON, C. M. & REISENAUER, H. M. (1961). Influence of cobalt on nitrogen fixation by *Medicago*. *Plant Physiol.* **36**, 73.

DINGER, R. (1895). De Els als stikstof verzamelaar. *Landb. Tijdschr.* **3**, 167.

FERGUSON, T. P. & BOND, G. (1953). Observations on the formation and function of the root nodules of *Alnus glutinosa*. *Ann. Bot., Lond.*, N.S. **17**, 175.

FLETCHER, W. W. (1955). The development and structure of the root-nodules of *Myrica gale* L. with special reference to the nature of the endophyte. *Ann. Bot., Lond.*, N.S. **19**, 501.

GARDNER, I. C. (1958a). Nitrogen fixation in *Elaeagnus* root nodules. *Nature, Lond.* **81**, 717.

GARDNER, I. C. (1958b). Aspects of symbiotic nitrogen fixation in non-leguminous plants. Ph.D. thesis, University of Glasgow.

GARDNER, I. C. & BOND, G. (1957). Observations on the root nodules of *Shepherdia*. *Canad. J. Bot.* **35**, 305.

HALLSWORTH, E. G., WILSON, S. B. & GREENWOOD, E. A. N. (1960). Copper and cobalt in nitrogen fixation. *Nature, Lond.* **187**, 79.

HARRIS, G. P. & MORRISON, T. M. (1958). Fixation of nitrogen by excised nodules of *Coriaria arborea* Lindsay. *Nature, Lond.* **182**, 1812.

HAWKER, L. E. & FRAYMOUTH, J. (1951). A re-investigation of the root-nodules of species of *Elaeagnus, Hippophaë, Alnus* and *Myrica*, with special reference to the morphology and life histories of the causative organisms. *J. gen. Microbiol.* **5**, 369.

HEWITT, E. J. & BOND, G. (1961). Molybdenum and the fixation of nitrogen in *Casuarina* and *Alnus* root nodules. *Plant & Soil*, **14**, 159.

HILTNER, L. (1896). Über die Bedeutung der Wurzelknöllchen von *Alnus glutinosa* für die Stickstoffernährung dieser Pflanze. *Landw. Versuchst.* **46**, 153.

HILTNER, L. (1898). Ueber Enstehung und physiologische Bedeutung der Wurzel-knöllchen. *Forst. Naturw. Z.* **7**, 415.

HILTNER, L. (1903). Über die biologische und physiologische Bedeutung der endo-trophen Mycorhiza. *Naturw. Z. Land- u. Forstw.* **1**, 9.

HUTCHINSON, J. (1959). *The Families of Flowering Plants*, Vol. I. Oxford: Clarendon Press.

KATAOKA, T. (1930). On the significance of the root-nodules of *Coriaria japonica* A.Gr. in the nitrogen nutrition of the plant. *Jap. J. Bot.* **5**, 209.

KREBBER, O. (1932). Untersuchungen über die Wurzelknöllchen der Erle. *Arch. Mikrobiol.* **3**, 588.

LAWRENCE, D. B. (1953). Development of vegetation and soil in Southeastern Alaska with special reference to the accumulation of nitrogen. *Final Report ONR Project Nr. 160–183*, Washington.

LEAF, G., GARDNER, I. C. & BOND, G. (1958). Observations on the composition and metabolism of the nitrogen-fixing root nodules of *Alnus*. *J. exp. Bot.* **9**, 320.

LEAF, G., GARDNER, I. C. & BOND, G. (1959). Observations on the composition and metabolism of the nitrogen-fixing root nodules of *Myrica*. *Biochem. J.* **72**, 662.

LIESKE, R. (1921). *Morphologie und Biologie der Strahlenpilze (Actinomyceten)*. Leipzig: Borntraeger.

MACCONNELL, J. T. & BOND, G. (1957). A comparison of the effect of combined nitrogen on nodulation in non-legumes and legumes. *Plant & Soil*, **8**, 378.

MIEHE, H. (1918). Anatomische Untersuchungen der Pilzsymbiose bei *Casuarina equisetifolia* nebst einigen Bemerkungen über das Mykorrhizenproblem. *Flora*, N.F. **11–12**, 431.

MORRISON, T. M. (1961). Fixation of nitrogen-15 by excised nodules of *Discaria toumatou*. *Nature, Lond.* **189**, 945.

MORRISON, T. M. & HARRIS, G. P. (1958). Root nodules in *Discaria toumatou* Raoul Choix. *Nature, Lond.* **182**, 1746.

MORRISON, T. M. & HARRIS, G. P. (1959). Root nodules in non-leguminous plants in New Zealand. *Proc. N.Z. ecol. Soc.* **6**, 23.

MOWRY, H. (1933). Symbiotic nitrogen fixation in the genus *Casuarina*. *Soil Sci.* **36**, 409.

NIEWIAROWSKA, J. (1959). Symbioza u rokitnika. *Acta Microbiol. Polonica*, **8**, 289.

NIEWIAROWSKA, J. (1961). Morphologie et physiologie des Actinomycetes sym-biotiques des *Hippophae*. *Acta Microbiol. Polonica*, **10**, 271.

NOBBE, F. & HILTNER, L. (1904). Ueber das Stickstoffsammlungsvermögen der Erlen und Elaeagnaceen. *Naturw. Z. Land- u. Forstw.* **2**, 366.

PANOSJAN, A. K. (1943). The biology of the nodules on the roots of *Elaeagnus angustifolia*. *Microbiol. Symp. Acad. Sci. U.S.S.R.* **1**, 147.

PARKER, C. A. (1957). Evolution of nitrogen-fixing symbioses in higher plants. *Nature, Lond.* **179**, 593.

PARKER C. A. & SCUTT, P. B. (1960). The effect of oxygen on nitrogen fixation by *Azotobacter*. *Biochim. Biophys. Acta*, **38**, 230.

PEKLO, J. (1910). Die pflanzlichen Aktinomykosen. *Zbl. Bakt.* (2), **27**, 451.

PETRY, E. J. (1925). Physiological studies on *Ceanothus americanus*. Ph.D. dissertation. Michigan State Coll. Agric. Appl. Sci.

VON PLOTHO, O. (1941). Die Synthese der Knöllchen an den Wurzeln der Erle. *Arch. Mikrobiol.* **12**, 1.

POMMER, E. H. (1956). Beiträge zur Anatomie und Biologie der Wurzelknöllchen von *Alnus glutinosa* Gaertn. *Flora*, **143**, 603.

POMMER, E. H. (1959). Über die Isolierung des Endophyten aus den Wurzelknöllchen *Alnus glutinosa* Gaertn. und über erfolgreiche Re-Infektionsversuche. *Ber. dtsch. bot. Ges.* **72**, 138.

QUISPEL, A. (1954a). Symbiotic nitrogen fixation in non-leguminous plants. I. Preliminary experiments on the root nodule symbiosis of *Alnus glutinosa*. *Acta Bot. Neerlandica*, **3**, 495.

QUISPEL, A. (1954b). Symbiotic nitrogen fixation in non-leguminous plants. II. The influence of the inoculation density and external factors on the nodulation of *Alnus glutinosa* and its importance to our understanding of the mechanism of the infection. *Acta Bot. Neerlandica*, **3**, 512.

QUISPEL, A. (1958). Symbiotic nitrogen fixation in non-leguminous plants. IV. The influence of some environmental conditions on different phases of the nodulation process in *Alnus glutinosa*. *Acta Bot. Neerlandica*, **7**, 191.

QUISPEL, A. (1960). Symbiotic nitrogen fixation in non-leguminous plants. V. The growth requirements of the endophyte. *Acta Bot. Neerlandica*, **9**, 380.

REISENAUER, H. M. (1960). Cobalt in nitrogen fixation by a legume. *Nature, Lond.* **186**, 375.

ROBERG, M. (1934a). Über den Erreger der Wurzelknöllchen von *Alnus* und den Elaeagnaceen *Elaeagnus* und *Hippophaë*. *Jb. wiss. Bot.* **79**, 472.

ROBERG, M. (1934b). Weitere Untersuchungen über die Stickstoffernährung der Erle. *Ber. dtsch. bot. Ges.* **52**, 54.

ROBERG, M. (1938). Über den Erreger der Wurzelknöllchen europäischer Erlen. *Jb. wiss. Bot.* **86**, 344.

SCHAEDE, R. (1948). *Die pflanzliche Symbiosen*. Jena: Gustav Fischer.

SCHWARTZ, W. (1959). Bakterien- und Actinomyceten-Symbiosen. In *Encyclopedia of Plant Physiology*, 11. Ed. W. Ruhland. Berlin: Springer-Verlag.

SERVETTAZ, C. (1909). Monographie des Eleagnacées. *Beih. bot. Zbl.* **25**, 1.

SHIBATA, K. (1902). Cytologische Studien über die endotrophen Mykorrhizen. *Jb. wiss. Bot.* **37**, 643.

SHIBATA, K. & TAHARA, M. (1917). Studien über die Wurzelknöllchen. *Bot. Mag. Tokyo*, **31**, 157.

STEVENSON, G. (1958). Nitrogen fixation by non-nodulated plants, and by nodulated *Coriaria arborea*. *Nature, Lond.* **182**, 1523.

STEWART, W. D. P. (1961). Studies in the biological fixation of nitrogen. Ph.D. thesis, University of Glasgow.

STEWART, W. D. P. (1962). A quantitative study of fixation and transfer of nitrogen in *Alnus*. *J. exp. Bot.* **13**, 250.

STEWART, W. D. P. & BOND, G. (1961). The effect of ammonium nitrogen on fixation of elemental nitrogen in *Alnus* and *Myrica*. *Plant & Soil*, **14**, 347.

TAUBERT, H. (1956). Über den Infektionsvorgang und die Entwicklung der Knöllchen bei *Alnus glutinosa* Gaertn. *Planta*, **48**, 135.

UEMURA, S. (1952). Studies on the root nodules of alders (*Alnus* spp.). IV. Experiments on isolation of actinomycetes from alder nodules. *Bull. (Jap.) Govt. Forest Exp. Sta.* **52**.

PLATE 1

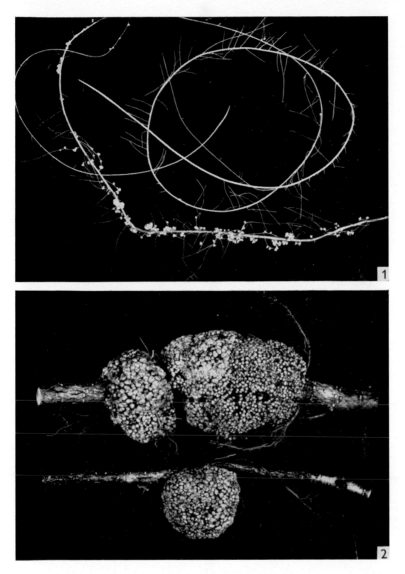

PLATE 2

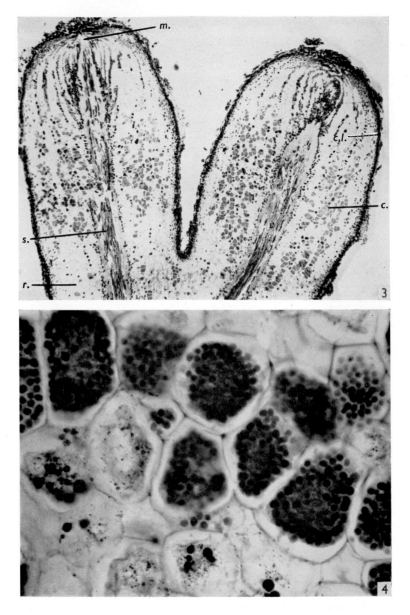

PLATE 3

PLATE 4

7

8

UEMURA, S. (1961). Studies on the *Streptomyces* isolated from alder root nodules. *Sci. Reps. Agric. Forest & Fish. Res. Co. Tokyo*, **7**.

VIRTANEN, A. I. & SAASTAMOINEN, S. (1936). Untersuchungen über die Stickstoffbindung bei der Erle. *Biochem. Z.* **284**, 72.

YOUNGKEN, H. W. (1919). The comparative morphology, taxonomy and distribution of the Myricaceae in the Eastern United States. *Contr. Bot. Lab. Univ. Pa.* **4**, 339.

EXPLANATION OF PLATES

PLATE 1

Fig. 1. Young nodules on the roots of a plant of *Hippophaë rhamnoides* grown in water culture. ($\times \frac{1}{2}$.) Photograph by Mr W. Anderson.

Fig. 2. Nodule clusters of *Alnus glutinosa*, field material. ($\times \frac{1}{2}$.) Photograph by Mr R. Cowper.

PLATE 2

Fig. 3. Younger part of a branched nodule of *Alnus glutinosa* in radial longitudinal section. *m.* = meristem; *s.* = stele; *c.* = cortex with darker stained infected cells; *r.* = region in which the contents of the infected cells have disappeared; *c.l.* = cork layer. ($\times 30$.) Section prepared by Miss E. Boyd, micrograph by Mr W. Anderson.

Fig. 4. Cortical region from a transverse section of a nodule lobe of *Alnus glutinosa* showing the dark-stained vesicles in the infected cells. ($\times 300$.) Acknowledgements as for Fig. 3.

PLATE 3

Fig. 5. Plants of *Myrica gale* after one season's growth in culture solution free of combined nitrogen; the plants on the left with nodules, those on the right without. ($\times \frac{1}{5}$.) Photograph by Mr R. Cowper.

Fig. 6. Plants of *Casuarina cunninghamiana* after one season's growth in culture solution free of combined nitrogen; the plants on the left with, those on the right without nodules. ($\times \frac{1}{7}$.) Photograph by Mr R. Cowper.

PLATE 4

Fig. 7. Plants of *Myrica gale* after 12 weeks' growth in water culture without combined nitrogen, no. 105 nodulated, molybdenum supplied, no. 102 nodulated, molybdenum not supplied, no. 109 without nodules, molybdenum supplied. ($\times \frac{1}{5}$.) Photograph by Mr R. Cowper.

Fig. 8. Nodulated plants of *Alnus glutinosa* after one season's growth in water culture without combined nitrogen, on the left with cobalt supplied, on the right without cobalt. $\times \frac{1}{4}$.) Photograph by Mr R. Cowper.

THE BIOCHEMISTRY OF NITROGEN FIXATION

D. J. D. NICHOLAS

Chemical Microbiology Department, Long Ashton Research Station, University of Bristol

Lieve God wat zynder al wonderen in soo een kleyn schepsel!
Leewenhoek's draughtsman (Letter 76, 15 *October* 1693)

INTRODUCTION

The biological fixation of nitrogen is, next to the photosynthetic assimilation of CO_2, one of the most fundamental processes in nature since it maintains the balance in the nitrogen economy of the world.

The early history of asymbiotic nitrogen fixation has been well reviewed by Burris & Wilson (1945) and Wilson (1958). Although Winogradsky is credited with the discovery of the first free-living micro-organism able to fix atmospheric nitrogen, Jodin as early as 1862 (Stephenson, 1939) noted good growth of 'mycoderms' in simple media without combined nitrogen and even more remarkable he established a loss of nitrogen from the atmosphere. Berthelot in 1885 showed an increase in nitrogen content of soil in sealed pots (Winogradsky, 1949). It is clear that the first pure culture of bacteria able to fix nitrogen was described by Winogradsky who named it *Clostridium pastorianum* to be changed later to *C. pasteurianum*.

The finding of the first *Azotobacter* species is accredited to Beijerinck who named them *chroococcum* and *agilis* (van Iterson, den Dooren de Jong & Kluyver, 1940; Winogradsky, 1949). *A. vinelandii* was isolated by Lipman (1903) from a soil in New Jersey, U.S.A.

As early as 1889 Frank (van Iterson *et al.* 1940) suggested that blue-green algae might fix nitrogen in soil, but the purity of his cultures was in dispute. Drewes (1928) and Allison & Morris (1930) demonstrated nitrogen fixation in *Nostoc muscorum* and in *Anabaena species* in un-adulterated cultures. Fogg & Wolfe (1954) have greatly extended these studies to include a range of blue-green algae. Photosynthetic bacteria including *Rhodospirillum rubrum, Chromatium, Chlorobium* and *Rhodomicrobium* also fix nitrogen.

Numerous other organisms incorporate nitrogen gas and these are listed in Table 1.

Not all these have been checked with the sensitive nitrogen-15 technique and for this reason it is doubtful whether the yeast *Rhodotorula* is able to incorporate atmospheric nitrogen. This list is not complete since claims, as yet unsubstantiated, have been put forward for actinomycetes, mycorrhiza, yeasts and some strange associations of bacteria with insects and goats (Toth, 1952, 1953).

Table 1. *Biological systems that fix nitrogen gas*

ASYMBIOTIC			
BACTERIA	Aerobe		*Azotobacter*
	Anaerobe	(Facultative)	*Aerobacter, Bacillus polymyxa, Achromobacter, Pseudomonas*
		(Obligate)	*Clostridium pasteurianum* (heterotrophic), *Desulphovibrio, Methanobacterium* (facultatively autotrophic?)
		(Photosynthetic)	*Rhodospirillum, Chromatium, Chlorobium, Rhodomicrobium*
YEASTS	Aerobe		*Rhodotorula*
BLUE-GREEN ALGAE	Aerobe		*Nostoc, Anabaena, Calothrix*, etc.
SYMBIOTIC			
LEGUMES			*Rhizobium*
NON-LEGUMES			*Actinomycetes* ?

One of the main obstacles in studying the biochemistry of nitrogen fixation has been the lack of success in preparing suitable extracts from the bacteria that will consistently incorporate the gas. Recently this was achieved, as is often the case, almost simultaneously and independently in a number of laboratories with a variety of micro-organisms. Thus active extracts were prepared from *Clostridium pasteurianum* by Carnahan, Mortenson, Mower & Castle (1960), from *Azotobacter vinelandii* by Nicholas & Fisher (1960), from blue-green algae and *Rhodospirillum rubrum* by Schneider *et al.* (1960), from *Bacillus polymyxa* by Grau (1961) and Grau & Wilson (1961), and from *Chromatium* by Arnon, Losada & Nozaki (1960). Thus for the first time it is possible to study nitrogen fixation at the enzyme level.

Progress in understanding the complex biochemical relations between host and symbiont in the root nodules has been slow. The main difficulty is that the rhizobia (legumes) or the actinomycetes (suggested as possible causal organisms in some of the non-legumes) do not fix nitrogen outside the host plant. Numerous claims that rhizobia grown in synthetic media are able to fix nitrogen have not been substantiated. Excised and crushed nodules readily lose their capacity to fix nitrogen, and bacteroids taken from the nodules are inactive (Aprison, Magee & Burris, 1954).

In this paper I will give an account of recent research, and hypotheses on the biochemistry of the nitrogen-fixing process. This will mostly concern asymbiotic systems but these have so many features linking them with symbiotic systems that their full discussion in this symposium can be justified on the grounds of comparative biochemistry. It will become clear that only preliminary studies have been attempted thus far and much remains to be done before the precise mechanism of nitrogen fixation is established.

METHODS FOR DETERMINING
BIOLOGICAL NITROGEN FIXATION

Stable isotope of nitrogen (^{15}N)

Burris & Wilson (1957) and Wilson (1958) have reviewed in detail the efficiency of the methods used to study nitrogen fixation. The limitations of the micro-Kjeldahl method led Burris, Eppling, Wahlin & Wilson (1942) to develop the use of the stable isotope of nitrogen (^{15}N) as a tracer. This technique has proved useful in determining the extent of fixation in whole cells, and more recently in extracts prepared from them. Values greater than 0·015 atom % ^{15}N excess are accepted as a reliable indication of fixation. At best, it is claimed, the method can detect a gain of nitrogen of 0·01–0·02 μg./ml. The method is relatively independent of both nitrogen present initially, and of sampling error, since the amount determined is the increase in percentage of the tracer and not the absolute content.

Radioactive nitrogen (^{13}N)

Since the ^{15}N method is time-consuming, some 2 days being required to complete an experiment from the initial exposure of the gas to the final analysis in the mass spectrometer, a technique employing ^{13}N has been developed (Nicholas, Silvester & Fowler, 1961).

Nitrogen-13 (half-life 10·05 min.) was produced continuously in a cyclotron. The ^{13}N which was carrier free, was swept from the target area in a stream of pure argon through small bore tubing to an adjacent laboratory where it was freed from ammonia and nitrogen oxides. For experiments with *Azotobacter vinelandii* the gas was mixed with the aid of rotameters with 10 % oxygen. This method is approximately 100 times as sensitive as the ^{15}N technique.

Microdiffusion method

Mortenson (1962) has used a Conway microdiffusion technique and a titrimetric assay to determine the production of ammonia in extracts of *Clostridium pasterianum*. This method is less sensitive than the ^{15}N procedure, and thus has limited use with cell-free preparations.

ASYMBIOTIC FIXATION

Extraction of enzymes

The methods employed in preparing cell-free extracts vary with the type of organism used.

The methods may be summarized as follows:

Clostridium pasteurianum. Carnahan *et al.* (1960) grew cells under anaerobic conditions for 16 hr. The cells were washed in phosphate buffer and disrupted in a Hughes press or preferably by autolysis of dried cells. The cells were dried at 40° under vacuum. Dried cells pulverized in a mortar were added to buffer and shaken under hydrogen at 30°. After centrifuging at 144,000g for 3 hr. all the nitrogen-fixing activity was in the supernatant fluid.

Schneider *et al.* (1960) autolysed the dried cells in phosphate buffer containing sodium molybdate and biotin under an atmosphere of hydrogen.

Azotobacter vinelandii (OP). Nicholas & Fisher (1960) grew cells in a well-aerated culture for 18 hr. The unwashed cells were suspended in culture medium in which the organism had been grown and the pH brought to 7·0. The suspension was subjected to ultrasonic treatment in the cold using a titanium probe.

Alternatively lysozyme has been used for disrupting the cells (Nicholas & Fisher, 1960). The cell suspension in the culture medium was adjusted to pH 8·0, and sodium pyrophosphate, disodium ethylenediaminetetra-acetic acid (EDTA) and egg-white lysozyme were added. By taking aliquots of this solution every 5 min., the progress of the lysis was followed in a spectrophotometer at 660 mμ. When the disruption of the cells was almost complete the pH was adjusted to 7·2.

Bacillus polymyxa. Grau (1961) and Grau & Wilson (1961) grew cells under anaerobic conditions. Cells were disrupted using a lysozyme and deoxyribonuclease treatment in phosphate buffer under H_2. The nitrogen-fixing system was found in the supernatant fluid after centrifuging at 10,000g for 15 min.

Rhodospirillum rubrum. This organism was grown in light at 30° in nitrogen-deficient medium. The cells were dried and extracted as described for *Clostridium pasteurianum* by Carnahan *et al.* (1960) and Schneider *et al.* (1960).

Chromatium. This organism was grown in the light in a medium containing thiosulphate (Arnon *et al.* 1960; Arnon, Losada, Nozaki & Tagawa, 1961). Details of conditions of growth and the way in which the cell-free extracts were prepared are not given.

Blue-green algae. Schneider *et al.* (1960) grew these organisms at 24° in light, in aerated mineral salt medium. Cell-free extracts were prepared by sonic disruption of freshly harvested cells. The nitrogen-fixing activity was not sedimented by centrifugal forces up to 45,000g for 45 min.

COMPONENTS OF THE NITROGEN-FIXING SYSTEM

Clostridium pasteurianum

Nitrogen fixation in cell-free extracts is dependent on the production of reducing substrates which probably serve as a source of electrons or hydrogen donors for the reduction of nitrogen. Some of these have been identified but others are not yet known. Thus in *Clostridium pasteurianum* the incorporation of nitrogen is coupled to the metabolism of pyruvate. The extracts appear to have all other factors required for fixation since the addition of a range of cofactors has no effect and only in some preparations did coenzyme A stimulate the reaction. The specific effect of pyruvate or α-ketobutyrate may reflect a requirement for some derivative of pyruvate metabolism produced during the phosphoroclastic reaction:

$$CH_3COCOOH + H_3PO_4 \rightarrow CH_3(CO)H_2PO_4 + CO_2 + H_2.$$

The final products of this reaction are acetyl phosphate, H_2 and CO_2. Extraordinarily large amounts of pyruvate are required for a relatively small amount of nitrogen fixation as shown in Fig. 1.

The time course for the incorporation of nitrogen and pyruvate consumption in Fig. 2 shows that 200–400 μmoles of pyruvate is consumed before any nitrogen is fixed. This lag period probably reflects the production of a reducing factor in the phosphoroclastic reaction required for nitrogen fixation. Recently Mortenson, Valentine & Carnahan (1962) isolated an electron-transferring protein that links pyruvate dehydrogenase with hydrogenase in the formation of H_2 from pyruvic acid by the phosphoroclastic reaction. This protein factor, named ferredoxin, contains 0·5 μmole Fe/mg. protein and has absorption

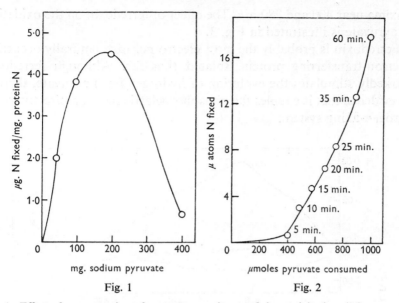

Fig. 1. Effect of concentration of pyruvate on nitrogen-fixing activity in cell-free extracts of *Clostridium pasteurianum*. (From Carnahan *et al*. 1960.)

Fig. 2. Time-course of nitrogen fixation and pyruvate consumption in extracts of *Clostridium pasteurianum*. (From Carnahan *et al*. 1960.)

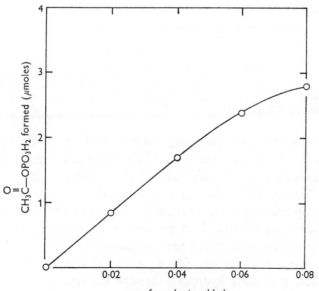

Fig. 3. Catalysis of pyruvate oxidation by ferredoxin in extracts of *Clostridium pasteurianum* as measured by acetyl phosphate formation. (From Mortenson, Valentine & Carnahan, 1962.)

maxima near 400 and 280 mμ. The effect of ferredoxin on the oxidation of pyruvate is illustrated in Fig. 3.

Ferredoxin is probably the most electro-negative naturally occurring electron-transferring protein isolated thus far. Although ferredoxin markedly stimulates the evolution of hydrogen from pyruvate, there is no evidence that it couples the phosphoroclastic reaction directly to the nitrogen-fixing system.

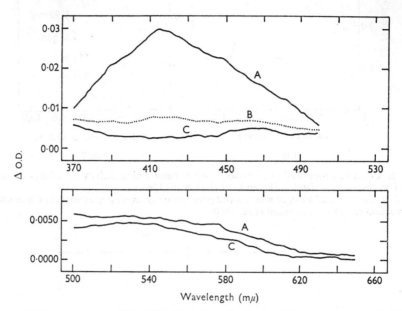

Fig. 4. Difference spectra (CO—H$_2$) of whole cells of *Clostridium pasteurianum* in 0·05 M phosphate buffer pH 6·8. A, CO passed through the extract; B, H$_2$ passed through the extract previously treated with CO; C, H$_2$ passed through the extract (control). (From Westlake, Shug & Wilson, 1961.)

It is of interest that Westlake, Shug & Wilson (1961) had earlier implicated a carbon monoxide-combining iron component in *Clostridium pasteurianum* which was not of the haemochromogen type as shown in Fig. 4. Difference spectra (CO—H$_2$) of crude extracts as well as of alkaline-pyridine treated preparations (flavin removed by adding trichloracetic acid) had an absorption maximum at 415 mμ but no appreciable absorption was noted in the visible region of the spectrum. Reversal of the reaction between CO and the iron compound was obtained by passing hydrogen through the reaction mixture.

Mortenson, Mower & Carnahan (1962) have prepared two fractions from their extracts which singly do not fix nitrogen, but when combined the activity is restored as shown in Table 2.

Table 2. *Nitrogen fixation by* Clostridium pasteurianum *enzymes.*
Determination of two protein requirements

(From Mortenson, Mower & Carnahan, 1962.)

Additions	mg. protein-N	Nitrogen fixed (μg. N/60 min.)
1. Cell extract	18·6	233
2. Nitrogen-activating fraction (NAS)	14·8	0
3. Hydrogen-donating fraction (HDS)	10·8	0
4. Addition of (2)+(3)	14·8+10·8	206

One fraction which contains the phosphoroclastic system and hydrogenase activity (HDS) is the source of reducing activity, and the other has a nitrogenase component (NAS). The phosphoroclastic activity is assayed by pyruvate consumption, and by acetyl phosphate, H_2 and CO_2 production. Hydrogenase was determined by hydrogen reduction of methylene blue. Heat treatment at 60° for 10 min. under H_2, as used by Shug, Hamilton & Wilson (1956) to purify hydrogenase, was employed by Mortenson, Mower & Carnahan (1962) to dispose of the nitrogen-fixing component. This treatment retains the HDS system. When the crude extract was treated with protamine sulphate the bulk of the reducing system was removed leaving the NAS part intact.

Cells grown with ammonia as the sole nitrogen source do not fix nitrogen but the pyruvate metabolism appears to be unimpaired. This was confirmed by coupling HDS and NAS fractions from NH_3-grown cells each with its complementary fraction from N_2-grown cells. The defect appears to be in the NAS fraction from the NH_3-grown cells. They suggest that nitrogen fixation may be repressed in the NAS fraction from the NH_3-grown cells. The addition of NAS fraction from the NH_3-grown cells to the combined NAS and HDS from the N_2-fixing system should show whether or not an inhibitor is present.

Shug *et al.* (1956) have shown that sonic extracts of *Clostridium pasteurianum* exposed to nitrogen undergo spectral changes between 390 and 450 mμ. They observed similar effects of nitrogen in sonic extracts of *Azotobacter* and soybean root nodules at 400 and 550 mμ (Hamilton, Shug & Wilson, 1957; Bergersen & Wilson, 1959). The extracts they examined did not, however, fix nitrogen. Carnahan *et al.* (1960) employed similar techniques to study changes in their active preparations from *C. pasteurianum*. Addition of sodium pyruvate or α-ketobutyrate resulted in increased absorption at 365 mμ as shown in Fig. 5.

The difference spectra were measured under N_2, A and O_2, relative to H_2. Responses to A and O_2 (decreased absorption at 365 mμ) were

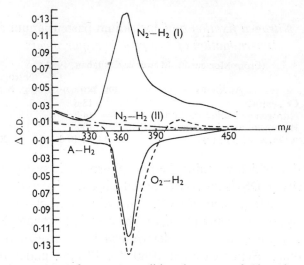

Fig. 5. Difference spectra of pyruvate-conditioned extracts of *Clostridium pasteurianum* exposed to N_2, A and O_2 relative to H_2. For curve II 20 mg., instead of 40 mg. sodium pyruvate was used. (From Carnahan *et al.* 1960.)

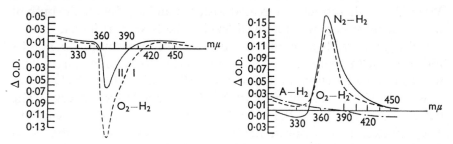

Fig. 6. Difference spectra of pyruvate-conditioned extracts of *Clostridium pasteurianum* exposed to O_2 reversed by H_2. Curve I is the difference spectrum after 15 min. exposure to O_2 at 0°. Curve II represents the difference spectrum as in I followed by exposure to H_2 for 10 min. (From Carnahan *et al.* 1960.)

also obtained and may have been caused by desorption of H_2 since they were reversed by replacing A or O_2 with H_2, but the N_2 response was not reversed as shown in Fig. 6.

The significance of these spectral changes in relation to N_2 fixation is difficult to evaluate since similar results were also reported by Shug *et al.* for bacterial extracts that did not fix nitrogen.

Azotobacter vinelandii

One of the first attempts to prepare cell-free extracts from *Azotobacter chroococcum* was reported by Bach, Jermoljeva & Stepanjan (1934), but Roberg (1936) was unable to reproduce the results. The Wisconsin group, in experiments from 1943 onwards, have tried a variety of techniques to prepare active extracts, but fixation was usually small and inconsistent. They found that these extracts synthesized amino acids from ammonia. The report by Nason, Takahashi, Hoch & Burris (1957) of fixation in sonicates of *Azotobacter* turned out to be due to whole cells in their preparations. Nitrogen is fixed by cell-free extracts prepared by the lysozyme treatment or by ultrasonic breakage of the cells in the medium in which they were grown (Nicholas & Fisher, 1960). Cell-free extracts were prepared by centrifuging at 25,000g for 30 min. Microscopic examination and plating experiments showed that there were no viable cells in the extracts. The results in Table 3 show that they incorporated appreciable amounts of the isotope.

Table 3. *Fixation of ^{15}N by cells and cell-free extracts of* Azotobacter vinelandii (*OP*)

(Atom % ^{15}N excess.)

Whole cells	Ultrasonic treatment	Supernatant after 25,000 g/ 30 min.	Pellet 25,000– 144,000 g/ 2 hr.	Supernatant 144,000 g
2·74	1·18	0·38	0·25	0·10

Table 4. *Uptake of ^{13}N by cell-free extracts of* Azotobacter vinelandii (*OP*) *during* 10 *min. exposure*

(Counts per sec. referred to standard time.)

Fraction	Normal extracts	Extracts boiled for 15 min. before exposure to ^{13}N
1. Supernatant solution (25,000 g/hr.)	348	110
2. Supernatant solution (144,000 g/hr.)	200	80
3. Solution as in fraction (1) flushed with O_2 or N_2 gas for 15 min. after exposure	250	10

After ultrasonic treatment of the cells the bulk of the activity is in the pellet collected between 25,000 and 144,000g. There is an enrichment in the supernatant fluid which may be due to small particles formed during the ultrasonic treatment, since centrifuging for a further 12 hr. sediments the activity. Hydrogenase activity is associated mainly with the particles.

Fixation of nitrogen by cell-free extracts was confirmed by exposing relatively dilute aliquots to ^{13}N at 30° (Nicholas, Silvester & Fowler, 1961). The results in Table 4 indicate that there were substantial counts in the boiled control samples. The tracer nitrogen dissolved in these samples could be removed by flushing the system for 10 min. with either N_2, O_2 or CO_2 after exposure to [13]N. When cells and extracts from micro-organisms that do not fix nitrogen were used as control samples, there was no incorporation of tracer.

Table 5. *Effect of composition of gas on the uptake of* [13]N *by cells of* Azotobacter vinelandii (*O strain*)

Gas mixtures as percentages		Counts per sec. referred to a standard time		
Argon containing carrier-free [13]N	Oxygen	Normal cells	Cells treated with 10^{-3}M KCN	Cells boiled for 15 min. before exposure to tracer
80	20	430	12	15
90	10	450	14	12
95	5	350	10	10
98	2	70	5	5

Extracts flushed with N_2 gas for 10 min. after exposure to the tracer.

Table 6. *Production of* [13]NH$_3$ *in cell-free extracts of* Azotobacter vinelandii (*OP*)

(Counts per sec. referred to standard time.)

Fraction	[13]N	[13]NH$_3$
1. Supernatant solution (25,000 *g*/hr.)	360	260
2. Supernatant solution (144,000 *g*/hr.)	140	130
3. Pellet (25,000–144,000 *g*)	250	210

The results in Table 5 show that 10 % oxygen is required for an active fixation of nitrogen in whole cells. Cyanide inhibited the reaction. The fixation was not increased by exposure times beyond 10 min. Similar results were obtained with cell-free extracts.

Extracts exposed to ^{13}N and flushed with ordinary N_2 gas for 10 min. were put in Conway units and carrier amounts of ammonia added, and after adding potassium carbonate, the ammonia was collected in the hydrochloric acid in the centre well. The results in Table 6 show that most of the counts can be recovered in the acid fraction.

The factor in the medium required for nitrogen fixation is not pyruvate but is probably a reducing substrate and will be discussed in a later section (Electron Paramagnetic Resonance Studies). Extracts of non-fixing micro-organisms did not incorporate tracer nitrogen.

Other micro-organisms

Washed cells of *Bacillus polymyxa* treated with lysozyme yield cell-free extracts that rapidly evolve hydrogen from reduced methyl viologen, formate and pyruvate (Grau & Wilson, 1961; Grau, 1961). Hydrogenase is particulate since 86 % was sedimented at $105,000g$ for 1 hr. The enzyme oxidizing pyruvate with methyl viologen as hydrogen acceptor is soluble. These extracts fix considerable amounts of N_2 when pyruvate is added, but the addition of formate or mannitol had no effect.

The distribution of the ^{15}N-fixation activity in cell fractions is shown in Table 7.

Table 7. *Distribution of nitrogen-fixing activity in lysed cells of*
Bacillus polymyxa

(After Grau, 1961.)

	Pyruvate (μmoles)	Atom % ^{15}N excess
Sediment (3 ml.)	228	0·001
	456	0·008
Supernatant (1·5)	228	0·308
+0·1 M phosphate buffer	342	0·192
pH 6·5 (1·5)	546	0·032
Supernatant (1·5)	228	0·205
+sediment (1·5)	342	0·225
	456	0·108

The fraction containing more of the membrane component (sediment plus supernatant solution) had a higher pyruvate requirement for maximum fixation of N_2 than the supernatant alone. The sediment did not incorporate N_2 despite the fact that the whole fraction was used, whereas only the trichloracetic acid soluble portion of the supernatant fluid was analysed.

There is a short time-lag in fixation after adding pyruvate, and a relatively large amount of pyruvate is required for maximum incorporation of the gas.

Extracts of *Rhodospirillum rubrum* incorporate appreciable amounts of N_2 and appear to have sufficient endogenous reducing substrates, since the addition of pyruvate had little effect on reduction in the light, or in the dark as shown in Table 8. The addition of α-ketoglutarate did, however, stimulate the fixation process (Schneider *et al.* 1960).

The nature of the 'reducing power' in these extracts has not been explored further.

Cell-free extracts of *Chromatium* supplied with a mixture of H_2 and N_2 in the dark absorbed more gas than did those given hydrogen only

(Arnon *et al.* 1960). The addition of NADH* also increased the fixation as shown in Table 9. The enrichment values are, however, low and more active fixation may now have been achieved in these extracts.

Table 8. *Nitrogen fixation by cell-free extracts of*
Rhodospirillum rubrum

(Adapted from Schneider *et al.* 1960.)

Substrate addition	Assay conditions	Atom % ^{15}N excess
None	Light	0·80, 1·15
	Dark	1·04, 0·85
10 mg. pyruvate	Light	1·17, 0·54
	Dark	1·23, 0·73
10 mg. α-keto-glutarate	Light	3·08
	Dark	1·58

Table 9. ^{15}N *fixation in the dark by cell-free extracts of* Chromatium

(Arnon *et al.* 1960.)

	^{15}N atom % excess	
Reductant added	Experiment A	Experiment B
None	0·0255	0·0274
NADH	0·0480	—
Hydrogen gas	0·0362	0·0430

Table 10. *Fixation of nitrogen by cell-free extracts of blue-green algae*

(Adapted from Schneider *et al.* 1960.)

Organism	Atom % ^{15}N excess
Mastigocladus laminosus	0·240
Nostoc muscorum	0·040
Anabaena cylindrica	0·037
Calothrix parietina	0·049

Extracts of a variety of blue-green algae incorporate N_2 in the light at 30° as shown in Table 10 (Schneider *et al.* 1960).

The activity was in the supernatant fluid left after centrifuging the broken cells at 45,000g for 45 min., but the values are low. The most active preparations were generally those from *Mastigocladus laminosus* incubated in the light. The addition of ATP or NADH to the extracts had no consistent effect on nitrogen fixation.

INTERMEDIATES IN FIXATION

Wilson & Burris (1953) defined the 'key intermediate' in nitrogen fixation as 'the inorganic compound at the end of the fixation reaction and the start of the assimilation of the fixed nitrogen into the carbon

---

* The following abbreviations are used: NADH, reduced nicotinamide-adenine dinucleotide; ATP, adenosine triphosphate; FADH, reduced flavin-adenine dinucleotide.

skeleton'. Both hydroxylamine and ammonia have been proposed as the key intermediate.

Blom (1931) suggested that hydroxylamine might be an intermediate although he did not produce substantial experimental evidence in support. Virtanen & Laine (1939) had earlier supported the hydroxylamine hypothesis, since they found oximes in the vicinity of root nodules. They claimed that the oxime of oxalacetic acid was then reduced to aspartic acid. Subsequently Virtanen (1947) conceded that 'it is not impossible that hydroxylamine is chiefly reduced to ammonia and the oximes are formed to a small extent through side reactions from hydroxylamine'. There is no conclusive evidence that oximes are utilized during metabolism of inorganic nitrogen and its compounds, but they probably serve as a non-toxic reservoir of hydroxylamine (Nicholas, 1959).

The following lines of evidence support the ammonia hypothesis proposed by Wilson & Burris (1953), and Newton, Wilson & Burris (1953): (a) ammonia is used without a lag period and in preference to other forms of nitrogen by N_2-fixing organisms, (b) after short exposures of N_2-fixing organisms to tracers, ammonia is more heavily labelled than any other compound, followed by glutamic acid which in most bacteria is formed by reductive amination of α-ketoglutaric acid—a result compatible with ammonia functioning as a key intermediate, (c) the kinetics of nitrogen fixation show that ammonia is the first free intermediate to become labelled, (d) *Clostridium pasteurianum* excretes about 50 % of ^{15}N-labelled N_2 fixed into the medium primarily in the form of highly enriched ammonia, and (e) recent work with cell-free extracts of *C. pasteurianum* and *Azotobacter vinelandii* using ^{15}N and ^{13}N respectively shows an impressive labelling in ammonia.

Since cell-free extracts have a limited capacity for converting ammonia to amino acids, the ammonia builds up to such high levels that it may often be determined by a micro-Conway diffusion technique (Mortenson, 1962).

Thus there is general agreement now that ammonia is the key intermediate in N_2 fixation but the problem of intermediates between N_2 and NH_3 remains to be solved.

A general scheme modified from Burris (1956) is shown in Fig. 7.

Three different ways may be postulated to activate nitrogen gas: reduction, oxidation or hydration.

The oxidation of N_2 is unlikely to occur since the first product would be nitrous oxide which is a competitive and specific inhibitor of N_2 fixation in *Azotobacter vinelandii* (Molnar, Burris & Wilson, 1948;

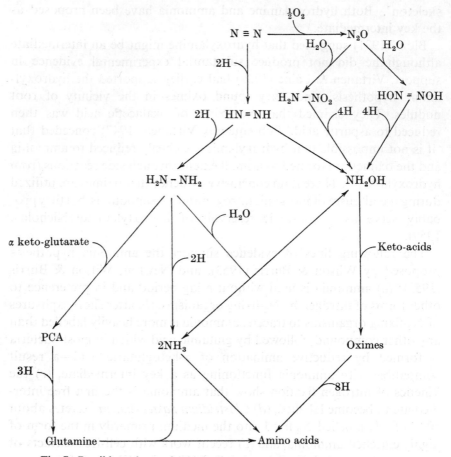

Fig. 7. Possible pathways for biological nitrogen fixation. (PCA = dihydro-pyridazinone-5-carboxylic acid.)

Table 11. *Nitrous oxide inhibition of nitrogen fixation in cell-free extracts of* Clostridium pasteurianum

(Adapted from Schneider, Bradbeer & Burris, 1962.)

pN_2O	mg. N_2 fixed	% of normal
0	27·7	100
0·01	28·0	100
0·1	29·8	89
0·5	3·1	11

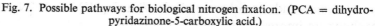

Repaske & Wilson, 1952; Wilson & Roberts, 1952; Virtanen & Lunbom, 1953) and in *Clostridium* sp. (Hino, 1955). Nitrous oxide labelled with ^{15}N is incorporated only slowly by *Azotobacter* and by sliced excised

nodules (Mozen & Burris, 1954). This inhibition has been confirmed recently with cell-free extracts by Schneider, Bradbeer & Burris (1962), as shown in Table 11.

The hydration of nitrous oxide results in a compound $(HNO)_2$ represented by the asymmetrical nitramide NH_2—NO_2 and the symmetrical hyponitrous acid HON=NOH. Neither of these compounds is utilized by *Azotobacter vinelandii* (Mozen & Burris, 1955; Chaudhary, Wilson & Roberts, 1954). Reduction of $(HNO)_2$ would produce hydroxylamine. It is unlikely that this oxidative process is involved in the fixation process and the present view is that it is primarily reductive. The possible intermediates in the reductive sequence are as follows:

$$
\begin{array}{ccccc}
 & N_2 & NH_2OH & NH_2.NH_2 & NH_3 \\
\text{oxidation/reduction} & 0 & -1 & -2 & -3 \\
\text{state of N atom} & & & &
\end{array}
$$

$$
N_2 \xrightarrow{\;6\text{ electrons}\;} 2NH_3
$$

Thus for the reduction of N_2 gas to ammonia, 6 electrons are involved. Should two electron transfers occur, the first product of reduction could be a hypothetical diimide HN=NH which does not necessitate a splitting of the N—N bond. Audrieth & Ogg (1951) were not successful in preparing this compound since it dismutes spontaneously to form hydrazine and nitrogen. It is unlikely that this free compound is an intermediate but it could be associated with organic compounds since a large number of organic compounds are known to contain the —N=N— group. Hydration of the diimide and splitting of the N—N bond would result in hydroxylamine, whereas its reduction would produce hydrazine. Hydration of hydrazine and splitting of the N—N bond could form hydroxylamine and ammonia. Hydrazine could react with alpha-keto-acids to form hydrazones or azines and likewise hydroxylamine would form oximes. The reaction of hydrazine with alpha-ketoglutaric acid to form dihydropyridazinone-5-carboxylic acid (PCA) and its reduction to glutamine was proposed as a mechanism by Bach (1957); see Fig. 8.

There is no evidence for hydroxylamine, hydrazine and PCA occurring as free intermediates in the fixation process. All are extremely toxic to *Azotobacter* at low concentrations (Novak & Wilson, 1948). Pethica, Roberts & Winter (1950, 1954) found that 1×10^{-5} M hydroxylamine inhibited N_2 fixation and ^{15}N-labelled hydroxylamine did not exchange with N_2. Hydrazine is even more toxic than hydroxylamine. Although Suzuki & Suzuki (1954) reported that suspensions of *A. vinelandii* oxidized hydrazine the products were not identified and it may well be that the loss of the compound was due to its adsorption on to the cell suspensions in organic combination.

In recent experiments with cell-free extracts of *Clostridium pasteurianum*, Garcia-Rivera (1961) and Garcia-Rivera & Burris (1962) showed that carrier amounts of hydroxylamine, added after exposure of the extracts to ^{15}N, were not enriched. Hydroxylamine was recovered as the copper salt of pyruvic oxime. Even hydrolysis of the extracts after exposure, to release any bound hydroxylamine that might be present, yielded no enrichment. They made similar experiments to determine whether hydrazine was labelled. Carrier amounts of hydrazine and benzaldehyde were added and the hydrazine was recovered from the preparations, but it was not enriched with ^{15}N. In these experiments the ammonia formed had 18 to 32 atom % ^{15}N excess, levels that are over 10,000 times the amounts detected in hydrazine or hydroxylamine. Negative results were also obtained with PCA.

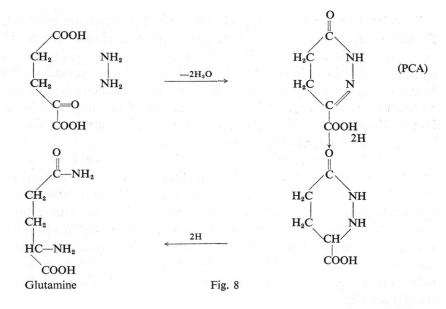

Glutamine   Fig. 8

Attempts have been made to solubilize and fractionate the nitrogen-fixing system in *Azotobacter* with the aid of tracer nitrogen. Preliminary results show that a nitrogenase fraction can be obtained which will incorporate ^{13}N provided a hydrogenase-containing fraction and media factor are added. This combined system incorporates ^{13}N in amounts in excess of ammonia or amino acids formed in the extract. There appears to be an impaired formation of ammonia which might mean either that compounds between N_2 and ammonia were accumulating, or that a labelled enzyme was formed, or both. Attempts were made to equilibrate this preparation with nitrite, hydroxylamine, hydrazine or PCA, but little or no labelling appeared in the exogenously supplied materials. Confirmation of labelled compounds in these extracts, other than ammonia, has come from very short-time exposure experiments with the radioactive tracer. Again no success was achieved with equilibration experiments.

The results therefore do not support the view that *free* hydroxylamine or hydrazine or PCA are intermediate in the fixation process. The results of these and other experiments suggest that either there are no *free* intermediates between N_2 and ammonia, or they are labile. It is feasible that after the initial adsorption of nitrogen on the enzyme surface it remains bound to the enzyme for the stepwise reduction, possible via diimide, until it is released as ammonia (Garcia-Rivera & Burris, 1962). These experiments do not rule out the possibility that enzyme-bound intermediates may be present, and indeed evidence for these has been obtained with ^{13}N experiments.

ELECTRON PARAMAGNETIC RESONANCE STUDIES

Electron paramagnetic resonance spectroscopy (EPR) is a relatively new technique used to identify free radicals and the valency state of transition metal ions and other systems containing unpaired electrons. This type of spectroscopy is concerned with electronic energy levels that do not exist in matter, but are created artificially by exposing the materials to an external magnetic field. Electrons by virtue of their spin possess a magnetic moment, and the unpaired ones, under the influence of an external magnetic field, will orientate themselves in one of two positions: parallel (stable) or antiparallel (unstable). The resultant moments are usually designated $g_{\parallel}$ (parallel) or $g_{\perp}$ (anti-parallel). Because of thermal equilibrium the stable state has only a slightly larger population than the unstable one. Exposure to radio-frequency produces transitions between both states resulting in absorption of energy (stable → unstable) or its emission (unstable → stable). The slightly larger population of the stable state results in a net absorption of energy.

A biological sample in a quartz tube (3 mm. bore) is assayed by freezing it in small volume (0·2 ml. sample containing about 40–50 mg. protein/ml.) in liquid nitrogen ($-175°$). It is then exposed to radiation at a fixed radio frequency whilst the magnetic field is varied. Knowing the strength of the magnetic field at which a given signal (indicating absorption of radiation) is produced, the magnetic moment (g.) of the electrons absorbing the radiation can be calculated. The sensitivity of the method is increased at low temperature and thus enzyme reactions can be stopped by freezing after short exposures to substrate, or any other reactant. This makes possible kinetic studies with enzyme reactions. A general account of the application of the EPR technique to biological problems is given in a recent symposium (Blois, Brown, Lemmon, Lindblom & Weissbluth, 1961).

The work to be reported in this section was carried out in collaboration with Dr Helmut Beinert, Institute for Enzyme Research, Wisconsin University, U.S.A.

EPR spectroscopy together with low-temperature optical reflectance spectroscopy (Beinert, Heinen & Palmer, 1962) have been used to study particles prepared from *Azotobacter vinelandii* (OP). Particles prepared from cells grown without combined nitrogen fix atmospheric nitrogen (Nicholas & Fisher, 1960), whereas those extracted from the same organism, grown with an ammonium salt as the source of nitrogen, do not. These are designated N_2 and NH_4^+ particles respectively. The N_2 and NH_4^+ particles contain a signal at $g_{\parallel} = 2 \cdot 00$, $g_{\perp} = 1 \cdot 94$ and $g_{\perp} = 4 \cdot 3$ superimposed on a broad underlying signal centred close to $g = 2 \cdot 0$. By analogy with spectra obtained with iron complexes in other materials (Beinert, *et al.* 1962), it is likely that the absorption at $g = 2 \cdot 0$ and $4 \cdot 3$ is associated with Fe^{3+}, and the $g = 1 \cdot 94$ signal, which increased markedly on reduction with sodium dithionite, is probably a reduced form of non-haem iron. The latter was first observed in materials of animal origin by Sands & Beinert (1960) and Beinert & Sands (1960). The non-haem iron functions near the flavin in the electron transfer chain (Beinert & Lee, 1961; Beinert, 1962). A sharp signal $g = 2 \cdot 0$ in both types of particles is probably associated with a flavin semiquinone. The N_2 particles had a signal at $g = 1 \cdot 97$ which varied in intensity with batch, but was always more intense in particles that were actively fixing nitrogen. This signal was not present in the NH_4^+ particles. A signal of this g value has been observed previously in nitrate reductase from *Pseudomonas aeruginosa* (Fewson & Nicholas, 1961 *a, b*), in xanthine oxidase (Bray, Pettersson & Ehrenberg, 1961) and in aldehyde oxidase (Rajagopalen *et al.* 1962), and is probably associated with Mo^{5+} in the three enzymes. The Mo^{5+} signal in these enzymes disappears on oxidation or on further reduction and these effects are also observed in the *Azotobacter* N_2 particles. The state of molybdenum after extensive reduction, when the Mo^{5+} signal disappears, is not known at the present time. Nicholas & Stevens (1956), using chromatographic techniques, had earlier implicated a valency change involving Mo^{6+} and Mo^{5+} in nitrate reductase from *Neurospora*, and showed that Mo^{3+} was unlikely to be involved since this reduced nitrate to nitrite non-enzymically.

Preliminary EPR results at $-175°$ together with those of optical reflectance spectroscopy at $-90°$, are shown in Table 12.

Table 12. *Electron paramagnetic resonance (EPR) spectra and reflectance spectroscopy of* N_2-*fixing particles prepared from* Azotobacter vinelandii *(OP)**

	EPR signals (relative heights)			Cytochrome (% reduction)		
		Non-haem				
Treatment	FADH	Fe	Mo^{5+}	b	c	a
1. Particles	36	6	1	30	27	31
2. H$_2$ gas to (1)	36	6	1	28	28	31
3. 10 μl. media to (1)	48	25	4	38	25	26
4. 10 μl. media to (2)	126	100	20	64	55	56
5. He gas to (4)	120	90	17	54	45	50
6. N$_2$ gas to (5)	80	32	4	26	25	20

* Particles prepared by the method of Nicholas & Fisher (1960) fixed 0·22 atom % excess [15]N.

When hydrogen gas was passed through 0·2 ml. suspension of the N_2 particles (30 μg. protein/ml.) in a quartz tube (3 mm. bore) there was little effect on the components of the respiratory chain. The addition of 10 μl. of culture medium (in which the bacteria had been grown) to a sample not treated with hydrogen resulted in small increases in signals for non-haem iron and Mo^{5+}, and little reduction in haem pigments. The effects were large when the hydrogen and media treatments were combined. Subsequent addition of helium gas had little effect, whereas nitrogen partially oxidized the haem components; the non-haem iron signal was diminished and the Mo^{5+} signal disappeared. This reversal with nitrogen was only achieved with particles containing Mo^{5+} that were actively fixing nitrogen. It is significant that the NH$_4^+$ particles did not contain appreciable amounts of Mo^{5+}, and the non-haem iron was only partially reduced (under hydrogen) by adding media from the NH$_4^+$ cultures as shown in Table 13. There was no oxidation when nitrogen was added, indicating the absence of the N_2-fixing system. The non-haem iron signal in the NH$_4^+$ particles was similar in amount to that in the N_2 particles after adding sodium dithionite. Pyruvate or succinate had little effect on the reduction of the electron transfer chain but NADH was an effective donor.

These preliminary results suggest that the electron transfer chain in the N_2-fixing particles of *Azotobacter* is reduced by hydrogen and an unidentified media factor, and at some point, presumably near non-haem iron and Mo, the electrons may be diverted to nitrogen gas as an alternative acceptor to oxygen. This would be analogous to the function of Mo^{6+} and Mo^{5+} in coupling the electron transfer chain to nitrate (Nicholas, 1959, 1961). One interesting feature of this coupled scheme would be the availability of the free energy change from H_2 to O_2 ($\Delta G = -57$ kcal.) for the endothermic reaction involving the first step reduction of nitrogen to diimide or hydroxylamine which requires a large amount of energy ($\Delta G + 102$ kcal.). Further studies are now required with purified fractions of the enzyme complex to decide whether this tentative interpretation of electron transfer during N_2 fixation is correct.

Table 13. *Electron paramagnetic resonance (EPR) spectra of particles** *prepared from* Azotobacter vinelandii *grown with ammonia as sole* N *source*

	EPR signals (relative heights)		
		Non-haem	
Treatment	FADH	Fe	Mo⁵⁺
1. Particles (25,000–144,000 g)	30	2	2
2. H_2 gas to (1)	29	2	2
3. 10 μl. media to (1)	20	4	2
4. H_2 gas to (3)	20	6	2

* Particles prepared by the method of Nicholas & Fisher (1960).

A number of other bacteria that fix nitrogen were examined by EPR spectroscopy. Thus, in crude extracts of *Clostridium pasteurianum*, a Mn^{2+} signal (six equally spaced components centred at $g = 2\cdot01$) is so strong that it obscures that of the non-haem iron ($g = 1\cdot94$–$1\cdot95$). Addition of hydrogen or pyruvate markedly increased the iron signal, and decreased the intensity of Mn^{2+} signal, presumably by the formation of Mn^{2+} complexes. The iron signal can also be differentiated by its decrease in size when the temperature is raised, since Mn^{2+} is unaffected. The separation of the 'nitrogenase' fraction from the crude extract of Mortenson, Mower & Carnahan (1962) eliminates practically all the Mn^{2+}, and a very clear signal of non-haem Fe appeared when the preparation was treated with hydrogen gas or pyruvate. A flavin semiquinone is also present. The Mn^{2+} sextet is recovered in the fraction concerned with the evolution of hydrogen and carbon dioxide from pyruvate after destruction of nitrogenase activity by heat treatment. The absence of a signal for Mo^{5+} in the nitrogenase fraction suggests that non-haem iron ($g = 1\cdot94$) may be an important metal for nitrogen fixation in this organism. Caution is necessary, however, in interpreting the EPR data, since valency states other than Mo^{5+} and Mo^{3+} would not produce signals, and substantial amounts of metals need to be present ($10\,M^{-4}$) before they are detected by this technique.

The non-haem iron ($g = 1\cdot94$) discussed here is quite distinct from the iron-containing protein ferredoxin isolated by Mortenson, Valentine & Carnahan (1962) from *Clostridium pasteurianum*. Ferredoxin is primarily involved in the metabolism of hydrogen in bacteria.

Bacteria that metabolize hydrogen, e.g. *Bacillus polymyxa*, *Aerobacter aerogenes*, *Micrococcus lactyliticus* and *Rhodospirillum rubrum* all exhibit strong Mn^{2+} signals. *Escherichia coli* has a hydrogenase activity when grown anaerobically but it does not fix nitrogen. It exhibits a pronounced Mn^{2+} sextet signal as well as a weaker one for non-haem iron. It would appear that Mn^{2+} is associated with the metabolism of hydrogen in a range of bacteria (Nicholas, Wilson, Heinen, Palmer & Beinert, 1962).

SYMBIOTIC NITROGEN FIXATION

Drs Nutman and Bond (pp. 51, 72) have already discussed certain morphological and physiological aspects of symbiotic fixation in legumes and non-legumes, so that I will confine my remarks to some of the more biochemical aspects of the process.

Legumes

It is established that the amount of N_2 fixed is dependent on the volume and stability of the central core of the nodule (Chen & Thornton, 1940), which is usually composed of enlarged host cells containing the symbiont bathed in haemoglobin (Smith, 1949). Bergersen & Briggs (1958) examined thin sections of soya bean nodules in the electron microscope, and found that the bacteroids were enclosed in double-layered membrane envelopes and were not free in the cytoplasm. They concluded, from a study of nodule development, that the envelopes were derived from membranes surrounding the original infection threads. The outer membranes of the infection threads, and the membrane envelopes enclosing the bacteria, may be homologous with the host-cell membrane (see Nutman, p. 53). Bergersen (1960 *a*, *b*) suggests that the membrane may be concerned with the formation of the enzymes required for the fixation process. It is estimated that the central nodule tissues contain 10^4 membrane envelopes, each containing bacteroids bathed in haemoglobin.

The first consistent fixation by excised nodules was achieved by Aprison *et al.* (1954) and Magee & Burris (1954) in Lincoln variety soya bean after 60 days' growth, in nodules of 5–6 mm. diameter. Crushed nodules were inactive but sliced nodules retained about a quarter of their N_2-fixing capacity which was stimulated by sucrose.

Bergersen (1960*b*), after exposing excised nodules of soya bean to ^{15}N for various periods, fractionated them in 0·15M phosphate buffer. The bacteroids, membrane fractions and supernatant fractions were separated and analysed as shown in Table 14.

Table 14. *Atom % excess ^{15}N in fractions of soybean nodules exposed to $^{15}N_2$ for 15 min.*

(From Bergersen, 1960*b*.)

Expt.	Soluble fraction	Membrane fraction	Bacteroid fraction
1	0·014	0·026	0·000
2	0·004	0·029	0·000
3	0·002	0·026	0·000
4	0·011	0·031	0·000
5	0·000	0·019	0·002
6	0·022	0·030	0·000

The membrane fraction contained significant amounts of the tracer after 15 min. exposure, and this was always higher than that in the soluble portion left after centrifuging at 23,000*g* for 20 min. The bacteroids were unlabelled even after 2 hr. exposure of the excised nodules. He concluded that fixation is associated with membrane fractions. It is difficult to interpret the data, however, since the enrichment values are low.

There is a close connexion between the haemoglobin content of the nodules and their capacity to fix N_2 (Virtanen, 1947, 1955, 1962). Keilin & Smith (1947) showed that it was unlikely that haemoglobin functioned as a catalyst in a mechanism of fixation involving hydroxylamine (Virtanen & Laine, 1939; Colter & Quastel, 1950). Smith (1949) showed that the haem-pigment was not involved in the transport of oxygen for nodule respiration, since amounts of carbon monoxide that inhibited N_2 fixation completely were also sufficient to complex all the haemoglobin, whilst respiration was unchanged. Hamilton *et al.* (1957) found that sonic extracts of soya bean nodules showed shifts of absorption spectra corresponding to the oxidation of haemoglobin from Fe^{2+} to Fe^{3+}, when the gas phase was changed from helium to nitrogen. Haemoglobin in the nodules was reduced in the absence of air, and they suggest that it was reduced by the bacteroids, and oxidized by N_2, although Keilin & Smith (1947) did not observe the oxidized form of haemoglobin in whole or in sliced nodules. It is of interest that Lowe & Evans (1961) have isolated an enzyme complex from bacteroids that will catalyse the reduction of nitrate to nitrite with NADH or succinate as an electron donor. The enzyme transfers electrons from succinate to oxygen or to oxyhaemoglobin isolated from the nodules. Thus there is good evidence that haemoglobin may be involved as a terminal acceptor in nodules, but its precise role in N_2 fixation, if any, remains obscure. Cheniae & Evans (1956) have postulated that nitrate reductase, found in the bacteroids of nodules from soya bean plants growing without nitrate nitrogen, may be induced by oxidized products of nitrogen fixation such as nitrates or nitrites. Bergersen (1961), however, showed that maximum nitrate reductase activity occurred in soya bean nodules 1 week before the nitrogen fixation process could be detected, when structural differentiation of the nodules was incomplete and haemoglobin absent. Thus the nature of the induction of nitrate reductase in the nodules is not clear and its function remains to be elucidated.

A hypothesis for N_2 fixation in the nodule proposed by Bergersen (1960a) is as follows: nitrogen is activated and reduced to ammonia in the membrane fraction. The activated nitrogen is the ultimate acceptor in an electron transport sequence that starts in the bacteroids and involves haemoglobin as a carrier. The host supplies carbon compounds which are partially oxidized by the bacteroids, and thus serve as a source of electrons for the reduction of the activated nitrogen. The products of the incomplete oxidation of the substrates then serve as acceptors of ammonia in the production of amino acids by the bacteroids, which then become available to the host plant. Future experiments will decide whether this hypothesis is valid.

Non-legumes

The present position in regard to physiological studies with nodules of non-legumes has been reviewed by Bond (1959). Leaf, Gardner & Bond (1958) exposed excised nodules of *Alnus* and *Myrica* to ^{15}N for periods ranging from 30 min. to 3 hr., and analysed the nitrogenous constituents. Glutamic acid and citrulline were enriched, and this is in agreement with the data of Miettinen & Virtanen (1952) for alder nodules. By degrading the citrulline molecule, Leaf *et al.* (1958) established that after a 30 min. exposure to ^{15}N, the carbamyl nitrogen atom was five times more highly labelled than the two remaining nitrogen atoms of the ornithine part of the citrulline molecule. The labelling of the carbamyl nitrogen exceeded that of glutamic acid in a given experiment. It is of interest that in animal tissues citrulline is formed from ornithine by condensation of ammonia and carbon dioxide, resulting in a carbamyl group which is attached to the ornithine molecule. The mechanism of citrulline formation in the nodules has not, however, been determined. In nodules of *Myrica*, glutamine had the highest labelling, followed by glutamic acid, but citrulline was not detected. The amide N of glutamine contained more ^{15}N than did the amino nitrogen and this was also true for asparagine, although it was enriched to a lesser degree than glutamine. These results strongly support the formation of ammonia during N_2 fixation, but give no clue to earlier intermediates. Davenport (1960) has found haemoglobin in root nodules of a non-legume (*Casuarina*).

METAL REQUIREMENTS

It is well established that molybdenum, iron and calcium are required by *Azotobacter* and other micro-organisms when fixing nitrogen (Bortels, 1930; Horner & Burk, 1934; Jensen, 1948; Wilson, 1958; Nicholas, Fisher, Redmond & Wright, 1960). Vanadium has a sparing action on molybdenum in some species of *Azotobacter* (Horner, Burk, Allison & Sherman, 1942; Nicholas, 1958; Nicholas *et al.* 1960), and tungstate is a competitive inhibitor of molybdate for N_2 fixation in *Azotobacter* (Keeler & Varner, 1957; Takahashi & Nason, 1957). Iron and molybdenum deficiencies restrict N_2 fixation in *Clostridium pasteurianum* (Nicholas, 1958), and Carnahan & Castle (1957) demonstrated a higher iron requirement when it is grown without combined nitrogen. Winfield (1951) has proposed that fixation occurs by hydrogenation at the site of the iron complex in hydrogenase. The consensus of opinion now is that hydrogenase is primarily an iron-dependent enzyme. Roberts (1959) has recently considered theoretical aspects of the role of metals in nitrogen fixation.

Holm-Hansen, Gerloff & Skoog (1954) showed that cobalt was essential for blue-green algae when fixing nitrogen. Thus the requirement for *Nostoc muscorum* was 0·4 μg. Co/l. Recent investigations have shown that cobalt is essential for certain leguminous species grown under symbiotic conditions, but no response was obtained when the plants were given combined nitrogen (Ahmed & Evans, 1959, 1960, 1961; Reisenauer, 1960; Hallsworth, Wilson & Greenwood, 1960; Delwiche, Johnson & Reisenauer, 1961). Since Lowe, Evans & Ahmed (1960) and Lowe & Evans (1962) have now shown that the micronutrient is essential for growth of *Rhizobium japonicum* in culture medium containing nitrate as the sole source of nitrogen, it is not established unequivocally that cobalt is specifically required for N_2 fixation in the symbiotic system. It is now known that cobalt deficiency markedly reduces nitrate reductase activity in *Rhizobium*, and in a range of micro-organisms grown with nitrate (Nicholas, Maruyama & Fisher, 1961; Nicholas & Wilson, 1962; Nicholas, Kobayashi & Wilson, 1962). A cobalt or vitamin B_{12} requirement has been established for *Azotobacter* when fixing N_2, since the requirement is less when it is grown with ammonium or DL-glutamate (Table 15). Vitamin B_{12} was more effective for growth at equivalent concentrations, and 5–6 dimethylbenzimidazolylcobamide coenzyme has been identified in *Azotobacter* by ionophoretic and enzymic analysis. Since only catalytic amounts of vitamin B_{12} are found, it is doubtful whether there is sufficient present to function directly in stoichiometric amounts as a cofactor in the fixation process (Nicholas *et al.* 1962).

Table 15. *Cobalt or vitamin* B_{12} *requirement for* Azotobacter vinelandii
(OP) shown by relative growth, and vitamin content of cells

	N source			
(μg./l.)	N_2	NO_3^-	NH_3	DL-Glutamate
0	55	51	70	80
0·05 Co	68	60	89	94
0·1 Co	100	100	100	100
0·05 B_{12}	100	100	100	100
Vitamin B_{12} mμg./g. lyophilized cells grown with 0·1 μg./l. Co.	25	15	8	5

It is of interest that Kliewer & Evans (1962) have now found the same cobamide coenzyme in nodules of a variety of legumes and in alder. Since this B_{12} coenzyme has been found necessary for nitrogen fixation in *Azotobacter*, it is likely that a similar role for it will be established in legumes and non-legumes. A further comparison concerns the vitamin B_{12} requirement for the formation of nitrate reductase in micro-organisms (Nicholas, Maruyama & Fisher, 1961; Nicholas *et al.* 1962), and the fact that this enzyme is also present in root nodules of legumes. Cheniae & Evans (private communication) have confirmed that *Rhizobium* from nodules of soya bean deficient in cobalt has a reduced nitrate reductase activity. Thus the B_{12} coenzyme may also have a dual function in the symbiotic system both for nitrate reductase activity and for N_2 fixation.

CONCLUDING REMARKS

The biochemistry of nitrogen fixation is now at an interesting stage since, at long last, cell-free extracts capable of fixing nitrogen gas have been prepared from numerous micro-organisms. It should now be possible to study the complete nitrogen-fixing process at the enzyme level.

Although some progress has been made in separating a hydrogen donating fraction from a nitrogen activating system in extracts, the work is not sufficiently advanced to propose a mechanism for fixation. Further purification will be necessary to elucidate the pathway of electron transfer to nitrogen, and to determine the components of the nitrogenase enzyme. In this connexion the newer techniques of electron paramagnetic resonance (EPR) and reflectance spectroscopy at low temperatures have proved useful.

  The question posed most frequently in discussions on fixation is 'How is N_2 gas, which is chemically inert, activated by a biological system at ordinary temperatures and pressures?' We do not know the answer. One can postulate that the diatomic gas is adsorbed tightly on to the enzyme surface either as a molecule, atom, or ion. The results of experiments with labelled nitrogen in cell-free extracts of *Clostridium pasteurianum* and *Azotobacter vinelandii* suggest that there are no free intermediates between N_2 and ammonia, since there is no exchange with exogenous supply of hydroxylamine, hydrazine, or dihydropyridazinone-5-carboxylic acid (PCA). There is evidence from ^{13}N experiments with partially purified fractions of *Azotobacter* that a labelled enzyme might be formed since there was an impaired incorporation of tracer into ammonia compared with the uptake of ^{13}N by the enzyme. It is likely, therefore, that nitrogen remains bound to the enzyme during the reduction process, and is released finally as two molecules of ammonia. Should the enzyme-intermediate complexes be tightly bound and labile, it will be exceedingly difficult to characterize the types of compounds involved.

  Although little is known about the biochemistry of the symbiotic system, since it involves a complex host-symbiont relationship, there is no doubt that its resolution will be made easier when the precise mechanism of N_2 fixation is established in the free-living microorganisms. The basic activation of N_2 gas may not differ markedly whether it occurs in asymbiotic, symbiotic, aerobic or anaerobic organisms.

  Since the industrial synthesis of nitrogen compounds requires a large expenditure of energy, estimated at about 5 tons of coal per ton of nitrogen, hope has been expressed in some quarters that by solving the mechanism of biological fixation of N_2, the process might then be used for the large-scale production of nitrogenous fertilizers, but this forecast is probably too optimistic. It is likely, however, that the biochemist will find in the N_2-fixing system some novel features of catalysis involving some of the transition metals, e.g. Fe–Mo and possibly cobalt as the 5-6-dimethyl-benzimidazolylcobamide coenzyme.

ACKNOWLEDGEMENTS

I wish to record my thanks to Dr P. W. Wilson, Department of Bacteriology, Dr R. H. Burris, Department of Biochemistry, Dr D. E. Green and Dr H. Beinert, Institute for Enzyme Research, for the hospitality and facilities accorded me in their laboratories during my sabbatical leave 1961–62 at the University of Wisconsin, Madison, U.S.A.

REFERENCES

AHMED, S. & EVANS, H. J. (1959). Effect of cobalt on the growth of soybeans in the absence of supplied nitrogen. *Biochem. biophys. Res. Comm.* **1**, 121.

AHMED, S. & EVANS, H. J. (1960). Cobalt: a micronutrient element for the growth of soybean plants under symbiotic conditions. *Soil Sci.* **90**, 205.

AHMED, S. & EVANS, H. J. (1961). The essentiality of cobalt for soybean plants grown under symbiotic conditions. *Proc. nat. Acad. Sci., Wash.* **47**, 24.

ALLISON, F. E. & MORRIS, H. J. (1930). Nitrogen fixation by blue-green algae. *Science (Lancaster, Pa.),* **71**, 221.

APRISON, M. H., MAGEE, W. E. & BURRIS, R. H. (1954). Nitrogen fixation by excised soybean root nodules. *J. biol. Chem.* **208**, 29.

ARNON, D. I., LOSADA, M. & NOZAKI, K. (1960). Photofixation of nitrogen and photoproduction of hydrogen by thiosulphate during bacterial photosynthesis. *Biochem. J.* **77**, 23.

ARNON, D. I., LOSADA, M., NOZAKI, K. & TAGAWA, K. (1961). Photoproduction of hydrogen, photofixation of nitrogen and a unified concept of photosynthesis. *Nature, Lond.* **190**, 601.

AUDRIETH, L. F. & OGG, B. A. (1951). *The Chemistry of Hydrazine.* New York: John Wiley and Sons.

BACH, A., JERMOLJEVA, Z. & STEPANJAN, M. (1934). Fixation de l'azote atmosphérique par l'intermédiaire d'enzymes estraites de cultures d'*Azotobacter chroococcum*. (Russian with French summary.) *C.R. Acad. Sci. U.R.S.S.* **1**, 22.

BACH, M. K. (1957). Hydrazine and biological nitrogen fixation. *Biochim. biophys. Acta,* **26**, 104.

BEINERT, H., HEINEN, W. & PALMER, G. (1962). Application of combined low temperature optical and electron paramagnetic resonance spectroscopy to the study of oxidative enzymes. In *Enzyme models and enzyme structure. Cold Spr. Harb. Symp. quant. Biol.* no. 14 (in the Press).

BEINERT, H. & LEE, W. (1961). Evidence for a new type of iron containing electron carrier in mitochondria. *Biochem. biophys. Res. Comm.* **5**, 40.

BEINERT, H. & LEE, W. (1962). Further studies on a new type of electron carrier in oxidative enzymes. *Fed. Proc.* **21** (2), 48.

BEINERT, H. & SANDS, R. H. (1960). Studies on succinic and DPNH dehydrogenase preparations by paramagnetic resonance (EPR) spectroscopy. *Biochem. biophys. Res. Comm.* **3**, 41.

BERGERSEN, F. J. (1960a). Biochemical pathways in legume root nodule nitrogen fixation. *Bact. Rev.* **24**, 246.

BERGERSEN, F. J. (1960b). Incorporation of $^{15}N_2$ into soybean nodule fractions. *J. gen. Microbiol.* **22**, 671.

BERGERSEN, F. J. (1961). Nitrate reductase in soybean root nodules. *Biochim. biophys. Acta,* **52**, 206.

BERGERSEN, F. J. & BRIGGS, M. J. (1958). Studies on the bacterial component of soybean root nodules: cytology and organization in the host tissues. *J. gen. Microbiol.* **19**, 482.

BERGERSEN, F. J. & WILSON, P. W. (1959). Spectrophotometric studies of the effects of nitrogen on soybean nodule extracts. *Proc. nat. Acad. Sci., Wash.* **45**, 1641.

BLOIS, M. S., BROWN, H. W., LEMMON, R. M., LINDBLOM, R. O. & WEISSBLUTH, M. (1961). *Free Radicals in Biological Systems.* New York: Academic Press.

BLOM, J. (1931). Ein Versuch die chemischen Vorgänge bei der Assimilation des molekularen Stickstoff durch Mikroorganismen zu erklären. *Zbl. Bakt.* **84**, 60.

BOND, G. (1959). In *Utilization of Nitrogen and its Compounds in Plants. Symp. Soc. exp. Biol.* no. 13. Ed. H. K. Porter. Cambridge University Press.

BORTELS, H. (1930). Molybdän als Katalysator bei der biologischen Stickstoffbindung. *Arch. Mikrobiol.* **1**, 333.

BRAY, R. C., PETTERSSON, R. & EHRENBERG, A. (1961). The chemistry of xanthine oxidase. The anaerobic reduction of xanthine oxidase studied by electron-spin and magnetic susceptibility. *Biochem. J.* **81**, 178.

BURRIS, R. H. (1956). Studies on the mechanism of biological nitrogen fixation. In *Inorganic Nitrogen Metabolism.* Ed. W. D. McElroy and B. Glass. Baltimore: Johns Hopkins Press.

BURRIS, R. H., EPPLING, F. J., WAHLIN, H. B. & WILSON, P. W. (1942). Studies of biological nitrogen fixation with isotopic nitrogen. *Proc. Soil Sci. Soc. Amer.* **7**, 258.

BURRIS, R. H. & WILSON, P. W. (1945). Biological nitrogen fixation. *Ann. Rev. Biochem.* **14**, 685.

BURRIS, R. H. & WILSON, P. W. (1957). Methods for measurement of nitrogen fixation. In *Methods in Enzymology*, **4**, 355. Eds. N. O. Kaplan and S. Colowick. New York: Academic Press.

CARNAHAN, J. E. & CASTLE, J. E. (1957). Some requirements of biological nitrogen fixation. *J. Bact.* **75**, 121.

CARNAHAN, J. E., MORTENSON, L. E., MOWER, H. F. & CASTLE, J. E. (1960). Nitrogen fixation in cell-free extracts of *Clostridium pasteurianum. Biochim. biophys. Acta*, **38**, 188; **44**, 520.

CHAUDHARY, M. T., WILSON, T. G. G. & ROBERTS, E. R. (1954). Studies in the biological fixation of nitrogen. II. Inhibition of *Azotobacter vinelandii* by hyponitrous acid. *Biochim. biophys. Acta*, **14**, 507.

CHEN, H. K. & THORNTON, H. G. (1940). The structure of 'ineffective' nodules and its influence on nitrogen fixation. *Proc. Roy. Soc.* B, **129**, 208.

CHENIAE, G. & EVANS, H. J. (1956). Nitrate reductase from nodules of leguminous plants. In *Inorganic Nitrogen Metabolism*, p. 184. Ed. W. D. McElroy and B. Glass. Baltimore: Johns Hopkins Press.

COLTER, J. S. & QUASTEL, J. H. (1950). Catalytic decomposition of hydroxylamine by haemoglobin. *Arch. Biochem.* **27**, 368.

DAVENPORT, H. E. (1960). Haemoglobin in root nodules of *Casuarina cunninghamiana. Nature, Lond.* **186**, 653.

DELWICHE, C. C., JOHNSON, C. M. & REISENAUER, H. M. (1961). Influence of cobalt on nitrogen fixation by *Medicago. Plant Phys.* **36**, 73.

DREWES, K. (1928). Über die Assimilation des Luftstickstoffs durch Blaualgen. *Zbl. Bakt.* **76**, 88.

FEWSON, C. A. & NICHOLAS, D. J. D. (1961a). Nitrate reductase from *Pseudomonas aeruginosa. Biochim. biophys. Acta*, **49**, 335.

FEWSON, C. A. & NICHOLAS, D. J. D. (1961b). Utilization of nitrate by microorganisms. *Nature, Lond.* **190**, 2.

FOGG, G. E. & WOLFE, M. (1954). The nitrogen metabolism of the blue-green algae (Myxophyceae). In *Autotrophic Micro-organisms*, p. 99. Ed. B. A. Fry and J. L. Peel. Cambridge University Press.

GARCIA-RIVERA, J. (1961). Intermediates in biological nitrogen fixation. Ph.D. thesis, University of Wisconsin, U.S.A.

GARCIA-RIVERA, J. & BURRIS, R. H. (1962). Intermediates in biological nitrogen fixation. Fed. Proc. 21, 399.

GRAU, F. H. (1961). Hydrogen metabolism and nitrogen fixation by cell-free extracts of Bacillus polymyxa. Ph.D. thesis, University of Wisconsin.

GRAU, F. H. & WILSON, P. W. (1961). Cell free nitrogen fixation by Bacillus polymyxa. Bact. Proc. p. 193.

HALLSWORTH, E. G., WILSON, S. B. & GREENWOOD, E. A. N. (1960). Copper and cobalt in nitrogen fixation. Nature, Lond. 187, 79.

HAMILTON, P. B., SHUG, A. L. & WILSON, P. W. (1957). Spectrophotometric examination of hydrogenase and nitrogenase in soybean nodules and Azotobacter. Proc. nat. Acad. Sci., Wash. 43, 297.

HINO, S. (1955). Studies on the inhibition by carbon monoxide and nitrous oxide of an anaerobic nitrogen fixer. Biochem. (Japan), 42, 775.

HOLM-HANSEN, O., GERLOFF, G. C. & SKOOG, F. (1954). Cobalt as an essential element for blue-green algae. Physiol. Plant. 7, 665.

HORNER, C. K. & BURK, D. (1934). Magnesium, calcium and iron requirements for growth of Azotobacter in free and fixed nitrogen. J. agric. Res. 48, 981.

HORNER, C. K., BURK, D. ALLISON, F. E. & SHERMAN, M. (1942). Nitrogen fixation by Azotobacter as influenced by molybdenum. J. agric. Res. 65, 173.

ITERSON, G. VAN, DEN DOOREN DE JONG, L. E. & KLUYVER, A. J. (1940). Martinus Willem Beijerinck, his Life and Work. The Hague: Martinus Nijhoff.

JENSEN, H. L. (1948). The influence of molybdenum, calcium and agar on nitrogen fixation by Azotobacter indicum. Proc. Linn. Soc. N.S.W. 72, 300.

KEELER, R. F. & VARNER, J. E. (1957). Tungstate as an antagonist of molybdate in Azotobacter vinelandii. Arch. Biochem. Biophys. 70, 585.

KEILIN, D. & SMITH, J. D. (1947). Haemoglobin and nitrogen fixation in the root nodules of leguminous plants. Nature, Lond. 159, 692.

KLIEWER, M. & EVANS, H. J. (1962). B_{12} coenzyme content of the nodules of alder and of Rhizobium meliloti. Nature, Lond. 194, 108.

LEAF, G., GARDNER, I. C. & BOND, G. (1958). Observations on the composition and metabolism of the nitrogen-fixing root nodules of Alnus. J. exp. Bot. 9, 320.

LIPMAN, J. G. (1903). Experiments on the transformation and fixation of nitrogen by bacteria. Rep. N.J. agric. Exp. Sta. no. 24, 217.

LOWE, R. H. & EVANS, H. J. (1961). Further studies on a particulate enzyme preparation from nodules of soybean plants. Plant Physiol. 36, 545.

LOWE, R. H. & EVANS, H. J. (1962). Cobalt requirement for the growth of Rhizobia. J. Bact. 83, 210.

LOWE, R. H., EVANS, H. J. & AHMED, S. (1960). The effect of cobalt on the growth of Rhizobium japonicum. Biochem. biophys. Res. Comm. 3, 675.

MAGEE, W. E. & BURRIS, R. H. (1954). Fixation of $^{15}N_2$ by excised nodules. Plant Physiol. 29, 199.

MIETTINEN, J. K. & VIRTANEN, A. I. (1952). The free amino acids in the leaves, roots and root nodules of the Alder (Alnus). Physiol. Plant. 5, 540.

MOLNAR, D. M., BURRIS, R. H. & WILSON, P. W. (1948). The effect of various gases on nitrogen fixation by Azotobacter. J. Amer. chem. Soc. 70, 1713.

MORTENSON, L. E. (1962). A simple method for measuring nitrogen fixation by cell-free enzyme preparations of Clostridium pasteurianum. Anal. Biochem. 2, 216.

MORTENSON, L. E., MOWER, H. F. & CARNAHAN, J. E. (1962). Nitrogen fixation by enzyme preparations. Bact. Rev. 26, 42.

MORTENSON, L. E., VALENTINE, R. C. & CARNAHAN, J. E. (1962). An electron transport factor from *Clostridium pasteurianum*. *Biochem. biophys. Res. Comm.* **7**, 448.

MOZEN, M. M. & BURRIS, R. H. (1954). The incorporation of ^{15}N-labelled nitrous oxide by nitrogen fixing agents. *Biochim. biophys. Acta*, **14**, 577.

MOZEN, M. M. & BURRIS, R. H. (1955). Experiments with nitramide as a possible intermediate in biological nitrogen fixation. *J. Bact.* **70**, 127.

NASON, A., TAKAHASHI, H., HOCH, G. & BURRIS, R. H. (1957). Nitrogen fixation in sonicates of *Azotobacter*. *Fed. Proc.* **16**, 224.

NEWTON, J. W., WILSON, P. W. & BURRIS, R. H. (1953). Direct demonstration of ammonia as an intermediate in nitrogen fixation by *Azotobacter*. *J. biol. Chem.* **204**, 445.

NICHOLAS, D. J. D. (1958). Some aspects of the biochemistry of nitrogen fixation. In *Nutrition of the Legumes*, p. 359. Ed. E. G. Hallsworth. London: Butterworth's Scientific Publications.

NICHOLAS, D. J. D. (1959). In *Utilization of Nitrogen and its Compounds in Plants*. Ed. H. K. Porter. *Symp. Soc. exp. Biol.* no. 13 p. 1. Cambridge University Press.

NICHOLAS, D. J. D. (1961). Minor mineral nutrients. *Annu. Rev. Pl. Physiol.* **12**, 63.

NICHOLAS, D. J. D. & FISHER, D. J. (1960). Nitrogen fixation in extracts of *Azotobacter vinelandii*. *Nature, Lond.* **186**, 735; *J. Sci. Fd Agric.* **11**, 603.

NICHOLAS, D. J. D., FISHER, D. J., REDMOND, W. J. & WRIGHT, M. A. (1960). Some aspects of hydrogenase activity and nitrogen fixation in *Azotobacter* spp. and in *Clostridium pasteurianum*. *J. gen. Microbiol.* **22**, 191.

NICHOLAS, D. J. D., KOBAYASHI, M. & WILSON, P. W. (1962). Cobalt requirement for inorganic nitrogen metabolism in microorganisms. *Bact. Proc.* p. 101; *Proc. nat. Acad. Sci. Wash.* **48**, 1537.

NICHOLAS, D. J. D., MARUYAMA, Y. & FISHER, D. J. (1961). The effect of cobalt deficiency on the utilization of nitrate nitrogen in *Rhizobium*. *Biochim. biophys. Acta*, **56**, 623.

NICHOLAS, D. J. D., SILVESTER, D. J. & FOWLER, J. F. (1961). Use of radioactive nitrogen in studying nitrogen fixation in bacterial cells and their extracts. *Nature, Lond.* **189**, 634.

NICHOLAS, D. J. D. & STEVENS, H. M. (1956). Role of molybdenum in oxidation-reduction processes in *Neurospora* and *Azotobacter*. In *Inorganic Nitrogen Metabolism*, p. 178. Ed. W. D. McElroy and B. Glass. Baltimore: Johns Hopkins Press.

NICHOLAS, D. J. D. & WILSON, P. W. (1962). Metabolism of inorganic nitrogen and its compounds in microorganisms. *Science*, **136**, 328.

NICHOLAS, D. J. D., WILSON, P. W., HEINEN, W., PALMER, G. & BEINERT, H. (1962). The use of electron paramagnetic resonance spectroscopy in studies on functional metal components in microorganisms. *Nature, Lond.* **196**, 433.

NOVAK, R. & WILSON, P. W. (1948). The utilization of nitrogen in hydroxylamine and oximes by *Azotobacter vinelandii*. *J. Bact.* **55**, 517.

PETHICA, B. A., ROBERTS, E. R. & WINTER, E. R. S. (1950). The exchange reaction of hydroxylamine and gaseous nitrogen. *J. chem. Phys.* **18**, 996.

PETHICA, B. A., ROBERTS, E. R. & WINTER, E. R. S. (1954). Studies in biological fixation of nitrogen. 1. Inhibition in *Azotobacter vinelandii* by hydroxylamine. *Biochim. biophys. Acta*, **14**, 85.

RAJAGOPALEN, R., ALEMAN, A., HANDLER, P., BEINERT, H., HEINEN, W. & PALMER, G. (1962). Electron paramagnetic resonance studies of iron reduction and semiquinone formation of metallo-flavoproteins. *Biochem. biophys. Res. Comm.* **8**, 220.

REISENAUER, H. M. (1960). Cobalt and nitrogen fixation by a legume. *Nature, Lond.* **186**, 375.

REPASKE, R. & WILSON, P. W. (1952). Nitrous oxide inhibition of nitrogen fixation by *Azotobacter*. *J. Amer. chem. Soc.* **74**, 3101.

ROBERG, M. (1936). Beiträge zur Biologie von *Azotobakter*. III. Zur Frage eines ausserhalb der Zelle den Stickstoff bindenden Enzymes. *Jb. wiss. Bot.* **83**, 567.

ROBERTS, E. R. (1959). In *Utilization of Nitrogen and its Compounds in Plants*. Ed. H. K. Porter. *Symp. Soc. exp. Biol.* no. 13, p. 15. Cambridge University Press.

SANDS, R. H. & BEINERT, H. (1960). Studies on mitochondria and submitochondrial particles by paramagnetic resonance (EPR) spectroscopy. *Biochem. biophys. Res. Comm.* **3**, 47.

SCHNEIDER, K. C., BRADBEER, C. & BURRIS, R. H. (1962). Effect of gaseous inhibitors on nitrogen fixation in cell-free extracts of *Clostridium pasteurianum*. (In the Press.)

SCHNEIDER, K. C., BRADBEER, C., SINGH, R. N., WANG, L. C., WILSON, P. W. & BURRIS, R. H. (1960). Nitrogen fixation by cell-free preparations from microorganisms. *Proc. nat. Acad. Sci., Wash.* **46**, 726.

SHUG, A. L., HAMILTON, P. B. & WILSON, P. W. (1956). Hydrogenase and nitrogen fixation. In *Inorganic Nitrogen Metabolism*, p. 344. Ed. W. D. McElroy and B. Glass. Baltimore: Johns Hopkins Press.

SMITH, J. D. (1949). Haemoglobin and the oxygen uptake of leguminous root nodules. *Biochem. J.* **44**, 591.

STEPHENSON, M. (1939). *Bacterial Metabolism*, 2nd ed., p. 391. London: Longmans, Green and Co.

SUZUKI, B. & SUZUKI, S. (1954). Hydroxylamine reduction and hydrazine oxidation. *Sci. Rep., Tohoku Univ. (Japan)*, **20**, 195.

TAKAHASHI, H. & NASON, A. (1957). Tungstate as a competitive inhibitor of molybdate in nitrate assimilation and N_2 fixation by *Azotobacter*. *Biochim. biophys. Acta*, **23**, 433.

TÓTH, L. (1952). The role of nitrogen-active microorganisms in the nitrogen metabolism of insects. *Tijdschr. Entomol.* **95**, 43.

TÓTH, L. (1953). Nitrogen active microorganisms living in symbiosis with animals and their role in the nitrogen metabolism of the host animal. *Arch. Mikrobiol.* **18**, 242.

VIRTANEN, A. I. (1947). The biology and chemistry of nitrogen fixation by legume bacteria. *Biol. Rev.* **22**, 239.

VIRTANEN, A. I. (1955). Biological nitrogen fixation. *Proc. 3rd int. Congr. Biochem., Brussels*, p. 425.

VIRTANEN, A. I. (1962). *Biological Nitrogen Fixation*. In *Plant Physiology*, vol. 3. Ed. F. C. Steward. New York: Academic Press. (In the Press.)

VIRTANEN, A. I. & LAINE, T. (1939). Investigations on the root nodule bacteria of leguminous plants. XXII. The excretion products of root nodules. The mechanism of *N*-fixation. *Biochem. J.* **33**, 412.

VIRTANEN, A. I. & LUNDBOM, S. (1953). Inhibition of nitrous oxide of biological nitrogen fixation and uptake of combined nitrogen. *Acta Chem. Scand.* **7**, 1223.

WESTLAKE, D. W. S., SHUG, A. L. & WILSON, P. W. (1961). The pyruvic dehydrogenase system of *Clostridium pasteurianum*. *Canad. J. Bact.* **7**, 515.

WILSON, P. W. (1940). *The Biochemistry of Symbiotic Nitrogen Fixation*. Madison: The University of Wisconsin Press.

WILSON, P. W. (1958). *Asymbiotic Nitrogen Fixation*. In *Handbuch der Pflanzenphysiologie*, p. 9. Ed. W. Ruhland. Berlin: Springer-Verlag.

124   D. J. D. NICHOLAS

WILSON, P. W. & BURRIS, R. H. (1953). Biological nitrogen fixation—a reappraisal. *Ann. Rev. Microbiol.* **7**, 415.
WILSON, T. G. G. & ROBERTS, E. R. (1952). Nitrous oxide as a specific competitive inhibitor of bacterial nitrogen fixation. *Chemistry and Industry,* **4**, 87.
WINFIELD, M. E. (1951). Adsorption and hydrogenation of gases on transition metals. *Aust. J. Sci. Res.* **4** (A), 385.
WINOGRADSKY, S. (1949). *Microbiologie du sol. Problèmes et méthodes.* Cinquante Années de recherches. Œuvres completes. Paris: Masson et Cie.

SOME EFFECTS OF FOREST TREE ROOTS
ON MYCORRHIZAL BASIDIOMYCETES

ELIAS MELIN

Institute of Physiological Botany of the University,
Uppsala, Sweden

INTRODUCTION

A large number of Basidiomycetes are known to live in symbiosis with forest trees such as members of Pinaceae, Betulaceae and Fagaceae, forming ectotrophic or ectendotrophic mycorrhizas with them. Among the Hymenomycetes, tree mycorrhizal fungi have been demonstrated experimentally in the following genera: *Amanita, Boletinus, Boletus, Cantharellus, Clitopilus, Clitocybe, Cortinarius, Entoloma, Lactarius, Lepiota, Paxillus, Russula* and *Tricholoma*, by Melin (1922, 1923 a, b, 1924, 1925 a, b, 1936), Melin & Nilsson (1953), Hatch & Hatch (1933), Doak (1934), Modess (1941), Rayner & Levisohn (1941), Santos (1941), Fries (1942), Norkrans (1949), Hacskaylo (1951, 1953), Hacskaylo & Palmer (1955), Bryan & Zak (1961) Vozzo & Hacskaylo (1961) and others. In addition, there are probably several other genera, for instance, *Gomphidius, Hebeloma, Hydnum, Hygrophorus* and *Inocybe* which contain mycorrhizal fungi. Among the Gasteromycetes only *Rhizopogon* and *Scleroderma* are known to contain mycorrhiza-forming species. Whether Ascomycetes such as *Elaphomyces* and *Tuber* form mycorrhizas with forest trees remains to be established (Melin, 1959 a). *Cenococcum graniforme*, the most unspecific of all known mycorrhizal fungi, is supposed to be the imperfect stage of an Ascomycete (Lihnell, 1942).

It is remarkable that many different Basidiomycetes are able to form mycorrhizas with the same tree species. More than forty species have been proved to form mycorrhiza with *Pinus silvestris*, which has been studied most thoroughly in this respect. The actual number may be many times larger. Even a single tree may be associated with many fungal species at one time.

On present evidence the mycorrhizal Basidiomycetes may be considered as root-inhabiting fungi, in the sense of Garrett (1950, 1956); the hyphae enter the rootlets and, under certain conditions, induce the characteristic mycorrhizal structures. According to this view the fungal symbionts may be primarily parasites which obtain from the host certain metabolites essential for their development. When, however, the symbiotic state of equilibrium is established, the higher partner generally also benefits from its fungal associate. As confirmed lately (see Harley, 1959; Lobanow, 1960), tree mycorrhizas act as nutrient-absorbing organs which, in most habitats, are more efficient than the uninfected roots.

Our knowledge of the mechanisms leading to the formation of normal mycorrhizas is still rather limited and we have therefore made extensive studies in this Institute in the last few years to obtain a better understanding of these fungus–root relationships. In the first instance we have attempted to throw some light upon host effects on the fungus which may be important for the establishment of symbiotic relations, and this aspect of ectotrophic mycorrhizal associations forms the main subject of this paper.

TRANSLOCATION OF CARBON-CONTAINING MATERIALS FROM THE HIGHER PLANT TO THE FUNGAL ASSOCIATE

By means of the isotope technique, Uppsala workers have recently proved that the basidiomycete associate, under pure culture conditions, obtains considerable amounts of carbon-containing materials from its higher partner (Melin & Nilsson, 1957). Pine seedlings were cultured aseptically in Erlenmeyer flasks and inoculated with mycelia of either *Boletus variegatus* or *Rhizopogon roseolus*. When mycorrhizas had formed, the intact seedlings were exposed for 0·5–1 hr. to an atmosphere containing C^{14}-labelled CO_2, at *c*. 70 % of daylight in clear weather. It was clearly demonstrated that carbon-containing compounds formed in photosynthesis are transported in considerable amounts from the root to the fungal associate. A similar flow of organic materials from the myco-trophic plant to its fungal associate may occur in nature, although it may vary quantitatively as well as qualitatively with the external and internal conditions. Among the transported substances there are no doubt some essential to the mycorrhizal fungi and their association with the root.

This finding of Melin & Nilsson is in line with observations of several workers that roots of many higher plants give off many substances when immersed in distilled water, or different solutions. Some root secretions have been identified as amino acids (Virtanen & von Hausen, 1951; Ratner, 1954; Linskens & Knapp, 1955; Rovira, 1956; and others), nucleic acid consituents (Lundegårdh & Stenlid, 1944; Stenlid, 1947; Fries & Forsman, 1951), B-group vitamins (West, 1939; Rovira & Harris, 1961), and different kinds of enzymes (Kouprevitch, 1954;

Ratner, 1954). Roots of mycotrophic plants such as pine have also been found, in this Institute and by Slankis (1958), to release vitamins and amino acids. According to Slankis (1958) they also exude sugars (glucose and arabinose) to some extent.

CARBON SUPPLIES

There is evidence that the mycorrhizal fungi, in their symbiosis with trees, obtain substances from the roots that serve as carbon and energy sources. Several workers (Melin, 1923a, 1936; Hatch & Hatch, 1933; Modess, 1941; Norkrans, 1949, and others) have shown that mycorrhizas are readily formed by several Basidiomycetes in purified sand in the presence of mineral nutrients and very small amounts of 'starter' glucose. No doubt the roots were the source of carbon and energy for the symbiotic fungus in these experiments as well as in other experiments of longer duration (Melin, 1925b).

Frank (1885) suggested that the tree mycorrhizal fungi obtain carbohydrates from their host. Björkman (1942, 1944, 1949), Harley & Waid (1955), and others (see Harley, 1959) have produced evidence supporting this view.

Physiological investigations on tree mycorrhizal Basidiomycetes have indicated that many prefer simple carbohydrates as carbon and energy sources (Melin, 1925b, 1953; How, 1940; Norkrans, 1950, and others). Generally they cannot utilize cellulose, or have only a slight capacity for producing adaptive enzymes for cellulose decomposition. However, there are exceptions to this rule. In a few cases litter-decomposing species such as *Boletus subtomentosus* and *Tricholoma fumosum* were able to form mycorrhizas with pine under pure culture conditions (Lindeberg, 1948; Modess, 1941; Norkrans, 1950). Thus tree mycorrhizal Basidiomycetes differ widely in their behaviour towards complex polysaccharides, even though the majority seem to utilize mainly simple carbohydrates.

It seems unlikely that mycorrhizal Basidiomycetes of this latter group are capable of satisfying their requirements for carbon-containing material from the humus of the forest soils in competition with saprophytic soil micro-organisms, since the humus contains only small amounts of soluble carbohydrates. In nature, tree roots may be the main carbon and energy source of these fungi. Cellulose-decomposing basidiomycete associates, on the other hand, may get their carbon requirements from the soil as well as from the roots. When they are in mycorrhizal association with trees, they probably do not produce cellulase, so long as soluble carbohydrates are available from the roots. However, when the roots no longer provide an excess of such substances, they utilize cellulose (Norkrans, 1950) which they may find in the roots as well as in the soil (Melin, 1962a).

Even though the carbohydrates in the host may be essential to 'sugar-requiring' basidiomycete symbionts, as is the case with many obligate parasites (Allen, 1954), this cannot explain all associations of mycorrhizal Basidiomycetes with tree roots. Neither can it explain the characteristic development of the fungal associates in the rootlets, or the balanced status of the mycorrhizas. This point of view accords with the observations of MacDougal & Dufrenoy (1944, 1946) that new mycorrhizas of normal appearance were formed continuously on detached segments of pine roots which had remained alive for several seasons, indicating that carbon-containing materials suitable for the particular fungal symbiont occurred in the soil used in these experiments. Apparently the mycorrhizal roots, in this case, obtained their carbon supplies from the soil through the fungus.

Harley (1959) also expressed the opinion that free sugars in the host root are not a factor leading to the production of ectotrophic mycorrhizas. Handley & Sanders (1962) suggested that in the search for factors controlling the formation of A- and B-type ectotrophic mycorrhizal associations, factors other than soluble reducing substances in the root should not be neglected.

SUPPLIES OF ESSENTIAL VITAMINS AND AMINO ACIDS

Tree mycorrhizal Basidiomycetes studied by Uppsala workers (Melin & Lindeberg, 1939; Melin & Norkrans, 1942; Melin & Nyman, 1940, 1941; Melin, 1953, and unpublished; Norkrans, 1950) were generally heterotrophic for one or more B-vitamins, when grown in synthetic nutrient medium in pure culture. All those tested were partially or completely dependent on thiamine or its constituent moieties, pyrimidine and/or thiazole (Melin, 1954). Most proved to be relatively thiamine-heterotrophic, and only a few were totally so. The degree of heterotrophy varied widely for different species and to some extent even within the same species (Melin & Nyman, 1941). Chudjakow & Woznjakowskaja (1951) have reported thiamine autotrophic strains of tree mycorrhizal Basidiomycetes, and Rawald (1962) reported such strains in *Tricholoma imbricatum* and *T. pessundatum*, but Swedish strains of these two species studied by Norkrans (1950) required thiamine. Some mycorrhizal

Basidiomycetes have one or more additional vitamin requirements *in vitro*. *Tricholoma imbricatum* is relatively heterotrophic for pantothenic acid (Norkrans, 1950), and *T. fumosum* and *Lactarius deliciosus* are partially heterotrophic for nicotinamide (Norkrans, 1950; Melin, 1953).

Table 1. *Effects of* L-*glutamic acid* (100 μmol./20 ml. *medium*) *and* α-*ketoglutaric acid* (100 μmol.) *on the growth rate of various strains of three pine mycorrhizal Basidiomycetes in buffered ammonium medium*

(In each case the total nitrogen was initially 3 mg./flask (20 ml. medium). Initial pH 5·6.)

	Incubation period (days)	Ammonium phosphate (mg. N/flask)	Glutamic acid (mg. N/flask)	α-Keto-glutaric acid (μmol./flask)	Dry weight (mg.)	Final pH
Boletus variegatus C	12	3·0	—	—	8·0	5·4
		1·6	1·4	—	21·4	5·6
		3·0	—	100	10·6	5·4
B. variegatus E	10	3·0	—	—	7·8	5·2
		1·6	1·4	—	18·7	5·3
		3·0	—	100	14·7	5·3
B. variegatus H	10	3·0	—	—	13·0	4·7
		1·6	1·4	—	22·8	4·9
		3·0	—	100	16·5	4·9
B. luteus E	7	3·0	—	—	12·1	5·1
		1·6	1·4	—	17·6	5·3
		3·0	—	100	13·4	5·2
B. luteus F	14	3·0	—	—	4·8	5·3
		1·6	1·4	—	24·6	5·2
		3·0	—	100	8·2	5·4
B. luteus G	7	3·0	—	—	8·7	5·3
		1·6	1·4	—	15·3	5·4
		3·0	—	100	8·8	5·4
Rhizopogon roseolus B	7	3·0	—	—	11·0	5·3
		1·6	1·4	—	22·9	5·3
		3·0	—	100	16·7	5·3

The majority of tree mycorrhizal Basidiomycetes studied were stimulated *in vitro* by one or more amino acids, particularly glutamic and aspartic acid, or their corresponding keto acids, in the presence of ammonium nitrogen (Norkrans, 1950, 1953; Melin, 1953). However, different species and even different strains have different demands (Melin, 1955, and unpublished). Most species responded positively to glutamic acid but the dose-response curves varied, probably due mainly to different sensitivity to its toxic action. Several species such as *Boletus luteus*, *B. variegatus* and *Rhizopogon roseolus*, reached their optimal growth at levels of 100–500 μmol./20 ml. medium. Among species most sensitive to the inhibitory action of glutamic acid were *B. versipellis* and *Lactarius rufus*, which showed positive growth response only at concentrations up to c. 1 μmol./20 ml. Above this concentration there was marked growth inhibition. Great differences in the response of various tree mycorrhizal Basidiomycetes to aspartic and some other amino acids were also observed.

In nature, tree mycorrhizal Basidiomycetes may obtain essential B-group vitamins and amino acids (or corresponding keto acids) from their hosts as well as from the forest soil (Melin, 1953), but the different response of various fungal associates to these metabolites seems to indicate that this does not offer an explanation for the formation of ectotrophic or ectendotrophic mycorrhizas.

EFFECTS OF OTHER ROOT METABOLITES

In numerous earlier experiments of Melin and his collaborators, many basidiomycete species—several suspected of forming mycorrhiza with trees—did not grow, or developed only very poorly in pure culture; the same experience has frequently been encountered elsewhere. The failure of growth was thought to be due, at least partly, to deficiencies of some complex substances occurring in the higher partner (Melin, 1953). I therefore decided to study the effects of living pine roots on the growth of different types of tree mycorrhizal Basidiomycetes in nutrient solutions (Melin, 1954, 1959 a, 1962).

(a) Qualitative experiments in 'maximum' nutrient medium

In the first group of experiments, aseptic pine roots obtained from root cultures or seedlings grown *in vitro* were tested. Later, growth effects of root exudates of pine seedlings grown in pot cultures were also studied.

(i) *Preliminary experiments.* Aseptic roots of *Pinus silvestris* were cultured in Erlenmeyer flasks according to the method of Slankis (1951) for about 5 months. Their dry weight was then 10–15 mg. The nutrient solution of these root cultures was then replaced by a medium found in this Institute to be especially favourable for the growth of many mycorrhizal Basidiomycetes. The basic solution was generally supplemented with a mixture of B-group vitamins and amino acids (Melin & Das, 1954). For convenience, this will be called the maximum nutrient solution. As inocula, mycelial suspensions (Wikén, Keller, Schelling & Stöckli, 1951) were generally superior to floating mycelia (Melin & Lindeberg, 1939; Norkrans, 1950).

The effects of pine roots on the growth rate of fourteen fungal species are illustrated in Table 2. Although all the responses were strongly positive, growth stimulation varied considerably for different fungi. The weakest response occurred in *Boletus subtomentosus*, the strongest in slow-growing species, such as *Russula xerampelina*, *Pholiota caperata* and *Cortinarius glaucopus*. In *R. xerampelina* practically no growth occurred in the control series in 38 days, whereas with addition of the roots there was an average mycelial yield of 25 mg. dry weight. It is noteworthy that the mycelia in this case developed almost entirely around the root, as illustrated in Pl. 1, fig. 1.

Table 2. *Growth-promoting effect of cultured roots of* Pinus silvestris *on various mycorrhizal Basidiomycetes added as mycelial suspensions to nutrient solution supplemented with* 19 *amino acids and* 10 *vitamins*

(Initial pH 5·1.)

Fungus	Days of incubation	Control		Added pine roots	
		Dry weight (mg.)	Final pH	Dry weight (mg.)	Final pH
Amanita muscaria (L. ex Fr.)	22	22·5	3·6	51·7	3·2
A. pantherina (DC. ex Fr.)	21	8·8	4·3	32·7	3·3
Boletus bovinus L. ex Fr.	13	14·4	3·8	34·2	3·5
B. edulis Bull. ex Fr.	14	5·9	4·9	23·0	4·0
B. Grevillei Klotzsch, *E*	13	14·9	4·7	43·7	3·3
B. luteus L. ex Fr., *A* *	10	33·0	3·4	69·5	3·2
B. subtomentosus Sw. ex Fr.	11	23·2	3·3	35·3	3·3
B. variegatus Sw. ex Fr., *D*	10	12·2	4·0	62·4	3·9
Cortinarius glaucopus (Schaeff. ex Fr.) Fr.	21	1·4	5·0	19·0	4·2
C. multiformis Fr.	14	4·7	5·0	12·2	4·4
Lactarius deliciosus (L. ex Fr.) Fr.	21	5·0	4·6	13·0	4·1
Pholiota caperata Pers. ex Fr.	22	1·2	5·0	21·4	4·3
Rhizopogon roseolus (Corda) Th. Fr., *B*	10	11·8	4·2	74·0	3·3
Russula xerampelina Schaeff. ex Fr.	38	0·1	4·9	24·9	3·4

* Inoculated as floating mycelia.

The pH of the culture media generally fell considerably during the experiment, although much more rapidly with, than without added roots. There was no correlation between these pH changes and the effect of roots on fungal growth.

It was assumed from these and similar experiments that the pine roots produce—besides vitamins and amino acids—one or more growth-promoting metabolites called factor M, which are essential to the growth of the fungal symbionts. *Russula xerampelina* could apparently utilize the nutrient solution only in the presence of this factor, indicating that it was totally, or almost totally heterotrophic for the M-factor. In most other fungi the M-factor increased the rate of growth but did not appreciably affect the maximum mycelial yields. This indicates that these species were themselves capable of synthesizing the M-factor but not as rapidly as needed for optimal growth. They may therefore be considered as partially heterotrophic for this factor.

The M-factor is not specific for pine roots and is also produced by the roots of many other species (Melin, 1954; Melin & Das, 1954). As tomato roots grow easily in pure culture, they were used to work out methods for further studies of root effects on mycorrhizal Basidio-mycetes.

(ii) *Diffusible and non-diffusible M-factor*. The M-factor seems to be composed of at least two components (Melin, 1955, 1959a, 1962). This was suggested by the following observation.

When the above-described experiment was discontinued, some rootlets which were surrounded by mycelium and sometimes somewhat swollen (Slankis, 1948), were fixed and embedded according to Jackson (1947). L. Orrhage (unpublished) found that hyphae of the mycorrhizal fungi had generally entered the cortex of these rootlets, mainly intercellularly but also intracellularly, although no mycorrhizal structures were observed. This entry into the rootlets may indicate that they contained not only diffusible but also non-diffusible material, essential to the fungi in a maximum nutrient solution.

Table 3. *Effects of cultured primary pine roots and their exudate on the growth-rate of* Boletus variegatus D *in maximum nutrient solution at 25°*

(Incubation period 6 days. Initial pH 5·2. Average of three observations with standard errors.)

Treatment	Dry weight of mycelium (mg.)	Final pH
A. Control	8·2 ± 1·5	4·2
B. Exudate (diffused out from one root in 6 days)	15·4 ± 1·2	3·8
C. Root after 6 days of exudation	24·5 ± 1·7	3·5
D. Root after 6 days of exudation and extraction at 100°	21·5 ± 2·5	3·6
E. Fresh specimen of cultured root	27·6 ± 2·0	3·5

Further evidence for this assumption was obtained from comparison of the growth of *Boletus variegatus* with added root exudates, and in the presence of roots with different pre-treatments (Table 3). Six-month-old cultured pine roots were placed in maximum nutrient solution for 6 days (treatment B), and then transferred to new flasks with the same medium. Half the roots were transferred directly (treatment C), the other half were extracted at 100° in redistilled water for 5 min. and water-washed repeatedly before re-suspension (treatment D). In a fifth treatment, fresh specimens of cultured roots were added to maximum nutrient solution. All flasks were inoculated simultaneously with measured amounts of a hyphal suspension of *Boletus variegatus*. The exudate caused a doubling of the mycelial yield after 6 days as compared with the control (Table 3), but the roots which had given off this exudate (treatment C), and also the extracted roots (treatment D) showed a considerably higher growth-promoting activity than the exudate.

These and several similar experiments support the view that the M-factor contains at least two substances, one diffusible through plasma membranes, the other bound within the cells. The latter is also available to the mycelia, presumably as a result of their enzymic activities. Treatment B contained only diffusible M-factor exuded into the medium before the fungal suspension was introduced, whereas treatment D contained only non-diffusible M-factor, i.e. substances bound in the root cells. In treatments C and E, on the other hand, there were available to the fungus the non-diffusible as well as the diffusible M-factor, the latter being liberated from the root continuously during the experimental period for 6 days. To judge from other experiments not reported here, the total amount of diffusible M-factor released in treatment E was somewhat larger than that in treatment C.

The relative amounts of the two kinds of substances could not be determined. Judging from the growth-promoting effects in treatments B and D, the bound M-factor had a somewhat higher activity than the diffusible one. However, the two treatments were not quite comparable, as the roots in treatment D had been heated.

The greater growth-promoting effect in treatment C compared with treatment B may be mainly a result of a synergistic action of diffusible and non-diffusible M-factor. It is noteworthy that both substances were inactivated in the presence of small amounts of adenine or its derivatives (Melin, 1959*b*), and this may indicate that they are chemically related.

(iii) *Exudate of aseptic attached pine roots.* Exudates from aseptic attached pine roots were obtained through the courtesy of Dr V. Slankis, who cultured seedlings of *Pinus silvestris* in closed glass cylinders (Slankis, 1951, pp. 44–6). In his experimental arrangement the roots grew aseptically in the medium, whereas shoots developed in the free air. The solution was aseptically changed as required, and was continuously aerated. Roots of 8-month-old seedlings were allowed to exude for 5 days in redistilled water, or in fresh nutrient solution. The solution containing the exudate was added aseptically to maximum nutrient solution in varying proportions, the final level of nutrients being adjusted to be the same in all treatments. The growth-promoting effect of the exudate on *Rhizopogon roseolus* is illustrated in Fig. 1; *Boletus luteus* responded similarly.

Experiments with attached roots of pine seedlings grown aseptically in purified sand or terralite in Erlenmeyer flasks for 1–2 years, gave similar effects to those obtained with cultured roots, and it was concluded that the diffusible M-factor was also released from attached roots.

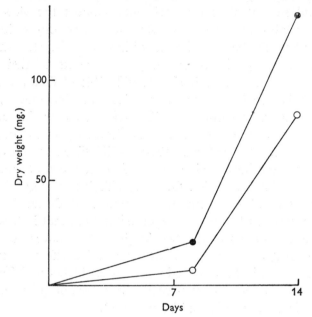

Fig. 1. Growth produced by floating inoculum of *Rhizopogon roseolus B.* in maximum nutrient solution, without supplement (hollow symbols) or supplemented with root exudate from aseptically cultured pine seedlings (solid symbols), suspended in redistilled water for 5 days.

(iv) *Auxanographic experiments*. The growth-promoting effect of the M-factor has been clearly demonstrated with several strains of *Boletus variegatus* and other mycorrhizal Basidiomycetes by means of the auxanographic method of Beijerinck (Melin, 1954, 1959*a*, 1962). The results of these and kindred experiments were reproducible only if washed, homogeneous mycelial suspensions were used as inocula. Measured amounts of mycelial suspensions, sieved aseptically through fine wire gauze, were thoroughly mixed with molten agar, and roots were then transferred, directly or in celluloid sacks, to the still unsolidified agar medium. If the mycelial portions had a proper size and density, species such as *B. variegatus* (for which nylon gauze with a hole size of

0·025 mm.² after repeated autoclaving and drying has been used success-fully) developed strongly in the first days around the roots on the plate, resulting in a distinct auxanogram demarcated by the limit of the diffusible M-factor (Pl. 1, fig. 2).

Slowly growing species such as *Pholiota caperata*—considered more deficient for the M-factor—have also been used in auxanographic experiments (Pl. 2, figs. 3, 4). For this species, however, suspensions of relatively large mycelial fragments had to be used to get an auxanogram

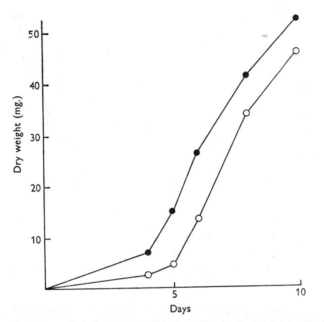

Fig. 2. Growth curves of *Boletus variegatus* (Sw.) Fr., strain *K*, in maximum nutrient solution without (hollow symbols) or with (solid symbols) cultured pine roots (3 months old). Inoculation with hyphal suspension.

of *Boletus variegatus* type (Pl. 2, fig. 3). Mycelial particles developed only in contact with the root (Pl. 2, fig. 4), which may indicate that they needed the synergistic action of diffusible and non-diffusible M-factor for development.

(v) *Growth response of various types of tree mycorrhizal Basidiomycetes.* The growth curves of several tree mycorrhizal Basidiomycetes were compared in maximum nutrient solution with and without cultured pine roots. Some results are illustrated in Figs. 2 and 3. Fig. 2 shows that *Boletus variegatus* responded to cultured pine roots as it did to cultured tomato roots (Melin & Das, 1954). Its growth rate was increased by the M-factor, particularly in the first stage of development, but the maximum yield reached the same value as the control. Several other tree mycorrhizal Basidiomycetes such as *Rhizopogon roseolus*, *Lactarius rufus* (Scop. ex Fr.) Fr. and *L. mitissimus* (Fr.) Fr., produced similar growth curves.

The behaviour of three *Russula* species, illustrated in Fig. 3, was quite different. *R. fragilis* Pers. ex Fr. is a mycorrhizal symbiont with *Pinus silvestris* and *P. montana* (Melin, 1924, 1925*b*), and *R. sardonia*

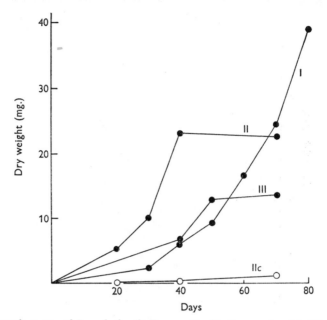

Fig. 3. Growth curves of *Russula fragilis* Pers. ex Fr. (I), *R. aeruginea* Lindbl. ap. Fr. (II) and *R. sardonia* Fr. (III) in maximum nutrient solution with supplement of cultured pine roots. Maximum mycelial yield of controls was 0·4 mg. for *R. fragilis*, 0·1 mg. for *R. sardonia*, and II*c* shows the control yield for *R. aeruginea*.

and *R. aeruginea* are probably also tree mycorrhizal symbionts (Lundell & Nannfeldt, 1934, 1936; Lange, 1940). *R. fragilis* produced a growth curve (I) similar to that of *R. xerampelina* (Melin & Das, 1954). In the control series very poor growth occurred, resulting in a mycelial yield of 0·4 mg. over a period of 80 days, and no control curve is therefore given in Fig. 3. With added pine roots, however, the average mycelial yield was 40 mg., the growth rate increasing continuously during the whole incubation period after a rather long lag phase. Thus *R. fragilis* was able to utilize the nutrient solution only in the presence of the M-factor, indicating that—in contrast to *Boletus variegatus*—it was totally heterotrophic for this factor under the experimental conditions. *Lactarius helvus* (Fr.) Fr. behaved in the same way. In *R. aeruginea* the pine root supplement also stimulated growth which, however, stopped suddenly after 40 days (Fig. 3, curve II). *R. sardonia* (Fig. 3, curve III) showed about half the growth stimulation of *R. aeruginea* and a very similar flattening of the growth curve after 50 days. These two species are therefore also totally or almost totally heterotrophic for the M-factor. At present it is not possible to say why they ceased growth after 40–50 days. One possible explanation is that they were more sensitive than *R. fragilis* to a diffusible inhibitor discussed below, which may have increased to a detrimental level in the culture flasks.

(b) Effects of various amounts of pine root exudate

It seemed desirable also to investigate the effect of root exudates quantitatively. Cultured pine roots as well as roots of pine seedlings grown in pots were used as test material. Amounts of exudate were measured in arbitrary units, one unit being the amount of active substance diffusing into redistilled water at 4° in 6 days from a quantity of living root equivalent to 1 mg. dry weight.

(i) *Exudate of cultured pine roots.* The response of *Boletus variegatus* to 0·3–40 units of exudate from aseptically grown roots of *Pinus silvestris* is shown in Fig. 4, curve I. Growth is expressed as percentage increase or decrease over controls (relative growth), which are represented by the horizontal line at 100. The final amount of nutrients was adjusted to be the same in all flasks. The greatest growth-promoting effect of the diffusible M-factor was obtained at levels between 5 and 10 units. At concentrations higher than 20 units the exudate exerted an inhibitory action. An amount of 40 units inhibited fungal growth by 70 % over a period of 8 days. The experiment thus confirms previous reports (Melin & Das, 1954; Melin, 1955; Melin, 1959a, 1962) that the exudate of cultured pine roots contains also a growth-inhibiting principle.

Two possible explanations of the inhibiting effect at higher concentrations may be suggested: (1) that cultured pine roots contain—in addition to the M-factor and other growth-promoting metabolites such as B-group vitamins—one or more growth-inhibiting substances that also diffuse from the intact root cells into the surrounding medium; (2) that the diffusible M-factor had a promoting or inhibiting effect according to its concentration.

Several observations favour the first alternative. When living primary pine roots were freeze-dried or lightly ground and placed aseptically in redistilled water, the diffusible M-factor was almost inactivated (presumably by some component of the protoplast released through the injured plasma membrane), while the activity of the inhibitor increased. This is shown by comparison of curves I and III in Fig. 4. If, however, the freeze-dried or ground pine roots were autoclaved before being put in distilled water, the growth-promoting activity of the released M-factor was almost as high as for untreated roots (Fig. 4, curves I and II). It therefore appears likely that the intact pine roots also released into the surrounding medium an inhibitor which, at certain concentrations, counteracted the growth-promoting effect of the M-factor. In comparison with this the inhibitor seems to have a rather low rate of dif-

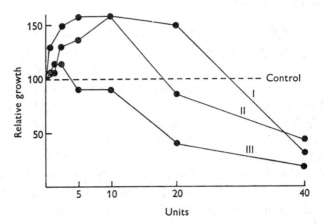

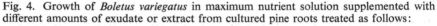

Fig. 4. Growth of *Boletus variegatus* in maximum nutrient solution supplemented with different amounts of exudate or extract from cultured pine roots treated as follows:

   Curve I = exudate collected from roots shaken lightly in redistilled water for 6 days at 4°.
   Curve II = extract from roots freeze-dried and then autoclaved before extraction.
   Curve III = extract from roots freeze-dried before extraction.

Incubation period 8 days. Dry weight of control = 6·8 mg.

fusion through the plasma membranes of intact root cells, as previously postulated by Melin (1955). When the membranes were injured—by freeze-drying or by light grinding of the roots—the amount of inhibitor liberated was very much increased. Inhibitory effects of root exudates have been reported also from other symbiotic plants such as legumes (Nutman, 1956).

   The amounts and proportions of the liberated M-factor and the inhibitor varied in different parts of the root, as was frequently demonstrated by means of the auxanographic method. The upper main axis of pine roots cultured for 6 months gave off the inhibitor in amounts sufficient to prevent completely the growth of *Boletus variegatus*, Pl. 3, fig. 5, whereas the M-factor was ineffective. On the other hand,

secondary rootlets arising from the same roots favoured fungal development (Pl. 3, fig. 6). This is in line with the finding of Lundegårdh &
Stenlid (1944) that the exudation of nucleotides is much greater from
parts of the root that are still growing than from somewhat older parts.

MacDougal & Dufrenoy (1944) provided some cytochemical evidence
that the pine root may contain a non-diffusible inhibitor affecting the
penetration of the fungal associate in the mycorrhizas. This may be
true, but there is as yet insufficient experimental evidence for acceptance
of the hypothesis.

(ii) *Exudates of roots of pine seedlings cultured in pots*. In 1960 seeds
of *Pinus silvestris* from the same tree as in the previous series of experiments were grown in a sand-terralite mixture (1:8) in pots, and irrigated
with a basic mineral nutrient solution (Melin, 1936) containing different
amounts of KH_2PO_4 and $(NH_4)_2HPO_4$. The seedlings did not appear to
have formed mycorrhizal associations.

Rinsed and washed roots of harvested seedlings were suspended in
redistilled water as described. As the roots were inhabited by saprophytic rhizosphere microbes, particularly bacteria, the dissolved exudates had to be sterilized; parallel experiments were performed with
filtered and autoclaved exudates. No attempt was made in these experiments to distinguish between exudation of intact root (and microbial)
cells, and the liberation of substances resulting from breakdown and
autolysis of dead root fragments, but it is thought that, as with aseptically
cultured roots, the released material mainly consisted of root exudates.

Dose-response curves for *Boletus variegatus* are given in Fig. 5. It
will be seen that the diffusible M-factor had its maximum activity at
very low levels. Inhibition began at levels of *c*. 10 units and with
20 units the fungal growth was inhibited about 90 % over a period of
10 days. Root exudates of pot-grown pine seedlings fed with nutrient
solutions containing respectively 1/20 and five times as much nitrogen
and phosphorus, produced essentially similar growth-response curves,
except that stimulation was greater from small amounts of autoclaved
exudate from roots grown at the higher nutrient level, and both filtered
and autoclaved exudates from such roots were already inhibitory at
lower concentrations (*c*. 7 units). As shown in Fig. 5 exudates were
always considerably more stimulatory after autoclaving than after
filtration.

It is interesting to compare the results of this series of experiments
with those presented in Fig. 4. In both series the exudates promoted or
inhibited growth of the fungus according to their concentrations.
However, the root exudates of the pot-grown seedlings became inhibitory

E. MELIN

at much lower dosage and it is therefore assumed that they contained much larger amounts of the inhibitor, than did the exudates of cultured roots.

The simplest explanation of these differences is that they may, at least in part, be connected with a somewhat different metabolism in cultured and attached roots, and the results indicate that exudation of diffusible inhibitor from the roots varies quantitatively with internal and external conditions.

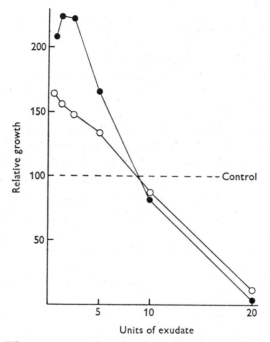

Fig. 5. Effect of different amounts of exudate from pot-grown seedlings of *Pinus silvestris* on *Boletus variegatus* F. Exudate was filtered (hollow symbols) or autoclaved (solid symbols) before adding to maximum nutrient solution. Incubation period 10 days. Dry weight of controls = 7·6 mg.

It seems surprising that the activity of the M-factor was regularly lower in the filtered than in the autoclaved exudate. A possible explanation may be a gradual inactivation of the M-factor in the filtered exudate by some heat-labile component of the protoplast, which had escaped from the root and/or the microbial cells.

There was evidence that amounts and proportions of released M-factor and inhibitor varied considerably in different parts of the root, and for the problem of mycorrhizal formation it may be of great importance to learn the behaviour in this respect of parts of the roots which are still growing.

Extensive studies have been made at this Institute to identify the substances concerned in the effects of the M-factor. H. Nilsson (1960, Report to the Swedish Council of Natural Science, unpublished), has found that the diffusible M-factor could be replaced in its effect on hyphal growth by diphosphopyridine nucleotide (DPN), which functions as coenzyme in many hydrogen transfer processes. The chemical nature of the diffusible inhibitor has not so far been identified.

CONCLUSION

The results obtained by the author and by other workers support the view that a complex of substances produced by the associated partners is involved in the formation of tree mycorrhizal symbiosis.

Some data from the present investigation suggest that excretion of growth-promoting and inhibiting root metabolites, such as the M-factor and the inhibitor, also occur under natural conditions, although the process of liberation of these substances in distilled water may not necessarily be identical with that in soil solutions. There is some indication that the exposure to distilled water may cause some injury of the plasma membranes (Laties, 1954; Fischer, 1956) and soil solutes may influence the permeability of the membranes differently. Thus, the root exudate may vary in amount and quality under different conditions.

In any case, it seems evident that, prior to the initial infection, the host greatly affects the fungal associate outside the rootlet by the liberation of material essential to germination (Melin, 1959a) and growth of the fungus. The most active growth-promoting constituent of the exudates seems to be the M-factor, although thiamine and occasionally some other B-group vitamins are also essential to the fungi concerned.

The stimulation of the fungus by the exudate appears to influence the fungal production of auxin, which was shown by Slankis to induce a morphogenetic effect on the rootlets of pine. He found that the mycorrhizal mycelia, by releasing this type of hormone, cause an increase in the width of the rootlets and a profuse dichotomous branching similar to that of pine mycorrhiza (Slankis, 1948, 1951, 1958).

The infection of the rootlets by the fungal associate may depend on the availability of energy material in them (Björkman, 1942, 1944), and on the bound M-factor. However, the mechanism of formation of such structures as the Hartig net, and the mycorrhizal sheath, whether attributable to a morphogenetic effect of the inhibitor or of some unknown hormone, still remains to be investigated.

The inhibiting principle produced by the root appears to play a most important part in the establishment of symbiotic relationships. The diffusible inhibitor may, at least in part, determine the susceptibility of the rootlet to infection. It may also, possibly together with a non-diffusible inhibitor (MacDougal & Dufrenoy, 1946), be the agent controlling the extension of the mycorrhizal hyphae in the root.

Thus, the symbiotic relations seem to be largely controlled by growth-promoting as well as growth-inhibiting substances.

It remains to be investigated whether the amounts of active substances in the rootlets are affected by environmental and/or internal factors known or presumed to influence ectotrophic mycorrhiza formation.

Finally, it must be emphasized that much more investigation is necessary before the mechanisms of ectotrophic mycorrhizal formation have been fully explored. As yet only a start has been made in our attempts to understand the relationships between the two symbiotic partners.

REFERENCES

ALLEN, P. J. (1954). Physiological aspects of fungus diseases of plants. *Annu. Rev. Pl. Physiol.* 3, 225.

BJÖRKMAN, E. (1942). Über die Bedingungen der Mykorrhizabildung bei Kiefer und Fichte. *Symb. bot. upsaliens.* 6, 1.

BJÖRKMAN, E. (1944). The effect of strangulation on the formation of mycorrhiza in pine. *Svensk Bot. Tidskr.* 38, 1.

BJÖRKMAN, E. (1949). The ecological significance of the ectotrophic mycorrhizal association in forest trees. *Svensk Bot. Tidskr.* 43, 223.

BRYAN, W. C. & ZAK, B. (1961). Synthetic culture of mycorrhizae of southern pines. *Forest Sci.* 7, 123.

CHUDJAKOW, J. M. & WOZNJAKOWSKAJA, J. M. (1951). [Reinkultur von Mykorrhizapilzen.] *Microbiologiya*, 20, 1. (In Russian.)

DOAK, K. D. (1934). Fungi that produce ectotrophic mycorrhizae of conifers. *Phytopathology*, 24, 7.

FISCHER, H. (1956). Ionenwirkungen. *Handb. Pflanzenphysiol.* 2, 706.

FRANK, A. B. (1885). Über die auf Wurzelsymbiose beruhende Ernährung gewisser Bäume durch unterirdische Pilze. *Ber. dtsch. bot. Ges.* 3, 128–45.

FRIES, N. (1942). Einspormyzelien einiger Basidiomyceten als Mykorrhizabildner von Kiefer und Fichte. *Svensk Bot. Tidskr.* 36, 151.

FRIES, N. & FORSMAN, B. (1951). Quantitative determination of certain nucleic acid derivatives in pea root exudate. *Physiol. Plant.* 4, 410.

GARRETT, S. D. (1950). Ecology of the root-inhabiting fungi. *Biol. Rev.* 25, 220.

GARRETT, S. D. (1956). *Biology of Root-Infecting Fungi*, pp. 1–293. Cambridge University Press.

HACSKAYLO, E. (1951). A study of roots in *Pinus virginiana* in relation to certain Hymenomycetes suspected of being mycorrhizal. *J. Wash. Acad. Sci.* 41, 399.

HACSKAYLO, E. (1953). Pure culture syntheses of pine mycorrhizae in terralite. *Mycologia*, 45, 971.

HACSKAYLO, E. & PALMER, J. G. (1955). Hymenomycetous species forming mycorrhizae with *Pinus virginiana*. *Mycologia*, 47, 145.

HANDLEY, W. R. C. & SANDERS, C. J. (1962). The concentration of easily soluble reducing substances in roots and the formation of ectotrophic mycorrhizal associations—a reexamination of Björkman's hypothesis. *Plant & Soil*, 16, 42.

HARLEY, J. L. (1959). *The Biology of Mycorrhiza*. In *Plant Science Monographs*. Ed. N. Polunin. London: Leonard Hill Ltd.

HARLEY, J. L. & WAID, J. S. (1955). The effect of light upon the roots of beech and its surface population. *Plant & Soil*, 7, 96.

HATCH, A. B. & HATCH, C. T. (1933). Some Hymenomycetes forming mycorrhizae with *Pinus strobus* L. *J. Arnold Arb.* 14, 324.

HOW, J. E. (1940). The mycorrhizal relations of larch. I. A study of *Boletus elegans* (Schum.) in pure culture. *Ann. Bot., Lond.* 4, 135.

JACKSON, L. W. R. (1947). Method for differential staining of mycorrhizal roots. *Science*, 105, 291.

KOUPREVITCH, V. F. (1954). Action des plantes phanérogames sur le substratum par les ferments dégagés de leurs racines. *Essais Bot.* 1, 100. (*Acad. Sci.* U.S.S.R.)

LANGE, J. E. (1940). *Flora agricina danica*, 5, 1.

LATIES, G. G. (1954). The osmotic inactivation *in situ* of plant mitochondrial enzymes. *J. exp. Bot.* 5, 49.

LIHNELL, D. (1942). *Cenococcum graniforme* als Mykorrhizabildner von Waldbäumen. *Symb. bot. upsaliens*, 5 (2), 1.

LINDEBERG, G. (1948). On the occurrence of polyphenol oxidases in soil-inhabiting Basidiomycetes. *Physiol. Plant.* 1, 196.

LINSKENS, H. F. & KNAPP, R. (1955). Über die Ausscheidung von Aminosäuren in reinen und gemischten Beständen verschiedener Pflanzenarten. *Planta*, 45, 106.

LOBANOW, N. W. (1960). *Mykotrophie der Holzpflanzen*, pp. 1–352. Berlin.

LUNDEGÅRDH, H. & STENLID, G. (1944). On the exudation of nucleotides and flavanones from living roots. *Ark. Bot.* 31 A, 1.

LUNDELL, S. & NANNFELDT, J. A. (1934, 1936). Fungi exsiccati suecici, praesentim Upsalienses. Fasc. I–II, 10; V–VI, 13.

MACDOUGAL, D. T. & DUFRENOY, J. (1944). Mycorrhizal symbiosis in *Aplectrum*, *Corallorrhiza* and *Pinus*. *Plant Physiol.* 19, 440.

MACDOUGAL, D. T. & DUFRENOY, J. (1946). Criteria of nutritive relations of fungi and seed plants in mycorrhizae. *Plant Physiol.* 21, 1.

MELIN, E. (1922). Untersuchungen über die *Larix*-Mykorrhiza. I. Synthese der Mycorrhiza in Reinkultur. *Svensk Bot. Tidskr.* 16, 161.

MELIN, E. (1923 a). Experimentelle Untersuchungen über die Konstitution und Ökologie der Mykorrhizen von *Pinus silvestris* L. und *Picea abies* (L.) Karst. In *Mykol. Unters. und Berichte*, 2, 73. Ed. R. Falck. Cassel.

MELIN, E. (1923 b). Experimentelle Untersuchungen über die Birken- und Espenmykorrhizen und ihre Pilzsymbioten. *Svensk Bot. Tidskr.* 17, 479.

MELIN, E. (1924). Zur Kenntnis der Mykorrhizapilze von *Pinus montana*. *Bot. Notiser*, 69.

MELIN, E. (1925 a). Untersuchungen über die *Larix*-Mykorrhiza. II. Zur weiteren Kenntnis der Pilzsymbionten. *Svensk Bot. Tidskr.* 19, 98.

MELIN, E. (1925 b). *Untersuchungen über die Baummykorrhiza. Eine ökologisch-physiologische Studie*, pp. 1–125. Jena: G. Fischer.

MELIN, E. (1936). Methoden der experimentellen Untersuchung mykotropher Pflanzen. In *Handbuch biol. Arbeitsmethoden*, Abt. XI. 1015. Ed. E. Abderhalden. Berlin: Urban and Schwarzenberg.

MELIN, E. (1953). Physiology of mycorrhizal relations in plants. *Annu. Rev. Pl. Physiol.* **4**, 325.

MELIN, E. (1954). Growth factor requirements of mycorrhizal fungi of forest trees. *Svensk Bot. Tidskr.* **48**, 86.

MELIN, E. (1955). Nyare undersökningar över skogsträdens mykorrhizasvampar och det fysiologiska växelspelet mellan dem och trädens rötter. *Acta Univ. Upsaliens*, **3**, 1–29.

MELIN, E. (1959a). Mycorrhiza. *Handb. Pflanzenphysiol.* **11**, 605.

MELIN, E. (1959b). Studies on the physiology of tree mycorrhizal Basidiomycetes. I. Growth response to nucleic acid constituents. *Svensk Bot. Tidskr.* **53**, 135.

MELIN, E. (1962). Physiological aspects of mycorrhizae of forest trees. In *Tree Growth*, p. 247. Ed. T. T. Kozlowski. New York: The Ronald Press Co.

MELIN, E. & DAS, V. S. R. (1954). Influence of root-metabolites on the growth of tree mycorrhizal fungi. *Physiol. Plant.* **7**, 851.

MELIN, E. & LINDEBERG, G. (1939). Über den Einfluss von Aneurin und Biotin auf das Wachstum einiger Mykorrhizenpilze. *Bot. Notiser*, 241.

MELIN, E. & NILSSON, H. (1953). Transport of labelled phosphorus to pine seedlings through the mycelium of *Cortinarius glaucopus* (Schaeff. ex Fr.) Fr. *Svensk Bot. Tisdkr.* **48**, 555.

MELIN, E. & NILSSON, H. (1957). Transport of C^{14}-labelled photosynthate to the fungal associate of pine mycorrhizae. *Svensk Bot. Tidskr.* **51**, 166.

MELIN, E. & NORKRANS, B. (1942). Über den Einfluss der Pyrimidin- und Thiazolkomponente des Aneurins auf das Wachstum von Wurzelpilzen. *Svensk Bot. Tidskr.* **36**, 271.

MELIN, E. & NYMAN, B. (1940). Weitere Untersuchungen über die Wirkung von Aneurin und Biotin auf das Wachstum von Wurzelpilzen. *Arch. Mikrobiol.* **11**, 318.

MELIN, E. & NYMAN, B. (1941). Über das Wuchsstoffbedürfnis von *Boletus granulatus* (L.) Fr. *Arch. Mikrobiol.* **12**, 254.

MODESS, O. (1941). Zur Kenntnis der Mykorrhizabildner von Kiefer und Fichte. *Symb. bot. upsaliens*, **5** (1), 1.

NORKRANS, B. (1949). Some mycorrhiza forming *Tricholoma* species. *Svensk Bot. Tidskr.* **43**, 485.

NORKRANS, B. (1950). Studies in growth and cellulolytic enzymes of *Tricholoma*. With special reference to mycorrhiza formation. *Symb. bot. upsaliens*, **11** (1), 1.

NORKRANS, B. (1953). The effect of glutamic acid, aspartic acid, and related compounds on the growth of certain *Tricholoma* species. *Physiol. Plant.* **6**, 584.

NUTMAN, P. S. (1956). The influence of the legume in root-nodule symbiosis. A comparative study of host determinants and functions. *Biol. Rev.* **31**, 109.

RATNER, F. I. (1954). Sur l'activité vitale des systèmes radiculaires dans ses relations avec la nutrition hétérotrophe des phanérogames et le rôle des microorganismes. *Essais de Bot.* **9**, 706 (Acad. Sci. U.S.S.R.).

RAYNER, M. C. & LEVISOHN, I. (1941). The mycorrhizal habit in relation to forestry. IV. Studies on mycorrhizal response in *Pinus* and other conifers. *Forestry*, **15**, 1.

RAWALD, W. (1962). Über die Bedeutung wasserlöslicher Vitamine für das Mycelwachstum höherer Pilze. *Arch. Mikrobiol.* **42**, 378.

ROVIRA, A. D. (1956). Plant root excretions in relation to the rhizosphere effect. I. The nature of the root exudate from oats and peas. *Plant & Soil*, **7**, 178.

ROVIRA, A. D. & HARRIS, J. R. (1961). Plant root excretions in relation to the rhizosphere effect. V. Exudation of B-group vitamins. *Plant & Soil*, **14**, 199.

SANTOS, N. F. (1941). Elementos para o estado des micorrizas ectendotroficas do *Pinus pinaster* Sol. *Publ. Dir. Geral. Serv. Florestais Aquicolas*, **8**, 65.

PLATE 1

Fig. 1

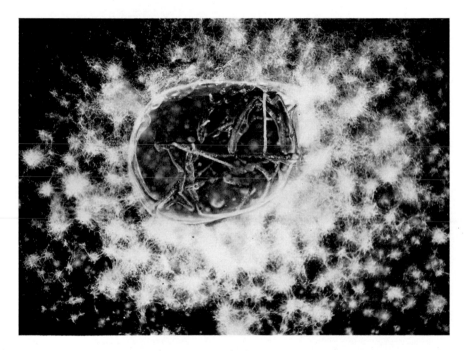

Fig. 2

(*Facing page* 144)

PLATE 2

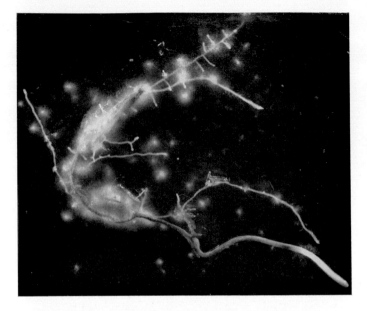

Fig. 3

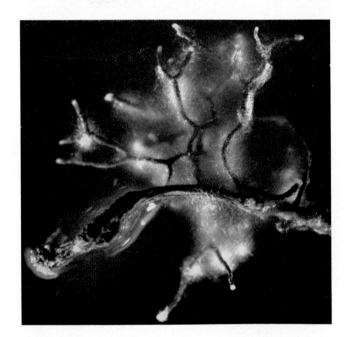

Fig. 4

PLATE 3

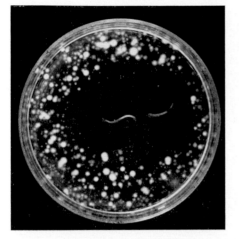

Fig. 5(a)

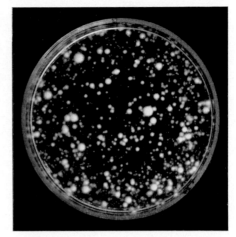

Fig. 5(b)

Fig. 6

SLANKIS, V. (1948). Einfluss von Exudaten von *Boletus variegatus* auf die dicho-tomische Verzweigung isolierter Kiefernwurzeln. *Physiol. Plant.* **1**, 390.

SLANKIS, V. (1951). Über den Einfluss von β-Indolylessigsäure und anderen Wuchsstoffen auf das Wachstum von Kiefernwurzeln. I. *Symb. bot. upsaliens*, **11** (2), 1.

SLANKIS, V. (1958). The role of auxin and other exudates in mycorrhizal symbiosis of forest trees. In *The Physiology of Forest Trees*, pp. 427–43. Ed. K. V. Thimann. New York: The Ronald Press Co.

STENLID, G. (1947). Exudation of excised pea roots as influenced by inorganic ions. *Ann. Roy. agric. Coll. Sweden*, **14**, 301.

WEST, P. M. (1939). Excretion of biotin and thiamine by roots of higher plants. *Nature, Lond.* **144**, 1050.

WIKÉN, T., KELLER, H. G., SCHELLING, C. L. & STÖCKLI, A. (1951). Über die Verwendung von Myzelsuspensionen als Impfmaterial in Wachstumsver-suchen mit Pilzen. *Experientia*, **7**, 237.

VIRTANEN, A. L. & VON HAUSEN, S. D. (1951). Dependence of nitrate reduction in green plants upon reducing substances. *Acta Chem. Scand.* **5**, 638.

VOZZO, J. A. & HACSKAYLO, E. (1961). Mycorrhizal fungi on *Pinus virginiana*. *Mycologia*, **53**, 538.

EXPLANATION OF PLATES

PLATE 1

Fig. 1. Response of mycelial suspension of *Russula xerampelina* Schaeff. ex Fr. to cultured pine roots in maximum nutrient solution. Fungal development occurred almost entirely around the root. From Melin (1954).

Fig. 2. Response of mycelial suspension of *Boletus variegatus* D. to cultured pine roots in a celluloid sack placed on maximum nutrient agar. Mycelial suspension was sieved through wire gauze with holes of 0·01 mm.^{2}. Incubation period 7 days. *c.* × 3·5.

PLATE 2

Fig. 3. Response of mycelial suspension of *Pholiota caperata* Pers. ex Fr. to cultured pine roots on maximum nutrient agar. Mycelial suspension sieved through wire gauze with holes of 0·1 mm.^{2}. Incubation period 14 days. × 2.

Fig. 4. Response of mycelial suspension of *Pholiota caperata* Pers. ex Fr. to cultured pine roots on maximum nutrient agar. Mycelial suspension sieved through wire gauze with holes of *c.* 0·01 mm.^{2}. Incubation period 20 days. × 4.

PLATE 3

Fig. 5. Growth produced by mycelial suspension of *Boletus variegatus* C. in maximum nutrient agar. *Right*, without supplement; *Left*, with the upper parts of the main axis of a cultured pine root (6 months old). Incubation period 10 days. *c.* × 0·8.

Fig. 6. Response of *Boletus variegatus* C. to secondary rootlets arising from the same pine root as in Pl. 3, fig. 5. Incubation period 7 days. × 2·5.

VESICULAR-ARBUSCULAR MYCORRHIZA: AN EXTREME FORM OF FUNGAL ADAPTATION

BARBARA MOSSE

Soil Microbiology Department, Rothamsted Experimental Station

INTRODUCTION

Among the many fungi that regularly inhabit plant roots or colonize the rhizosphere, those causing vesicular-arbuscular infections are some of the commonest. The interactions between host and endophyte in this association are as yet very imperfectly understood, largely because it is difficult to isolate the endophytes, and to find soils that do not contain them. Progress made during the last decade in the isolation of these fungi has led to the establishment of typical infections under micro-biologically controlled conditions, but the advantages this offers for future investigations have yet to be exploited. In the meantime the inclusion of vesicular-arbuscular infections within the concept of mycorrhizal association—implying as it does some mutualistic relation-ship—rests largely on anatomical evidence, i.e. the extensive and inti-mate intracellular association between host and endophyte, the evidence of fungal digestion in the host cells, the restriction of infection to certain relatively impermanent parts of the root system and the absence of pathogenic disturbances in the infected plant tissue. In addition, some experimental evidence which will be presented suggests that, in sterilized soils at least, inoculated plants may grow better than uninoculated ones. While the benefit to the plant remains somewhat controversial, it is widely agreed that the fungus in its endophytic state has become highly adapted to life within the plant root, and having in most instances lost the capacity for independent saprophytic existence, derives obvious advantages from the association. It is the intention of this article to show how far such adaptation has proceeded, and to what extent avail-able evidence supports the view that vesicular-arbuscular infection is a labile condition, in which environmental factors influence the balanced relationship between a plant root and an invading fungus.

Before considering particular aspects of the association, the prevalence of such infections should be noted; they are the rule rather than the exception in perennial plants, are only slightly less common in annual flowering plants and occur also in many ferns and liverworts. Vesicular-arbuscular infections are world-wide in their distribution, both under natural conditions and in cultivation. There have been many detailed descriptions of such infections, referred to in reviews by Rayner (1927), Butler (1939), Harley (1950, 1959) and Mosse (1962b) and in the literature indices of Kelley (1937), Strzemska (1954) and Hepden (1960). The evidence for their occurrence in fossil plants has been reviewed by Butler (1939).

THE ANATOMY OF
VESICULAR-ARBUSCULAR MYCORRHIZA

Vesicular-arbuscular mycorrhiza consist of three organically connected components: the host plant, the endophyte within the root, and the endophyte in the rhizosphere. In this they differ from most non-mycorrhizal symbiotic associations in which the micro-symbiont does not usually have a continuous free-living and endophytic phase.

The endophyte within the root

The characteristics of vesicular-arbuscular infections have been extensively described by Peyronel (1924), McLennan (1926), Lihnell (1939), Butler (1939) and many others. Two fungal structures, arbuscules and vesicles serve as diagnostic features.

The arbuscules (Pl. 1, fig. 1), structurally similar to haustoria, develop by repeated dichotomous branching from one, occasionally several, intracellular hyphae that arise as short lateral branches from coarser intercellular distributive hyphae. At their extremities the fine arbuscular branches may pass beyond the resolution of the light microscope, and sometimes each arbuscular tip appears to be surrounded by a cloud of granular material staining like fungal protoplasm. Such structures have been called sporangioles, but they are now interpreted as extrusions of naked fungal protoplasm into the host cell. Well-defined arbuscules are relatively rare, and the 'sporangiole' stage, or a dense mass of granular protoplasm containing some hyphal remains are more common. Both are considered to be stages in the breakdown of arbuscules. Other intermediate stages are well illustrated by Harley (1959) after drawings by McLuckie & Burges (1932). Eventually, only a small region around the cell nucleus retains affinity for fungal stains, and remains as evidence of earlier intracellular infection (Pl. 1, fig. 2).

A much enlarged nucleus with prominent nucleolus is usually attached to the centre of the arbuscule. The increase in nuclear diameter in three kinds of mycorrhizal root is shown by the measurements in Table 1. The figures also show that the nucleus enlarges appreciably before actual cell penetration, in response to the (near) presence of the fungus within the root.

Lihnell's (1939) measurements show a similar, statistically significant, though numerically smaller effect on nuclear size in arbuscule-containing cells of juniper. After disintegration of the arbuscule the nucleus contracts again. Very similar nuclear changes occur during the 'digestion' of the pelotons in orchids (Bernard, 1909; Burgeff, 1909). In *Platanthera chlorantha* Burgeff (1909) found several nucleoli in the

Table 1. *Mean nuclear diameter (μ) in cortical cells with and without arbuscular infection*

Host	Uninfected root	Uninfected cell in infected root	Cell with arbuscules
Onion	4·6	5·8	9·0
Strawberry	3·4	4·2	8·4
Apple	3·4	4·6	9·6

For each host, differences between columns are significant ($P < 0.001$).

enlarged nucleus, which sometimes became 'fragmented' giving rise to several nuclei. The causes of increased nuclear size have not been studied. It could be induced by polyploidy, as suggested by Burgeff's observation, or by osmotic effects following a dilution of the cytoplasm. Goldstein & Harding (1950) and Harding & Feldherr (1958) experimentally induced swelling in nuclei of frog oocytes, and concluded that the nuclear membrane was impermeable to large molecules (the size of albumin) but permeable to smaller ones. That the nuclear swelling has some special relationship with the digestion process is suggested by Bernard's (1909) observations in *Odontoglossum* where nuclei on the dorsal side of the protocorm remain of normal size and shape, and there is no digestion of the pelotons, whilst those on the ventral side where digestion occurs become enlarged and lobed, even before cell penetration. Burgeff (1909) considered that infection and fungal digestion can occur more than once in the same cell if starch accumulates again after digestion, the nucleus then passing for a second time through the same cycle of enlargement, nucleolar changes and eventual contraction. The disappearance of starch from arbuscular cells has been regularly reported, and forms another parallel with orchid mycorrhiza. Again the biochemical implications are little understood. Khrushcheva (1960) reported more sugar in mycorrhizal than in uninfected wheat roots, possibly as a result of starch hydrolysis. Such hydrolysis together with increased respiration are fairly common reactions to invasion by pathogenic fungi, and are usually regarded as a protective mechanism. Bernard so regarded the digestion of pelotons, but in vesicular-arbuscular mycorrhiza where an active external mycelium exists in the soil, fungal digestion may confer some additional benefit on the host.

Vesicles (Pl. 2, figs. 3, 4) are sack-like swellings at the tip or in the middle of the distributive hyphae. Their development was studied in detail by Lihnell (1939), Skeppstedt (1952), Schrader (1958), Otto (1959) and Koch (1961). Vesicles are often intercellular and usually multinuclear, having open connexions with their parent hyphae. When young they have thin walls and contain a homogeneous protoplasm. Later the walls thicken, the protoplasm becomes vacuolated and numerous oil droplets develop. These tend to coalesce and mature vesicles often contain one very large oil droplet surrounded by a thin peripheral cytoplasm. Vesicle size and shape depend on the host and nutritional conditions. When infections were produced under sterile conditions in clover seedlings grown in an agar medium lacking nitrogen, the first vesicles always appeared near the entry point of the fungus, i.e. in the oldest parts of the infection, whereas arbuscules characteristically developed towards the root tip. As the clover seedlings declined, production of vesicles increased greatly, but all traces of arbuscules disappeared; no new ones were formed and old ones disintegrated without leaving any trace. Eventually vesicles became so prominent (Pl. 2, fig. 4) that they protruded from the root. Similar hypertrophied vesicles developed in very large numbers in strawberries suffering from red core (O'Brien & M'Naughton, 1928), in semi-moribund rootlets of citrus (Neill, 1944), and in strawberries 18 months after partial soil sterilization with chlorobromopropene (Wilhelm, 1959). As was suggested by Wilhelm, and is supported by the observations in agar cultures, such extensive development of vesicles may reflect an unhealthy state in the host, induced by external conditions, by the endophyte, or by both.

Two functions have been ascribed to the vesicles, storage and reproduction. Storage functions are suggested by the large accumulation of oil in them. McLennan (1926) thought that some of this oil, which also occurs in the hyphae, might become available to the host when the arbuscules break down. The reproductive functions attributed to vesicles are of two kinds: to act as reproductive propagules when the host roots decay, or as sporangia from which numerous smaller endogenously produced spores are set free. The most convincing evidence for their function as propagules is provided by Schrader's (1958) observation that some vesicles in detached and decaying pea roots buried in a sterilized soil and sand mixture can germinate, and by Koch's (1961) infection experiments with single excised vesicles. Small lemon-shaped spores within vesicles were first seen by Peyronel (1923) in decaying wheat roots, and have since been illustrated by Otto (1959) in apple roots, and by Koch (1961), who found up to 25 % of such spore-containing vesicles in various roots collected during the winter. As these workers pointed out, the small spores could have belonged to another fungus living parasitically in the vesicles, and the functions of vesicles as sporangia must remain in question. Larger round spores were pictured in vesicles of *Dryopteris* by Skeppstedt (1952) and these bear some resemblance to the proliferating vesicles illustrated by Butler (1939), and also observed by the writer. Both Otto (1959) and Koch (1961) thought it likely that vesicles served a dual purpose, as storage organs while they were young, and as occasional overwintering and reproductive structures when the root decays.

The spread of infection within the roots of angiosperms is strictly limited to the primary cortex, an impermanent tissue with a limited life span. There is no penetration of the endodermis, and neither the vascular tissue nor older parts of the root system are ever invaded after the primary cortex has sloughed off. All investigators are agreed that infection does not spread internally from parent to branch roots, but that every new rootlet is re-invaded from the soil. Entry into the root is commonly through epidermal cells, and is preceded by the formation of a massive appressorium. Entry through root hairs is rare, but it occurs regularly in oats (Strzemska, 1955; Meloh, 1961) and has been described as common in some species under particular conditions (Schrader, 1958; Nicolson, 1960). Liverworts and fern prothalli are infected through their rhizoids.

It is difficult to draw any final conclusions about the frequency of entry points and the number of functional hyphae connecting the mycelium in the soil with that in the root. Some workers have concluded that connexions are few (Winter, 1951), and often non-viable (Neill, 1944). Nicolson (1959) found that in grass roots they depended on soil type and habitat, and Mosse (1959b) that they depended on season and manurial treatments. Lihnell (1939) found 4–16 infection points/mm. root length in juniper; McLuckie & Burges (1932) counted 2·5–4·1 mm.^{2} root surface in *Eriostemum*; Mosse (1959b) recorded 2·6–21·2/mm. root length in strawberries and 4·6–10·7/mm. in apples. Counts of entry points in serial sections have been criticized because other fungi may be included, and entry and exit points may be difficult to distinguish. More valid is the objection that main hyphae growing towards the root often branch shortly before entering it, and the branches individually penetrate the root (Pl. 2, fig. 5). Because of this the number of entry points does not necessarily provide a good index of connexion between the internal and external mycelium in translocation studies. It is fairly clear that the number of entry points will depend, at least in part, on the amount of active fungal mycelium near the root surface and on the rate of root growth, both being strongly influenced by water availability and other environmental conditions. In agar cultures where the course of infection can be more easily followed, roots with an established infection tend to be invaded a second and third time. This may reflect the invigorating effect which penetration has on the external mycelium and its increased growth along the root surface, or it could indicate changes in root metabolism that render subsequent penetration easier once the root has become infected. A mechanism analogous to the increased polygalacturonase activity induced in legume roots by effective strains of *Rhizobium* (Fåhraeus & Ljunggren, 1959) could be involved. In agar cultures the connecting hyphae remain viable and aseptate for weeks after entry.

The endophyte in the soil

The all but universal occurrence of vesicular-arbuscular infections indicates that the fungi concerned must be very widely distributed in soils, but as they cannot be cultured there is no way of testing this other than by visual evidence of their presence in the rhizosphere. Fortunately the external mycelium, well described by Peyronel (1924) and Butler (1939), has some very characteristic features, thought by experienced mycologists to be sufficient for identification. Gerdemann (1955b) has described features of wound healing in the external mycelium (also noticed by Peyronel, 1924) which he regards as highly characteristic. The soil mycelium has a limited spread, perhaps 1 cm. from the root surface, and it is often possible to trace it into the root. Careful study shows that the soil mycelium with its characteristic vesicle-like spores, and fine, thin-walled, irregularly septate branches, arising from angular projections of the permanent thick-walled hyphae, resembles the vesicles and arbuscules of the endophytic phase. The fine branches that may function as adsorptive structures are most profuse when attached to particles of organic matter or root surfaces on which they assume remarkable amoeboid shapes. In a re-examination of the soil mycelium Nicolson (1959) has emphasized its dimorphic character, the ephemeral nature of the fine absorbing branches and their special relationship to the angular projections of the permanent hyphae. The dimorphic nature of the external mycelium can be seen in Pl. 2, fig. 5, and Pl. 3, fig. 6 illustrates the growth form of the endophyte on a piece of filter-paper, used instead of agar, as a supporting medium in a two-membered culture. The structure reminiscent of an arbuscule appears to be a response to the filter-paper medium, as nothing like it was ever noticed in otherwise comparable agar cultures. Similar structures were found in peat particles in potting composts where the successful establishment of mycorrhizal infection after addition of an impure inoculum was accompanied by the disintegration of the peat particles (Mosse, 1959b). No such disintegration occurred if, for some reason, the inoculum did not establish infections. The interest of these observations lies in their bearing on possible saprophytic activity of the endophyte in the soil. Although of limited spread, the density of mycelium in the root vicinity is high, and it may act as an extended absorption system for the root. Rapid protoplasmic streaming, according to Schütte (1956) an indication of the ability to translocate, can be observed in the external mycelium, and anastomoses are frequent. The endophyte can survive in the soil without a plant but it does not spread. Koch (1961) found no loss of infectivity in soil or dry roots stored for a year, but infectivity of stored soil declined after 2 years (Meloh, 1961). I have found that there was no spread from an inoculum of infected roots or fungal fructifications buried in sterilized soil, unless living plant roots were also present. In agar cultures there was a sudden outgrowth of hyphae from infected clover roots which were becoming moribund (Mosse, 1962a), and this may be the most usual way in which infection spreads. Soil below the root zone (40–160 cm.) was not infective (Schrader, 1958).

The large spores (100–300 μ) recently described by Gerdemann & Nicolson (1962a) as a soil phase of some vesicular-arbuscular endophytes are probably very resistant structures. This is suggested by their very thick walls, sometimes double, as well as by their similarity to certain resting spores of an *Endogone* sp. with which the writer is familiar. These spores often survive overgrowth by other microorganisms as well as prolonged dessication in the laboratory.

Clearly the vesicular-arbuscular association cannot be adequately understood without a consideration of the soil mycelium. Whilst the endophyte without its plant associate appears to remain inert in the soil, observations on the mycelium growing from the root suggest active saprophytic functions, which in their turn may have direct, or indirect effects on the host.

THE IDENTITY AND CULTURE OF THE ENDOPHYTES

The name *Rhizophagus* was given by Dangeard (1900) to a fungus in poplar roots which produced infections like those here described. The fungus was never isolated, and the name is no more than a convenient label, giving no clue to identity. Numerous attempts to isolate the causal fungi from infected root fragments resulted either in limited fungal growth which then ceased (Magrou, 1946), or in the isolation of common soil fungi, often *Rhizoctonia* spp., which on re-inoculation did not produce typical infections. Harley (1950) has reviewed this phase of the work.

On morphological evidence it has long been considered that the endophyte is a Phycomycete, possibly belonging to the Mucorales, and Peyronel (1923, 1937) first demonstrated hyphal connexion between the mycorrhizal roots of various alpine plants and the fructifications of three *Endogone* spp. (*E. vesiculifera, E. fuegiana* and one other). Because of the frequent isolation of a *Rhizoctonia* sp. and the common occurrence of regularly septate hyphae in later stages of infection, Peyronel (1924) put forward a theory of 'double infection', in which he attributed the characteristic appearance of mature infections to the presence of two fungi; this theory is fully reviewed by Rayner (1927). Butler (1939) re-emphasized the similarity of the endophyte to known species of *Endogone*, and reviewed the evidence that one or more species of this genus were responsible for vesicular-arbuscular infection in most plants.

The genus *Endogone* is not well known. It was established by Link (1809), and has been reviewed by Thaxter (1922). It is distinguished on spore character, the structure of the fruiting body (sporocarp) and general resemblances in habit and habitat; both its systematic position and some of its constituent species (many identified from fragments of fruiting bodies in herbarium collections) are in doubt. According to Thaxter (1922) the genus *Endogone* includes both zygospore- and chlamydospore-forming species, and at least one species produces both together in the same sporocarp. Some species in which sporangia occur in the sporocarps are also included in the Endogonaceae. Germination of the zygospores has not so far been observed, but if endogenous spores were formed the genus would probably be transferred to the Ascomycetes, with which it is considered to have close phylogenetic relationship. Godfrey (1957a, b), who made a detailed study of some British species and their spore germination (Godfrey, 1957c), has also reviewed the literature relating to their classification.

Rhizophagus

A new phase in the identification of vesicular-arbuscular endophytes started with Barrett's (1947) report of the successful isolation of a fungus from infected roots, using fragments of hemp seed as an intermediate substrate. After a few weeks the fungus grew from the hemp seed fragments into standard fungal media such as malt-, Czapek- and hemp-seed agar. Once growing it could then be freed from contaminants. The technique, now described in detail (Barrett, 1961), has recently been used by J. W. Gerdemann (private communication) to establish a culture differing only in minor details from the eleven slightly different isolates made by Barrett. In culture this fungus, a Phycomycete, produces round, vesicle-like vegetative spores, similar to those of the external mycelium of the endophyte, but as no reproductive structures have yet been produced it cannot be assigned to a genus, and is referred to as *Rhizophagus*. The cultures do not resemble those of any known soil fungus, but the vegetative spores are very like those formed in dual cultures by an *Endogone* sp., and J. T. Barrett (private communication) thinks that it may belong to this genus.

There is, unfortunately, no published record of the details of Barrett's inoculation experiments with *Rhizophagus*, but, at two scientific meetings (Barrett, 1958, 1961), pictures were shown of typical vesicular-arbuscular infections produced by inoculation with *Rhizophagus* in various plants grown in open pots in sterilized sand, and in a sand/soil mixture. J. W. Gerdemann's (private communication) inoculation experiments have so far failed; mine once succeeded (Mosse, 1961), but in subsequent tests plants in open pots in a mixed greenhouse had to be maintained for so long, and infected plants were so few, that the possibility of chance infection could not be excluded; in more adequately protected plants inoculation failed. Possibly significant is the fact (J. T. Barrett, private communication) that after successful inoculation, the fungus can only be re-isolated on hemp seed, and cannot grow directly from the infected roots into the agar medium in which the inoculum had originally grown. That hemp may have some peculiar property for growth of the endophyte is also suggested by some abnormal infections in hemp roots inoculated with another *Endogone* sp. (Mosse, 1961). In some roots, deeply staining, tightly coiled hyphae were formed instead of arbuscules, and a dense fungal mat developed along the endodermis (Pl. 3, fig. 7) which was occasionally penetrated, followed by discoloration and lesions in the cortex. There was also an abnormal concentration of hyphae around some emerging branch roots (Pl. 3, fig. 8), a region normally immune from infection.

Pythium

Culture of another endophytic fungus was reported by Hawker, Harrison, Nicholls & Ham (1957) who repeatedly isolated a *Pythium* sp., closely resembling *P. ultimum*, from thoroughly washed root sections of various plants, using a hanging-drop technique elaborated by Harrison (1955). The fungus had no special growth requirements. Isolations from roots with predominantly arbuscular infection were rarely successful, and more recently Hawker (1962) added that *Pythium* was isolated only from a relatively small proportion of the plants examined. There have been previous claims, especially in the earlier literature (fully reviewed by Hawker, 1962), that *Pythium* spp. caused vesicular-arbuscular infections.

As many *Pythium* species habitually invade plant roots, inoculation experiments and the establishment of typical infections assume particular importance. Proof rests on papers by Hawker *et al.* (1957), Hawker & Ham (1957), Hepden (1960), Carré & Harrison (1961) and Ham (1962). Hepden established infections in sporophytes and prothalli of *Phyllitis scolopendrium* grown respectively in an agar medium, and in sterile sand in pots, and illustrated typical infections in the latter. Neither Carré & Harrison (1961) nor Ham (1962) have illustrated the infections produced, and those shown by Hawker *et al.* (1957) and Hawker & Ham (1957) are not altogether convincing that typical vesicular-arbuscular infections resulted from inoculation. Ham's (1962) paper is of special interest because it sets out in some detail conditions under which such infections were obtained. In an agar medium containing both nitrate and glucose, two isolates of *Pythium* tended to destroy the test seedlings, but in a medium lacking both these nutrients, the fungi grew less vigorously, and one strain formed typical mycorrhiza. In pot culture both strains caused typical infections at pH above neutrality (presumably non-optimal for the fungus), whereas at lower pH other more pathogenic kinds of root invasion occurred. This is in line with my experience with two *Pythium* isolates supplied by the above workers. In apples the roots were not penetrated, but onion roots frequently were, the infection being at first non-arbuscular with occasional penetration of the vascular tissue; later typical vesicular-arbuscular infections sometimes became established. In anatomical detail they differed slightly from those caused by inoculation with an *Endogone* sp., and Hawker (1962) also noted some anatomical differences between infections of plants from which *Pythium* was isolated and those from which it was not. Such differences, however, are insufficient for a reliable identification of the endophytes, especially if mixed infections also occur.

Endogone

Other evidence concerning the identity of endophytes is based on spore inoculations.

Endogone sporocarps organically connected with mycorrhizal roots of a pot-grown strawberry plant (Mosse, 1953) were used in inoculation experiments. The sporocarps, approximately 1 mm. in diameter, contain up to ten thick-walled yellow resting spores which at maturity are extruded from the sporocarp, each leaving behind a transparent papery envelope, the remains of the original spore wall. This *Endogone* sp. differs from those described by Thaxter (1922), and is also unlike any previously described British species (Hawker, 1954). Inoculation with such sporocarps invariably produced typical infections in various plants grown in sterilized soil in open pots, and many new sporocarps formed in the rhizosphere of test plants (Mosse, 1956). The sporocarps which are without peridium and cannot be adequately sterilized are not a pure inoculum. Inoculations with excised, surface-sterilized resting spores occasionally succeeded (Mosse, 1956), but results were unpredictable until a method was found (Mosse, 1959 a) to overcome the difficulties of erratic spore germination and internal parasitization by other fungi with septate mycelia. Briefly, the method consists of the visual selection of non-parasitized spores, their surface sterilization, and exposure, under aseptic conditions, to some water-soluble substances produced by a mixed population of soil micro-organisms in a soil/agar plate. By this method a high proportion of the *Endogone* spores is stimulated to germinate rapidly. Typical infections can regularly be established with such an inoculum in test seedlings grown aseptically in an agar medium (Mosse, 1962 a); the conditions required will be discussed below. This *Endogone* sp. cannot be grown in culture alone, although it makes considerable growth through an agar or filter-paper medium in dual cultures, once an infection is established in a host root. Growth studies using the limited hyphal systems produced by germinated spores showed that all the simple sugars were inhibitory, that a range of nitrogen sources, both organic and inorganic, did not affect growth, but that tartaric and some other low-carbon organic acids could be used as carbohydrate source (Mosse, 1959 a). The near-obligate condition of this fungus seems to be connected with a highly specialized nitrogen requirement. Growth occurred only over a narrow pH range around 6·3. The limited growth made by the germ tubes entirely depended on some connexion with a parent spore, although such connexion could be provided by anastomosing hyphae with other spores, or with the external mycelium attached to an infected root.

In addition to fully formed fruit-bodies, the soil contains a varied population of more or less naked *Endogone*-type spores, occurring either singly or aggregated in groups (Gerdemann, 1955*a*, 1961; Ohms, 1956, 1957; Dowding, 1955, 1959; Nicolson, 1958, 1959; Kubíková, 1961; Gerdemann & Nicolson, 1962*a*). Many of these can cause vesicular-arbuscular infections. Nicolson (1958) and Dowding (1959) found spore aggregations organically connected with mycorrhizal roots. Gerdemann (1955*a*) produced arbuscules but not vesicles in various seedlings growing in sterilized sand, by inoculation with very large spores, up to 800 μ diameter, which had a characteristic bulbous swelling at the top of the supporting hypha. After considering the possibility that these might be the sexual spores of an Oomycete, Gerdemann now thinks that they are zygo- or chlamydospores of an *Endogone*. This species also forms echinulate spores borne in clusters on the external mycelium; such spores have not previously been described in association with arbuscular endophytes. Gerdemann (1961) then tested the infectivity of organic particles obtained by a wet-sieving technique from soil adjacent to the roots of maize plants. The particles, divided into six size categories, all produced typical infections, but the differences in external mycelium—presence or absence of sporocarps, of large spores with swollen hyphal bases, and size differences in the chlamydospores—led Gerdemann to conclude that the soil contained three or more distinct species of phycomycetous mycorrhizal fungi. Using the same wet-sieving technique Gerdemann & Nicolson (1962*a*) recently demonstrated large and varied populations of *Endogone* spores in some cultivated soils in Scotland. Having tentatively divided them into six distinct types on morphological criteria, particularly the structure of the spore base (Gerdemann & Nicolson, 1962*b*), they have now confirmed the validity of such a grouping by inoculation experiments; four of the six types produced vesicular-arbuscular infection, and reproduced their particular spore type on the external mycelium. Using the same technique, I have found *Endogone* spores, some similar to the types found in Scotland, in a wheat field at Rothamsted. Some, though not all, of these spores have also produced typical infections and reproduced their kind.

These recent developments clearly require more study; spores used in inoculation experiments were washed but not sterilized, and little is known of their development, their germination or their ability to grow in culture. The absence of Endogones from soil dilution plates and other estimates of soil microbial populations, suggests that they are not easily cultured. There is one report of the isolation of an *Endogone* sp., since lost (Kanouse, 1936), and others have been grown in culture for limited periods. The conditions leading to their chance isolation, as well as their relation to cultures of *Rhizophagus*, that so much resemble *Endogone*, require investigation. The demonstration of the widespread occurrence of *Endogone* spores in the soil, and the ability of many to cause vesicular-arbuscular infections, may stimulate a much needed re-investigation of the genus.

'ENDOGONE' INFECTIONS UNDER ASEPTIC CONDITIONS

Highly specialized conditions were required to establish vesicular-arbuscular infections under aseptic conditions, using sterilized spores of a sporocarpic strain of *Endogone* as inoculum (Mosse, 1962*a*). The germ tubes of this fungus did not penetrate the roots of various test seedlings grown on different agar media, even when the roots were cut or damaged mechanically. Germ tubes were not attracted towards the root, but when contact was established by chance, hyphal growth was stimulated, and continued along and around the root surface until spore reserves were exhausted; inability to form appressoria prevented infection. Appressorium formation, followed by infection, could be induced in the absence of soluble nitrogen, by some factor provided by a species of *Pseudomonas*, originally present as a contaminant. Addition of nitrogen, in excess of 30 μg./10 ml., delayed infection until the added nitrogen was used up in plant growth. The *Pseudomonas* sp. can be replaced, though less effectively, by other soil bacteria including *Bacillus subtilis* and species of *Arthrobacter, Agrobacterium, Alcaligenes* and even *Cytophaga*, but not by either *Azotobacter* or an ineffective strain of *Rhizobium*. Infection can also be promoted in the absence of bacteria by certain concentrations of EDTA (ethylenediaminetetra-acetic acid), by 'Pectinol' (a commercial pectinase enzyme preparation), and by cell-free filtrates of culture solutions in which plants and *Pseudomonas* have been grown together; all these substitutes were less consistent in their effects than the *Pseudomonas*. Although most of the inoculation experiments were made with a small-seeded clover species, infections were also established in cucumber, *Dactylis glomerata*, wheat and onions, using the conditions that were optimal for clover. While these conditions had some general applicability to other hosts, they are unlikely to be the only ones under which *Endogone* infections can develop; they did not lead to infection by either *Rhizophagus* or *Pythium*.

With *Endogone* inoculations root penetration is invariably followed by the establishment of typical infections, and this is also true of *Rhizophagus* infections in open pots. With *Pythium* by contrast, fungal entry rarely presents a problem, and vesicle-like spores are usually formed both in aseptic culture and in open pot inoculations, but atypical (non-arbuscular) infections often result (Ham, 1962). If entry is the main obstacle to infection by *Endogone*, it is all the more curious that after entry the hyphae apparently pass through the internal walls of the cortex readily.

When *Endogone* infections were established in clover seedlings growing in a mineral salt medium that also lacked a carbon source, no effect on shoot growth was evident, but individual infected roots were longer and more branched than neighbouring uninfected roots, or similarly placed roots on uninfected seedlings. This observation first made on a random sample of clover seedlings (Mosse, 1962*a*) has since been repeated (Table 2); the mean length of mycorrhizal roots differed significantly ($P < 0.05$) from that of non-mycorrhizal roots.

Table 2. *Length and percentage branching of mycorrhizal and non-mycorrhizal secondary rootlets of* Trifolium parviflorum *infected under aseptic conditions*

	No. of roots examined	Mean root length (mm.)	Percentage rootlets with laterals
Inoculated seedlings that did not become mycorrhizal	218	13·4	1·4
Uninfected roots on inoculated seedlings	226	13·1	1·3
Mycorrhizal roots on inoculated seedlings	47	21·6	16·3

Under aseptic conditions internal spread of the infection from the entry point was strongly directional towards the root apex. It thus seems that, in a medium lacking nitrogen, infection stimulates apical growth of the infected root, and the fungus is in turn stimulated by some substance produced near the growing root tip. There is no indication of such directional spread in pot or field infections where the soil provides a source of nitrogen.

SPECIFICITY

There is very little evidence of host specificity among vesicular-arbuscular endophytes. Stahl (1949) demonstrated cross-infection between different species and genera of liverworts by planting infected and uninfected thalli 10 cm. apart in sterile quartz sand. In control experiments in which the infected thalli were replaced by some infected soil, test thalli remained free from infection, and this was taken as evidence that the infection had arisen from the endophyte in the thallus, rather than from adhering soil particles. Koch (1961) transmitted infection from both surface-sterilized and non-sterilized roots of *Atropa bella-donna* to forty-five plant species belonging to twenty-two different families, growing in sterile sand in pots. Of the six species that remained uninfected, only one, *Ruta angustifolia* normally contains a vesicular-arbuscular endophyte. In a similar type of experiment, Meloh (1961) demonstrated cross-infection between oat, barley, rye, wheat and maize. Gerdemann (1955*a*) infected corn, strawberry and sweet clover with large spores collected from the rhizosphere of red clover, and Mosse (1962*a*) infected wheat, *Dactylis*, onion, cucumber and clover seedlings in aseptic culture with *Endogone* spores formed in association with onion, apple, strawberry and hemp roots. Barrett (1961) infected a range of plants with four different isolates of *Rhizophagus*, and Ham (1962) infected both onion and lettuce seedlings with *Pythium* isolates from *Allium ursinum* and lettuce.

The only evidence of specificity is provided by experiments of Tolle (1958). Using thin sections of surface-sterilized roots as inoculum she found that of oats, barley, rye and wheat grown in sand culture only rye and wheat exchanged endophytes. Strzemska (1955) noted that infections in oats consistently differed in appearance from those in other grain crops, and this may indicate a preferential association of some plant species with a particular endophyte, as is also suggested by Hawker (1962). Barrett (1961), by contrast, considered that minor anatomical differences in infection depended on host influence which overshadowed effects attributable to strain differences in *Rhizophagus*.

It is perhaps surprising that a fungus which is almost an obligate symbiont should show so little host specificity. No doubt this accounts for its wide distribution in soils. It also indicates that the obligate relationship rests on some fundamental property of root metabolism common to many plants.

INTERACTIONS BETWEEN HOST AND ENDOPHYTE

The effect on plant growth

Because of the economic importance of many host species there have been numerous attempts to assess the effects of vesicular-arbuscular infection on plant growth despite obvious difficulties, i.e. the need to use mixed inocula, and to sterilize soils, or to use some soil substitute in which the endophyte may be unable to perform all its functions. Although these limitations make the relevance of results difficult to assess, such experiments, more fully described elsewhere (Mosse, 1962b), have given fairly consistent results. Clear benefits from inoculation have been shown in experiments using test plants grown in autoclaved soil (Asai, 1943; Mosse, 1957), in steam-sterilized soil (Baylis, 1959; J. W. Gerdemann, private communication), in unsterilized subsoil (Peuss, 1958), in sand culture (Winter & Meloh, 1958; Meloh, 1961), and in water culture (Peuss, 1958). Infections were established by the addition of small amounts of garden soil (Asai, 1943), by pre-growth of test seedlings in an infected soil, followed by transplantation (Bayliss, 1959), by inoculation with infected roots (Winter & Meloh, 1958; Meloh, 1961), by implantation of washed sections of infected roots into the root system of test plants (Peuss, 1958), and by addition of washed spores or sporocarps of *Endogone* spp. (Mosse, 1957; J. W. Gerdemann, private communication). Growth improvements, often statistically secured, have been of the order of 20–100 % increase in dry weight. In Asai's (1943) experiments control seedlings grew badly, and the improved growth could have come from microbial destruction of deleterious substances produced by autoclaving. There were, however, some plant species, normally without vesicular-arbuscular infection, which grew normally in the autoclaved soil and neither benefited from, nor became infected by, the garden soil inoculum. Asai also noted a direct relationship between improvements in growth and intensity of infection, where several seedlings were grown together in the same pot. Although such a relationship is quite often observed (Schrader, 1958; Peuss, 1958; Meloh, 1961), the improved growth is not necessarily caused by the mycorrhizal infection. In Baylis's (1959, 1961) experiments the mycorrhiza were more definitely implicated in the improved growth. *Griselinia* seedlings grown in a poor soil deficient in phosphorus showed both an increase in dry weight, and an increase in phosphorus per unit dry weight, indicating that infected seedlings had taken up 3–5 times as much phosphorus as uninfected. Increased uptake of phosphorus by infected plants is also suggested by preliminary results of experiments with maize seedlings (J. W. Gerdemann, private communication), in which leachings from sporocarps were added to the controls in an attempt to introduce at least some of the micro-organisms which might be present as contaminants of the sporocarp inoculum. Differences in concentration and total uptake of other mineral components were also found by Mosse (1957) between infected and uninfected apple cuttings. The best evidence that vesicular-arbuscular infection can benefit the host was provided by Peuss (1958), who grew tobacco seedlings in water culture at three different pH levels. By an elegant technique of implanting sections of infected tissue into the root, she demonstrated an increased dry weight of both root and shoot at all pH levels, with the greatest relative improvement occurring in the least favourable medium. Sections of uninfected roots were grafted into control seedlings. Infected seedlings grew better at pH 4 than uninfected seedlings at the most favourable pH 7. These findings were supported by experiments in which the section implantation technique was supplemented by a root inoculum, and plants were grown in an unsterilized subsoil with fertilizer additions at three levels. The greatest relative improvement—over 100 % increase in dry weight—again occurred at the lowest nutrient level. The observation that maximum benefits occur under conditions least favourable for the uninfected plant is also supported by Johnson's (1949) experiments with Sea Island cotton, and by the experiments of Meloh (1961) and Koch (1961), and is in line with the general observation of maximum infection under conditions of low fertility, and only sporadic infection in rich garden soils.

It remains arguable that the beneficial effects obtained in the experiments discussed could come from some contaminating organism preferentially associated with infected roots, and some evidence for such a preferential association in ectotrophic mycorrhiza of yellow birch has recently been found by Katznelson, Rouatt & Peterson (1962). In some respects, however, such an argument may represent too narrow a viewpoint, because if the endophyte is regularly associated with a specialized population of rhizosphere organisms not occurring on uninfected roots, and if both together exert a beneficial effect on plant growth, then it is immaterial from a practical point of view which component of the association exerts the effect. For this reason an inoculum of live uninfected roots may represent the best control if root inocula have to be used in growth experiments.

Nevertheless it does not follow that every vesicular-arbuscular infection confers benefit on the host. Results of some inoculation experiments have been inconsistent (Laycock, 1935), in others no growth effects were observed (*Ann. Rep. Rothamsted Exp. Sta. for* 1951, p. 58). The same applied to *Rhizophagus* infections in different hosts (Barrett, 1961), and to *Pythium* infections in *Phyllitis* (Hepden, 1960), although in endophytic lettuce seedlings A. M. Ham (private communication) observed some improved growth after *Pythium* inoculation. Deleterious effects occurred in some experiments by Meloh (1961), and these

are of particular interest because they provide the first experimental
evidence that different endophytes may have different effects. Meloh
distinguished two types of infections, a normal type in which both
vesicles and arbuscules but no sporocarps occurred (these he called
'Vesikelmycorrhiza'), and a much less common type (called 'Sporokarpienmycorrhiza') in which no normal vesicles were formed inside
the root, but instead there were very large spores occurring singly or in
groups. Because they were associated with a variable sheath of en-
veloping hyphae Meloh regarded these spores as sporocarps; more fully
developed sporocarps also occurred outside the root. While the relationship between the 'Sporokarpienmycorrhiza' and the sporocarp-
forming *Endogone* sp. studied by Mosse is not clear—Meloh regards
the two as different, and they also seem to differ in their effects on plant
growth—the interest of the former lies in the consistently adverse effects
it had on growth. Whereas inoculation with roots containing 'Vesikelmycorrhiza' caused statistically significant increases in dry weight of
maize seedlings in both sand and water culture, and also induced an
increased uptake of dihydrostreptomycin from liquid culture, inoculation with roots containing 'Sporokarpienmycorrhiza' caused a smaller,
but significant decrease, and had no effect on uptake of the antibiotic.
Oat seedlings reacted in the same way as maize to the two different
inocula.

The effects of environment on the host endophyte association

Two factors, light and manurial treatments, have been examined in
some detail. Both Peyronel (1940) and Stahl (1949) experienced some
difficulty in maintaining vigorous infections in plants growing in a
greenhouse, and Stahl found that infection rapidly died out in liverworts
growing in plugged glass vessels. Peyronel attributed these difficulties
to the effects of insufficient light. In ecological studies of different hosts
and habitats he found that in some plant species, e.g. *Fegatella* (now
Conocephalum) *conica* and *Viola biflora* illumination and intensity of
infection were closely correlated, whereas in the humus-loving species
Oxalis acetosella and *Circaea alpina* a good supply of organic matter
counteracted some of the adverse effects of poor illumination on
infection. Direct experimental evidence of the effects of light on infection in pea seedlings was obtained by Schrader (1958). Peuss (1958)
showed that at 50 % normal light intensity infection decreased by two-
thirds, and was further reduced when the mature leaves were also
removed. Most workers have concluded that infection is reduced
because the supply of assimilation products to the roots is deficient.

There is some evidence that manurial treatments can affect the development of the endophyte in the root. An increase, particularly in the arbuscular phase of infection, has followed applications of farmyard manure to citrus (Reed & Frémont, 1935; Sabet, 1946), strawberries (Mosse, 1954), Sea Island cotton (Johnson, 1949), and maize (Khrushcheva, 1960). Responses to the addition of mineral fertilizers have been less consistent and have included decrease in total infection and/or the arbuscular phase, and increase in the number and size of vesicles. Whether such effects were caused by a direct action on the endophyte, by action on the host, or by a preferential stimulation of a particular endophyte species or strain in a mixed population, is unknown. The value of these observations lies in the demonstration that the vesicular-arbuscular infection is labile and can be affected by agricultural practices.

DISCUSSION

The investigations of the last decade have advanced knowledge concerning the identity and isolation of vesicular-arbuscular endophytes, and have also brought improvements in techniques for studying their effect. Search of the rhizosphere has revealed several different endophytes, and others probably remain to be found. The frequency of anastomosis in the soil mycelium and its aseptate multinucleate character provide extensive opportunity for adaptive changes, and for the exchange of genetic material, and both may have been important in the evolution of the endophytic habit, which is evidently of great age; Endogonaceae have even been regarded as the common ancestors of Phyco- and Mycomycetes. Among vesicular-arbuscular fungi as a group (as in most symbiotic relationships involving fungi) there appear to be trends towards a purely saprophytic mode of life as in *Rhizophagus*, or towards parasitism as in *Pythium* and in the endophyte causing 'Sporokarpienmycorrhiza'. Similar trends can apparently be induced in the more typical endophytes by environmental conditions as was shown by inoculation of hemp roots.

The typical endophyte of the *Endogone* type has become so highly adapted to association with a higher plant, that it has lost the capacity for independent existence on synthetic media. In the *Endogone* species studied by the writer adaptation has occurred not only towards the host, but also to the microbial environment, so that the endophyte now depends on it for the vital functions of spore germination and entry into the root. It is not known how far this may apply also to other endophytic Endogones, but the spores of many are notoriously difficult to germinate. The relative frequency of the different species or strains and the extent to which they are distinct are unknown, but there may be considerable overlap. The tendency to hyphal anastomosis can be used in a preliminary classification of endophytes, both to test the identity of apparently different spores, and of the endophytic hyphae that can at times be grown from the cut ends of infected roots.

The establishment of vesicular-arbuscular infections in agar culture has provided information about some aspects of their physiology, and the arbuscular phase is emerging as the more truly representative of a symbiotic condition. Vesicles developed to an abnormal extent when seedlings were declining from lack of nitrogen, and this links up with field observations, with the adverse effects of inoculation with 'Sporo-karpienmycorrhiza' (essentially a type of infection producing abnormally enlarged vesicles), and with the difficulty of obtaining growth of endo-phytes from roots with predominantly arbuscular infections.

In a medium lacking nitrogen infected roots grew after uninfected ones had stopped growing. This suggests that the infected roots either used nitrogen more efficiently or obtained it preferentially, or that they can use some other source. This could be some compound present as an impurity in the agar, or it could be atmospheric nitrogen. Experiments are in progress to investigate the reason for the improved growth.

Although the specialized conditions required for synthesis in culture may inadvertently have demonstrated a previously unsuspected pro-perty of vesicular-arbuscular associations, they are probably restrictive in other respects. There was no improvement in the shoot growth of infected seedlings comparable to that produced by impure inocula in pot- and water-culture experiments. The agar medium used to establish infections contained all plant nutrients except nitrogen in a readily available form, so that any mycorrhizal functions that improve nutrient availability or rate of uptake would not be shown. At the same time the medium, lacking a carbon source, may be unsuitable for the normal functioning of the external mycelium, which may be needed to demon-strate the mutualistic features of the association. Experiments of a transitional kind, perhaps involving the transplantation of seedlings from agar into soil, preferably of low nutrient status, may bridge the gap between the two kinds of investigation.

Finally there is evidence, from inoculation experiments under aseptic conditions, from growth experiments at different nutrient levels, and from field observations, that the vesicular-arbuscular condition is both established more readily, and benefits the host and perhaps also the endophyte more, under conditions unfavourable to the growth of the

166   B. MOSSE

higher plant. Examples of a similar trend in other symbiotic associations come to mind. Ectotrophic mycorrhiza of forest trees are more highly developed on poor soil, and stable mycorrhizal associations in orchids can sometimes be established by growing the plants in poor nutrient media or using an attenuated form of the endophyte. Establishment and maintenance of the lichen *Acarospora fuscata* was recently shown to depend on starvation conditions (Ahmadjian, 1962), and under aseptic conditions nodule formation in legumes is depressed by soluble nitrogen. Other examples of a similar trend may come to light in this symposium and recognition of the general principle may help in the choice of appropriate media, where the establishment of symbiotic associations presents difficulty.

REFERENCES

AHMADJIAN, V. (1962). Investigations on lichen synthesis. *Amer. J. Bot.* **49**, 277.
ASAI, T. (1943). Bedeutung der Mycorrhiza für das Pflanzenleben. *Japan J. Bot.* **12**, 359.
BARRETT, J. T. (1947). Observations on the root endophyte *Rhizophagus* in culture. *Phytopathology*, **37**, 359.
BARRETT, J. T. (1958). Synthesis of mycorrhiza with pure cultures of *Rhizophagus*. *Phytopathology*, **48**, 391.
BARRETT, J. T. (1961). Isolation, culture and host relation of the phycomycetoid vesicular-arbuscular mycorrhizal endophyte *Rhizophagus*. *Recent Advances in Botany*, **2**, 1725. University of Toronto Press.
BAYLIS, G. T. S. (1959). Effect of vesicular-arbuscular mycorrhizas on growth of *Griselinia littoralis* (Cornaceae). *New Phytol.* **58**, 274.
BAYLIS, G. T. S. (1961). The significance of mycorrhizas and root nodules in New Zealand vegetation. *Proc. Roy. Soc. N.Z.* **89**, 45.
BERNARD, N. (1909). L'évolution dans la symbiose. Les orchidées et leurs *champignons commenseaux*. *Ann. Sci. Nat. Bot.* **9**, 1.
BURGEFF, H. (1909). Die Wurzelpilze der Orchideen—ihre Kultur und ihr Leben in der Pflanze. Jena: G. Fischer.
BUTLER, E. J. (1939). The occurrences and systematic position of the vesicular-arbuscular type of mycorrhizal fungi. *Trans. Brit. mycol. Soc.* **22**, 274.
CARRÉ, C. G. & HARRISON, R. W. (1961). Studies on vesicular-arbuscular endophytes. III. An endophyte of *Conocephalum conicum* (L.) Dum identified with a strain of *Pythium*. *Trans. Brit. mycol. Soc.* **44**, 565.
DANGEARD, P. A. (1900). Le '*Rhizophagus populinus*' Dangeard. *Botaniste*, sér. VII, 285.
DOWDING, E. S. (1955). *Endogone* in Canadian rodents. *Mycologia*, **47**, 51.
DOWDING, E. S. (1959). Ecology of *Endogone*. *Trans. Brit. mycol. Soc.* **42**, 449.
FÅHRAEUS, G. & LJUNGGREN, H. (1959). The possible significance of pectic enzymes in root hair infection by nodule bacteria. *Physiol. Plant.* **12**, 145.
GERDEMANN, J. W. (1955a). Relation of a large soil-borne spore to phycomycetous mycorrhizal infections. *Mycologia*, **47**, 619.
GERDEMANN, J. W. (1955b). Wound-healing of hyphae in a phycomycetous mycorrhizal fungus. *Mycologia*, **47**, 916.
GERDEMANN, J. W. (1961). A species of *Endogone* from corn causing vesicular-arbuscular mycorrhiza. *Mycologia*, **53**, 254.

GERDEMANN, J. W. & NICOLSON, T. H. (1962*a*). *Endogone* spores in cultivated soils. *Nature, Lond.* **195**, 308.

GERDEMANN, J. W. & NICOLSON, T. H. (1962*b*). Spores of mycorrhizal *Endogone* species extracted from soil by wet sieving and decanting. *Trans. Brit. mycol. Soc.* (in the Press).

GODFREY, R. M. (1957*a*). Studies of British species of *Endogone*. I. Morphology and taxonomy. *Trans. Brit. mycol. Soc.* **40**, 117.

GODFREY, R. M. (1957*b*). Studies of British species of *Endogone*. II. Fungal parasites. *Trans. Brit. mycol. Soc.* **40**, 136.

GODFREY, R. M. (1957*c*). Studies of British species of *Endogone*. III. Germination of spores. *Trans. Brit. mycol. Soc.* **40**, 203.

GOLDSTEIN, L. & HARDING, C. V. (1950). Osmotic behaviour of isolated nuclei. *Fed. Proc.* **9**, 48.

HAM, A. M. (1962). Studies on vesicular-arbuscular endophytes. IV. Inoculation of species of *Allium* and *Lactuca sativa* with *Pythium* isolates. *Trans. Brit. mycol. Soc.* **45**, 179.

HARDING, C. V. & FELDHERR, L. (1958). Semipermeability of the nuclear membrane. *Nature, Lond.* **182**, 676.

HARLEY, J. L. (1950). Recent progress in the study of endotrophic mycorrhiza. *New Phytol.* **49**, 213.

HARLEY, J. L. (1959). *The Biology of Mycorrhiza.* London: Leonard Hill Ltd.

HARRISON, R. W. (1955). A method of isolating vesicular-arbuscular endophytes from roots. *Nature, Lond.* **175**, 432.

HAWKER, L. E. (1954). British hypogeous fungi. *Phil. Trans.* B, **237**, 429.

HAWKER, L. E. (1962). Studies on vesicular-arbuscular endophytes. V. A review of the evidence relating to identity of the causal fungi. *Trans. Brit. mycol. Soc.* **45**, 190.

HAWKER, L. E. & HAM, A. M. (1957). Vesicular-arbuscular mycorrhizas in apple seedlings. *Nature, Lond.* **180**, 998.

HAWKER, L. E., HARRISON, R. W., NICHOLLS, V. O. & HAM, A. M. (1957). Studies on vesicular-arbuscular endophytes. I. A strain of *Pythium ultimum* Trow. in roots of *Allium ursinum* L. and other plants. *Trans. Brit. mycol. Soc.* **40**, 375.

HEPDEN, P. M. (1960). Studies on vesicular-arbuscular endophytes. II. Endophytes in the Pteridophyta with special reference to leptosporangiate ferns. *Trans. Brit. mycol. Soc.* **43**, 559.

JOHNSON, A. (1949). Vesicular-arbuscular mycorrhiza in Sea Island cotton and other tropical plants. *Trop. Agric.* **26**, 118.

KANOUSE, B. B. (1936). Studies of two species of *Endogone* in culture. *Mycologia*, **28**, 47.

KATZNELSON, H., ROUATT, J. W. & PETERSON, E. A. (1962). The rhizosphere effect of mycorrhizal and non-mycorrhizal roots of yellow birch seedlings. *Canad. J. Bot.* **40**, 377.

KELLEY, A. P. (1937). *The Literature of Mycorrhizae.* Landenberg Laboratory, Landenberg, Pen.

KOCH, H. (1961). Untersuchungen über die Mykorrhiza der Kulturpflanzen unter besonderer Berücksichtigung von *Althaea officinalis* L., *Atropa belladonna* L., *Helianthus annuus* L., und *Solanum lycopersicum* L. *Gartenbauwiss.* **26**, 5.

KHRUSHCHEVA, E. P. (1960). [The mycorrhiza of wheat and its importance for the growth and development of the plant.] *Izv. Akad. Nauk.* (*Ser. biol.*), no. 2, 230.

KUBÍKOVÁ, J. (1961). *Endogone* sp. in association with vesicular-arbuscular mycorrhiza of ash (*Fraxinus excelsior* L.). *Ceská Mykol.* **15**, 161.

LAYCOCK, D. H. (1936). Preliminary investigations into the function of endotrophic mycorrhiza of *Theobroma cacao*. *Trop. Agric.* **22**, 77.

168   B. MOSSE

LIHNELL, D. (1939). Untersuchungen über die Mykorrhizen und die Wurzelpilze von *Juniperus communis*. *Symb. bot. upsaliens*, **3**, 1.

LINK, H. F. (1809). Observationes in Ordines Plantarum naturales. *Ges. naturf. Freunde Berl.* Magazin f. d. neuesten *Entdeckungen in der gesammten Naturkunde*, **3**, 33.

McLENNAN, E. I. (1926). The endophytic fungus of *Lolium*. II. The mycorrhiza on the roots of *Lolium temulentum* L. with a discussion of the physiological relationships of the organism concerned. *Ann. Bot., Lond.* **40**, 43.

McLUCKIE, J. & BURGES, A. (1932). Mycotrophism in the Rutaceae. I. The mycorrhiza of *Eriostemon crowei*. *Proc. Linn. Soc. N.S.W.* **57**, 291.

MAGROU, J. (1946). Sur la culture de quelques champignons de micorrhizes à arbuscule et à vésicules. *Rev. gén. Bot.* **53**, 79.

MELOH, K. A. (1961). Untersuchungen zur Biologie und Bedeutung der endotrophen Mycorrhiza bei *Zea mays* L. und *Avena sativa* L. Dissertationsschrift der Universität Köln.

MOSSE, B. (1953). Fructifications associated with mycorrhizal strawberry roots. *Nature, Lond.* **171**, 974.

MOSSE, B. (1954). Studies on the endotrophic mycorrhiza of some fruit plants. Ph.D. thesis, London.

MOSSE, B. (1956). Fructifications of an *Endogone* sp. causing endotrophic mycorrhiza in fruit plants. *Ann. Bot.* **20**, 349.

MOSSE, B. (1957). Growth and chemical composition of mycorrhizal and non-mycorrhizal apples. *Nature, Lond.* **179**, 922.

MOSSE, B. (1959*a*). The regular germination of resting spores and some observations on the growth requirements of an *Endogone* sp. causing vesicular-arbuscular mycorrhiza. *Trans. Brit. mycol. Soc.* **42**, 273.

MOSSE, B. (1959*b*). Observations on the extra-matrical mycelium of a vesicular-arbuscular endophyte. *Trans. Brit. mycol. Soc.* **42**, 439.

MOSSE, B. (1961). Experimental techniques for obtaining a pure inoculum of an *Endogone* sp. and some observations on the vesicular-arbuscular infections caused by it and by other fungi. *Recent Advances in Botany*, **2**, 1728. University of Toronto Press.

MOSSE, B. (1962*a*). The establishment of vesicular-arbuscular mycorrhiza under aseptic conditions. *J. gen. Microbiol.* **27**, 509.

MOSSE, B. (1962*b*). Le micorize vescicolari-arbuscolari ed alcuni problemi associati al loro studio. *Nuovo G. bot. Ital.* (in the Press).

NEILL, J. C. (1944). Rhizophagus in citrus. *N.Z. J. Sci. Tech.* A, **25**, 191.

NICOLSON, T. H. (1958). Vesicular-arbuscular mycorrhiza in the Gramineae. *Nature, Lond.* **181**, 718.

NICOLSON, T. H. (1959). Mycorrhiza in the Gramineae. I. Vesicular-arbuscular endophytes with special reference to the external phase. *Trans. Brit. mycol. Soc.* **42**, 421.

NICOLSON, T. H. (1960). Mycorrhiza of the Gramineae. II. Development in different habitats, particularly sand dunes. *Trans. Brit. mycol. Soc.* **43**, 132.

O'BRIEN, D. G. & M'NAUGHTON, E. J. (1928). Endotrophic mycorrhiza of strawberries and its significance. *Bull. W. Scot. agric. Coll.* **1**, 1.

OHMS, R. E. (1956). A phycomycetous mycorrhiza on barley roots in South Dakota. *Plant Dis. Reptr*, **40**, 507.

OHMS, R. E. (1957). A flotation method for collecting spores of a phycomycetous mycorrhizal parasite from soil. *Phytopathology*, **47**, 751.

OTTO, G. (1959). Beitrag zur Frage der funktionellen Bedeutung der Vesikel der endotrophen Mycorrhiza an Sämlingen von *Malus communis* L. *Arch. Mikrobiol.* **32**, 373.

PEUSS, H. (1958). Untersuchungen zur Ökologie und Bedeutung der Tabakmycor-rhiza. *Arch. Mikrobiol.* **29**, 112.

PEYRONEL, B. (1923). Fructification de l'endophyte à arbuscules et à vésicules des mycorrhizes endotrophes. *Bull. Soc. mycol. Fr.* **39**, 119.

PEYRONEL, B. (1924). Prime ricerche sulle micorize endotrofiche e sulla microflora radicicola normale delle Fanerogame. *Riv. Biol.* **5**, 463; **6**, 17.

PEYRONEL, B. (1937). Le 'Endogone' quali produttrici di micorrhize endotrofiche nelle Fanerogame alpestri. *Nuovo G. bot. Ital.* **44**, 584.

PEYRONEL, B. (1940). Prime osservazioni sui rapporti tra luce e simbiosi micor-rizica. *Estr. Ann. No. 4. Labor. della Chanousia Giardino Botanico dell'Ordine Mauriziano Picc. San Bernardo,* **4**, 1.

RAYNER, M. C. (1927). *Mycorrhiza.* Cambridge University Press. Also *New Phytol.* **25**.

REED, H. S. & FRÉMONT, TH. (1935). Étude physiologique de la cellule à mycor-rhizes dans les racines de citrus. *Rév. cytol. cytophysiol. vég.* **1**, 327.

SABET, Y. S. (1946). Reactions of citrus mycorrhiza to manurial treatment. *Proc. Egypt. Acad. Sci.* **1**, 21.

SCHRADER, R. (1958). Untersuchungen zur Biologie der Erbsenmycorrhiza. *Arch. Mikrobiol.* **32**, 81.

SCHÜTTE, K. H. (1956). Translocation in the fungi. *New Phytol.* **55**, 164.

SKEPPSTEDT, A. (1952). An investigation of the *Dryopteris Linnaeana* mycorrhiza with special reference to the vesicle problem. *Svensk Bot. Tidsk.* **46**, 454.

STAHL, M. (1949). Die Mycorrhiza der Lebermoose mit besonderer Berücksichti-gung der thallosen Formen. *Planta,* **37**, 103.

STRZEMSKA, J. (1954). Bibliografia prac z zakresu mikoryzy lata 1758–1953. *Acta microbiol. Polon.* **3**, 155.

STRZEMSKA, J. (1955). [Investigation on the mycorrhiza in corn plants.] *Acta microbiol. Polon.* **4**, 191.

THAXTER, R. (1922). A revision of the Endogoneae. *Proc. Amer. Acad. Arts Sci.* **57**, 291.

TOLLE, R. (1958). Untersuchungen über die Pseudomycorrhiza von Gramineen. *Arch. Mikrobiol.* **30**, 285.

WILHELM, S. (1959). Parasitism and pathogenesis of root-disease fungi. In *Plant Pathology,* 1908–58, p. 356. Ed. C. S. Holton and others. University of Wisconsin Press.

WINTER, A. G. (1951). Untersuchungen über die Verbreitung und Bedeutung der Mycorrhiza bei kultivierten Gramineen und einigen anderen landwirtschaft-lichen Nutzpflanzen. *Phytopathologische Z.* **17**, 421.

WINTER, A. G. & MELOH, K. A. (1958). Untersuchungen über den Einfluss der endotrophen Mycorrhiza auf die Entwicklung von *Zea mays* L. *Naturwissen-schaften,* **45**, 319.

EXPLANATION OF PLATES

PLATE 1

Fig. 1. Young arbuscule with host nucleus (N) near its centre. L.S. onion root.

Fig. 2. L.S. apple root showing different stages in arbuscule development; at centre mature arbuscules with enlarged cell nuclei (N) containing prominent nucleoli, and above two arbuscules in advanced stages of disintegration (d.a.).

PLATE 2

Fig. 3. Young vesicles containing oil globules in a longitudinal section of an apple rootlet.

Fig. 4. Vesicle development in an aseptically grown clover root of a seedling declining from lack of nitrogen. Some vesicles are protruding from the cortex. The root was cleared by immersing in boiling lactophenol.

Fig. 5. Entry of the endophyte (*Endogone* sp.) into an aseptically grown clover rootlet. The dimorphic character of the external mycelium, and multiple entry points from one main hypha can be clearly seen. Several vesicles have developed near the entry point.

PLATE 3

Fig. 6. Growth habit of the external mycelium of an endophytic *Endogone* sp. based in a clover root and growing aseptically on filter paper. The structure on the right resembles the intracellular arbuscules.

Fig. 7. Abnormal development of an endophytic *Endogone* sp. in a hemp root. The fungus is forming a mat along the endodermis, prior to penetration of the vascular tissue. The cortical cells of the root are collapsing.

Fig. 8. Abnormal development of an endophytic *Endogone* sp. in a hemp root. The fungus is penetrating the emerging branch root.

I thank Mr E. Yoxall Jones and Miss R. J. Dolling for the photographs in figs. 1–3, 7 and 8, which are reproduced by courtesy of East Malling Research Station.

PLATE 1

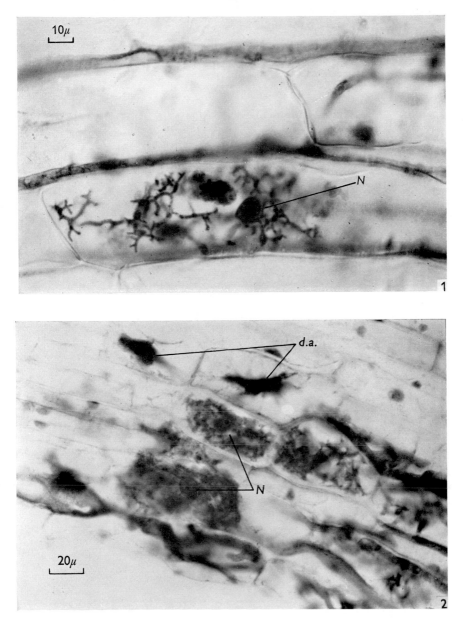

PLATE 2

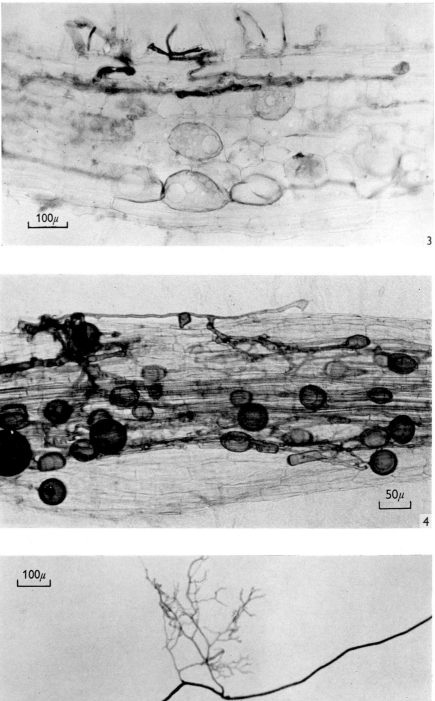

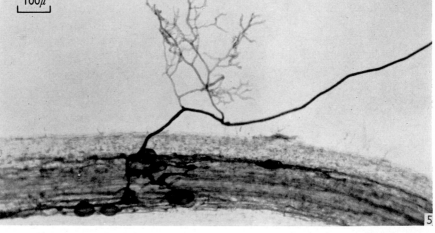

PLATE 3

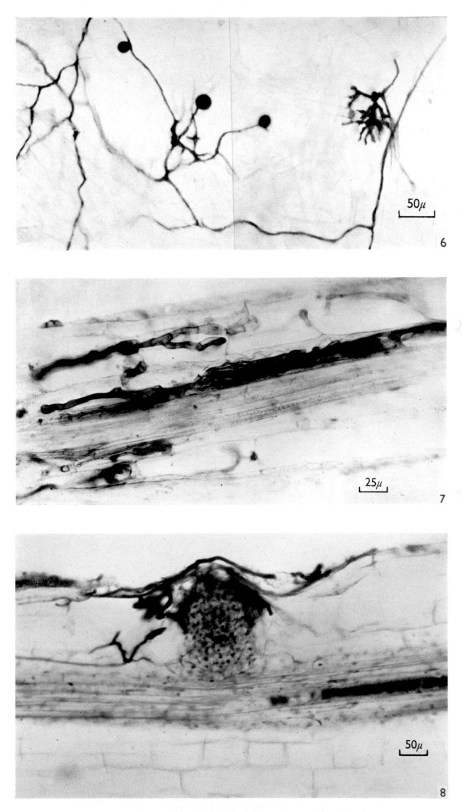

6

7

8

50μ

25μ

50μ

ALGAE AND INVERTEBRATES
IN SYMBIOSIS

M. R. DROOP

The Marine Station, Millport, Scotland

This universal nitrogen-hunger is a misery
which makes strange bed-fellows.
(Keeble: *Plant-animals*, 1910)

Symbiosis, *die Erscheinung des Zusammenlebens ungleichnamiger Organismen*, as de Bary (1879) has defined it in reference to lichens, is also to be seen between unicellular algae and aquatic invertebrates. Some associations, like the common green hydra, need no introduction; reef corals are hardly less familiar, although it is difficult for any one unacquainted with tropical faunas to comprehend the extent to which these symbiotic coelenterates are deployed in the warm oceans of the world. Indeed, corals form as remarkable a feature of tropical seascape as lichens do of arctic tundra.

Underlying every inquiry into the origin and significance of symbiosis is the question of function. Fundamentally, the relationship between endo-symbiont and its host is a nutritional one, and a degree of syntrophy exists between the two partners. This syntrophy may be slight or deep, lop-sided or symmetrical; it is potentially most perfect when one partner is a chemotroph and the other a phototroph, when the products of respiration of the one are exchanged for those of photosynthesis of the other, and there is a net short-circuiting of the cycle of nutrient elements. Any subtler nutritional interdependence is more likely to be a consequence rather than a cause of association, but once established it could become a major factor in the evolutionary persistence of the habit. On the other hand, no symbiont is there for the benefit of the other partner. The idea of antagonism expressed by Elenkine (1906) and Bernard (1909) serves to emphasize that primarily one of the partners must stand to gain by the association, regardless of the other. In algal/animal symbioses the physical relationship is such that the primary benefit must always be to the alga, although the well-being of the animal will often assume greater prominence when a symbiosis is viewed from an ecological standpoint.

Not even the most cursory introduction to algal/invertebrate symbiosis would be complete without reference to the reviews of Buchner

(1930) and Yonge (1944, 1958). By way of apology, the pages that follow are a botanist's interpretation to microbiologists of a preponderantly zoological field of endeavour.

OCCURRENCE

The animals

The variety of animals harbouring symbiotic algae is impressive: it includes Protozoa, Coelenterata, Porifera, Platyhelminthes (Turbellaria), Mollusca, in addition to isolated instances in five other phyla. Protozoa and Coelenterata furnish the most examples (Table 1). It is remarkable that digestion in very nearly all these animals is intracellular (Yonge, 1937a), the few exceptions having invariably been loose associations, with the alga never in the tissues or cells. No doubt the reason for the correspondence is to be found in the way in which the symbiosis is established and transmitted. Mostly the symbionts are situated within the cells of the host: in the outermost region of the endoplasm in ciliates (Dangeard, 1900), in amoebocytes in sponges, in carrier cells of the endodermal epithelium in coelenterates (in the epithelium itself in *Chlorohydra*), and in phagocytic blood cells in the giant clams (Tridacnidae). All these tissues are either concerned with digestion or with food transport. In Turbellaria, on the other hand, the algae are mostly extracellular in the spaces of the subepidermal parenchyma, as they are also in the nudibranch *Tridachia*, but in Turbellaria they occur also in the wandering parenchyma cells. The earlier (1930) edition of Buchner's monograph very fully summarizes the descriptive literature prior to that date.

The algae

The algal partners in these associations are, by contrast, very stereotyped. In size they are, of course, much the minor partner. They are almost invariably coccoid and are mainly drawn from only three classes of algae: Cyanophyceae, Chlorophyceae and Dinophyceae. It is customary to refer to them as Cyanellae (Pascher, 1929a), Zoochlorellae or Zooxanthellae (Brandt, 1883a), according to whether they are blue-green, green or yellow in colour. No precise taxonomic connotation is now attached to these terms.

Just as the green groups of free-living algae are most widely distributed in inland waters and the brown groups in the sea, so in general zoochlorellae associate with fresh water, and zooxanthellae with marine animals. The green alga in the marine worm *Convoluta roscoffensis* is a notable exception, and *Myrionema amboinense*, a marine hydroid, harbours either chlorellae or xanthellae depending on its geographical location (Fraser, 1931). Cyanellae are confined to a few remarkable freshwater protozoa and algae (Pascher, 1929a). The zooxanthella of *Anemonia sulcata* is shown in Fig. 1.

Taxonomy

There is little doubt that most zoochlorellae can be assigned to the genus *Chlorella* or, possibly in certain cases, to *Pleurococcus* (Beyerinck, 1890; Limberger, 1918; van Trigt, 1919; Chodat, 1924; Genevois, 1924;

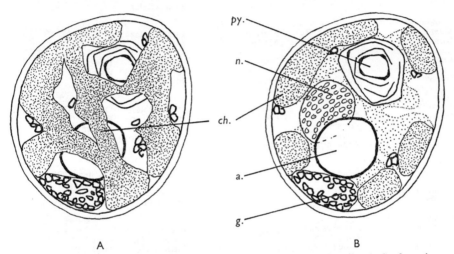

A   B

Fig. 1. Zooxanthella of *Anemonia sulcata* (from Clachan Sound, Argyll). A, Surface view; B, in optical section; showing reticulate parietal chromatophore (*ch.*), nucleus (*n.*), pyrenoid with starch sheath (*py.*), 'assimilatory product' of unknown composition (*a.*) and vacuole with refractive granules (*g.*). The shorter diameter of this cell is 10 μ.

Haffner, 1925; Loefer, 1936a). But the alga in *Convoluta roscoffensis* is a member of the Volvocales, a *Carteria*, according to Keeble & Gamble (1907), although, judging by their figures, more probably a *Tetraselmis* or *Prasinocladus*, both common marine genera.

Zooxanthellae were originally placed in either the Cryptophyceae or Dinophyceae, but all recent evidence (drawn from symbionts of marine coelenterates) points unequivocally to the latter. Thus, Hovasse & Teissier (1923) demonstrated a dinoflagellate type of nucleus, and Hovasse (1937), by silver impregnation, a typical dinoflagellate sculpturing of the cell wall. Gymnodinioid swarmers were first observed by Kawaguti (1944a), while their connexion with the zooxanthellae has now been amply confirmed by McLaughlin & Zahl (1959), who have

Table 1. *Some genera of invertebrates in which symbiosis with algae is recorded and some alga/alga associations*

(References in Buchner, 1930.)

ost	Algal symbiont	Habitat	Position of alga	Principal references
PROTOZOA CILIATA				
Frontonia	Zoochlorella	Fresh water	Intracellular	—
Ophrydium	Zoochlorella	Fresh water	Intracellular	—
Paraeuplotes	Zooxanthella	Marine	Intracellular	Wichterman (1942)
Paramoecium	Zoochlorella	Fresh water	Intracellular	Dangeard (1900), Pringsheim (1928), Siegel (1960)
Stentor	Zoochlorella	—	—	—
Tintinnus	Chaetoceras (Bacillariophyceae)	Marine	Extracellular	—
Trichodina	—	Marine	Intracellular	—
PROTOZOA SARCODINIA				
Acanthometra	Zooxanthella	Marine	Intracellular	—
Acanthocystis	Zooxanthella	Marine	Intracellular	—
Actinosphaerium	—	—	—	—
Amoeba	Zoochlorella	Fresh water	Intracellular	Pascher (1930)
Collozoum	Zooxanthella	Marine	Intracellular	Brandt (1885)
Difflugia	Zoochlorella	Fresh water	Intracellular	—
Globigerina	Zooxanthella	Marine	—	—
Heliophrys	—	—	—	—
Orbitolites	Zooxanthella	Marine	Intracellular	Doyle & Doyle (1940)
Paulinella	Cyanella	Fresh water	Intracellular	Pascher (1929 a, b)
Peneroplis	Zooxanthella	Marine	Intracellular	Winter (1907)
Sphaerozoum	Zooxanthella	Marine	Intracellular	Brandt (1885)
Trichosphaerium	Zooxanthella	Marine	Intracellular	—
PROTOZOA MASTIGOPHORA				
Cryptella	Cyanella	Fresh water	Intracellular	Pascher (1929 a)
Cyanophora	Cyanella	Fresh water	Intracellular	Korschikoff (1924)
Noctiluca	—	Marine	Intracellular	—
Peliaina	Cyanella	Fresh water	Intracellular	Pascher (1929 a)
PORIFERA				
Carterius	Scenedesmus	Fresh water	Extracellular	—
Ephidatia	Zoochlorella	Fresh water	Intracellular	Trigt (1919)
Euspongilla	Zoochlorella	Fresh water	Intracellular	Trigt (1919)
PLATYHELMINTHES TURBELLARIA				
Amphiscolops	Zooxanthella	Marine	—	Welsh (1936)
Castrada	Zoochlorella	Fresh water	—	Limberger (1918)
Convoluta paradoxa (= C. convoluta)	Zooxanthella	Marine, temperate	Extracellular	Keeble (1908)
C. roscoffensis	Carteria	Marine, temperate	Extracellular	Keeble & Gamble (1907)
Dalyellia	Zoochlorella	Fresh water	Intra- and extracellular	Haffner (1925)
Phaenocora	Zoochlorella	Fresh water	—	Genevois (1924)
Typhoplana	Zoochlorella	Fresh water	—	Genevois (1924)
COELENTERATA HYDROZOA				
Chlorohydra	Zoochlorella	Fresh water	Intracellular	Goetsch (1924), Haffner (1925)
Halecium	Zooxanthella	Marine	Intracellular	Hadzi (1911)
Hydra	Zoochlorella	Fresh water	Intracellular	Goetsch (1924)
Millepora	Zooxanthella	Marine	Intracellular	Mangan (1909)
Myrionema	Zoochlorella or Zooxanthella	Marine	Intracellular	Fraser (1931)
Porpita	Zooxanthella	Marine	Intracellular	—
Velella	Zooxanthella	Marine	Intracellular	—
COELENTERATA SCYPHOZOA				
Cassiopeia	Zooxanthella	Marine	Intracellular	Smith (1936)
Cotylorhiza	Zooxanthella	Marine	—	—
Crambessa	Zooxanthella	Marine	—	—
Linuche	Zooxanthella	Marine	—	—
Mastigias	Zooxanthella	Marine	—	—

Host	Algal symbiont	Habitat	Position of alga	Principal references
COELENTERATA ACTINOZOA				
Actinaria:				
Actinia	Zooxanthella	Marine	Intracellular	—
Aiptasia	Zooxanthella	Marine	Intracellular	Trendelenberg (1909), Pütter (1911)
Anemonia	Zooxanthella	Marine, temperate	Intracellular	Smith (1939)
Anthopleura	Zooxanthella	Marine	Intracellular	McLaughlin & Zahl (1959)
Condylactis	Zooxanthella	Marine	Intracellular	McLaughlin & Zahl (1959)
Heliactis	Zooxanthella	Marine, temperate	Intracellular	Naville (1926)
Alcyonaria:				
Alcyonium	Zooxanthella	Marine	Intracellular	Pratt (1906)
Clavularia	Zooxanthella	Marine	Intracellular	—
Heliopora	Zooxanthella	Marine	Intracellular	—
Heteroxenia	Zooxanthella	Marine	Intracellular	Gohar (1940)
Lobophytum	Zooxanthella	Marine	Intracellular	Pratt (1906)
Sarcophytum	Zooxanthella	Marine	Intracellular	Pratt (1906)
Sclerophytum	Zooxanthella	Marine	Intracellular	Pratt (1906) Gohar (1940)
Tubipora	Zooxanthella	Marine	Intracellular	
Xenia	Zooxanthella	Marine	Intracellular	Gohar (1940)
And several other genera —	—	—	—	—
Madreporaria:				
Acropora	Zooxanthella	Marine	Intracellular	—
Cyphastrea	Zooxanthella	Marine	Intracellular	—
Favia	Zooxanthella	Marine	Intracellular	—
Fungia	Zooxanthella	Marine	Intracellular	—
Loboghyllia	Zooxanthella	Marine	Intracellular	—
Porites	Zooxanthella	Marine	Intracellular	—
Seriatopora	Zooxanthella	Marine	Intracellular	—
And 40 other genera of corals —	—	—	—	Boschma (1924), Yonge & Nicholls (1931a)
MOLLUSCA LAMELLIBRANCHIA				
Anodonta	Chlorella	Fresh water	Extracellular	Goetsch & Scheuring (1926)
Corculum	Zooxanthella	Marine	—	Kawaguti (1950)
Hippopus	Zooxanthella	Marine	Intracellular	Yonge (1936)
Tridacna	Zooxanthella	Marine	Intracellular	Yonge (1936)
Unio	Chlorella	Fresh water	Extracellular	Goetsch & Scheuring (1926)
MOLLUSCA NUDIBRANCHIA				
Aeolidiella	Zooxanthella	Marine, temperate	Intracellular	Naville (1926)
Aeolis	Zooxanthella	Marine	—	Yonge & Nicholas (1940)
Doridoeides	Zooxanthella	Marine	—	Yonge & Nicholas (1940)
Favonius	Zooxanthella	Marine	—	Yonge & Nicholas (1940)
Melibe	Zooxanthella	Marine	—	Yonge & Nicholas (1940)
Phyllirhoe	Zooxanthella	Marine	—	Yonge & Nicholas (1940)
Spurilla	Zooxanthella	Marine	—	Yonge & Nicholas (1940)
Tridachia	Zooxanthella	Marine	Extracellular	Yonge & Nicholas (1940)
OTHER GROUPS				
Ctenophora:				
Beroe	'Red flagellate'	Marine	—	—
Rotatoria:				
Cephalodella	Chlorella	Fresh water	Extracellular	—
Mollusca Pulmonata:				
Limnaea	Chlorella	Fresh water	Extracellular	—
Echinodermata:				
Ophioglypha	Coccomyxa	Marine	In epidermal plates	—
Ascidiacea:				
Didemnum	—	Marine	Extracellular	Smith (1935)
Diplosoma	—	Marine	Extracellular	Smith (1935)
Trididemnum	—	Marine	Extracellular	Smith (1935)
ALGAE				
Cyanoptyche	Cyanella	Fresh water	Intracellular	Pascher (1929a), Geitler (1959)
Glaucocystis	Cyanella	Fresh water	Intracellular	Pascher (1929a), Geitler (1959)
Glaeochaete	Cyanella	Fresh water	Intracellular	Pascher (1929a), Geitler (1959)
Rhizosolenia	Richelia (Nostocales)	Marine	Intracellular	Pascher (1929a), Geitler (1959)

bacteria-free cultures of a number of strains from different coelenterates. Freudenthal (1962) has just published a full account of the organism isolated by McLaughlin & Zahl from the Scyphozoon *Cassiopeia*, under the name *Symbiodinium microadriaticum*.

On the other hand, it is by no means certain that all zooxanthellae are dinoflagellate. Brandt (1885) and Winter (1907) figured *Cryptomonas-*

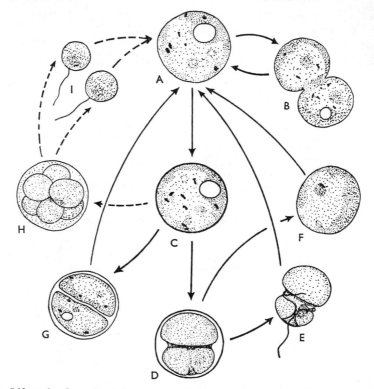

Fig. 2. Life cycle of *Symbiodinium microadriaticum*. A, Vegetative cell; B, vegetative cell undergoing binary fission; C, vegetative cyst with thick cell wall; D, mature zoosporangium; E, gymnodinioid zoospore; F, aplanospore; G, cyst containing two autospores; H, cyst containing developing isogametes (?); I, liberated isogametes (?). Diagrammatic, after Freudenthal (1962).

like swarmers in connexion with the symbionts of some Radiolaria and Foraminifera, and the yellow cells of *Convoluta paradoxa* were thought by Hovasse (1937) not to be dinoflagellates. The differences observed by Yonge (1936) and Yonge & Nicholas (1940) between the cells in *Tridacna* and *Tridachia* and most coelenterate zooxanthellae were mainly a matter of cell-wall thickness and reserve product. One notes, however, that the life history of *Symbiodinium* as described by Freudenthal (1962) and illustrated in Fig. 2 includes propagation in a non-motile naked state as well as autospore formation by thick-walled cells. Furthermore, the reserve product of free-living dinoflagellates, though plentiful, may never give an iodine reaction for starch unless the cells are encysted, when it may be very strong. Possibly the observed differences between zooxanthellae are of only minor taxonomic import.

MECHANISM

Transmission

For a symbiotic association to persist there must be an efficient method of passing the symbiont from one animal generation to the next. The method of transmission varies widely. In some animals there is no transmission *sensu stricto*, since each succeeding generation needs to be re-infected by algae, which necessarily also have a free-living existence. One can find sharp seasonal variation in the incidence of symbiosis which may be correlated with the seasonal abundance of the infecting alga (Wesenberg-Lund, 1909; Haffner, 1925). Re-infection by phago-cytosis occurs incidentally in Protozoa and Porifera having a rather casual relationship with their algae; it is the only way in many Turbellaria, e.g. *Convoluta*, and the freshwater *Dalyellia*, as it is also in Foraminifera with a reproductive phase involving very small cells (Winter, 1907; Keeble & Gamble, 1907; Haffner, 1925). Some wholly carnivorous animals also acquire a complement of algae with their food, but in this instance they receive them second-hand, so to speak, being unable to digest the algae contained in the tissues of their prey. An example is the nudibranch *Aeolidiella* feeding exclusively on *Heliactis*, an anemone with zooxanthellae (Naville, 1926). Undigested chlorellae in the gut of *Daphnia* and other small Crustacea might be responsible for the spontaneous infection of *Hydra attenuata* recorded by Goetsch (1924).

In *Paramoecium* and most Protozoa the symbionts are divided between the daughter cells at cell division, thus ensuring continuance of the infection. An analogous refinement is found in some asexually reproducing multicellular animals, where the buds (in Coelenterata) or gemmules (in sponges) contain their complement of infected cells. Sponge gemmules are reminiscent of the soredia of lichens.

Finally, in sexually reproducing animals transmission is surest when the egg is infected with algae before fertilization, as in Coelenterata generally (e.g. Mangan, 1909; Hadzi, 1911; Goetsch, 1924; Haffner, 1925). *Hydra attenuata* is apparently an exception to this rule, for in this species sexual reproduction of green individuals (= *H. viridescens*) gives rise to the normal colourless form which may or may not become infected later (Goetsch, 1924).

It is interesting that the strictest routine is found in coelenterates, for which symbiosis is not generally so vital, whereas in *Convoluta*, for which it is, the method of transmission is haphazard by comparison. But this is a matter of feeding habits: the Turbellarian is omnivorous in the early part of its life and has every chance of picking up a suitable alga, especially so in the case of *C. roscoffensis* where the alga is motile and attracted chemotactically to the egg capsules. Coelenterates, on the other hand, are generally held not to ingest algal cells, so that infection would be uncertain, even if their symbionts were numerous in the phytoplankton.

Regulation

Once established in the host, the algae increase in numbers, but only until they reach the population density appropriate to the situation, for example from 50 to several hundred per paramoecium or 7500 per coral planula $1\cdot0 \times 0\cdot5$ mm. (Marshall, 1932; Siegel, 1960). Thereafter the numbers are maintained at this level by some process of regulation, dynamic stability being one of the characteristic features of symbiosis.

One can see in a general way how control is achieved, for in so far as the alga is dependent upon the excretory products of the host's metabolism, the ratio of algal to animal tissue will be determined primarily by the metabolic rate of the animal. It is a necessary condition that the alga should have as high a potential specific growth rate as the animal tissue it inhabits. This may be an important factor in associations with Protozoa.

Control may also be effected by digestion in those animals which regularly kill and digest their algae, as Tridacnidae do. Van Trigt (1919) noted a regularly high proportion of dead algal cells in sponge tissues, and although it was not clear whether this was due to starvation (including the effect of low light intensity) or to an active reaction on the part of the animal on possibly weakened symbionts, nevertheless mortality was a major factor in the dynamics of the symbiosis. One is reminded of the way coelenterates eject moribund zooxanthellae.

There are indications of a rather more intimate adjustment as, for instance, in the number of symbionts, never more than two, held by individual carrier cells in marine coelenterates and lamellibranchs. Although mere *Lebensraum* may account for this, there is no indication that the adjustment is effected by migration of symbionts from one carrier cell to another, or alternatively by synchronization of cell-division between carrier and symbiont. The nicest examples of such control are provided by two protozoans, *Peliaina cyanea* and *Paulinella chromatophora* (Hoogenraad, 1927; Pascher, 1929 *a*, *b*).

Peliaina, a colourless flagellate, contains from one to six *Synechococcus* cells. Generally, both alga and flagellate divide at more or less the same rate and both daughter flagellates receive the same number of algae, but occasionally this is not so and when the number received by any daughter is reduced to one, cell-division of the alga is impaired, synchronization is lost and colourless viable flagellates are produced. Colourless individuals of *Paulinella*, on the other hand, are not known. Synchronization between cell-division of this rhizopod and the two *Synechococcus* cells is perfect, though out of phase: first the animal divides and the daughters receive a cyanella each, then the algae divide, bringing the number once more to two per cell. The phasing of the divisions suggests a nutritional relationship with control exerted by the animal, although from the precision achieved one might infer that rather more is involved.

Prerequisites

One essential condition for symbiosis—tolerance of the other partner—seems to be more difficult for the alga because apparently so few algal types are in symbiosis with such a wide variety of animals. However, the little experimental evidence, mostly from Protozoa and Hydrida, though confirming the specialization of endozoic algae reveals that much also depends on the animal species.

One recalls that, although Oehler (1922) found it easier to infect *Paramoecium bursaria* with chlorellae from other individuals than with free-living strains, this was the only species among a number of ciliates in which he was able to establish a permanent association. Similarly, Goetsch (1924), in a very well-known study, found the various species of Hydrida differed widely in their liability to experimental infection with the parasitic chlorella from *Anodonta*. *Pelmatohydra*, for instance, retained no infection and, moreover, when grafted on to *Hydra attenuata* actively inhibited infection of the latter; whereas both *H. vulgaris* and especially *H. attenuata* were easily infected, and *Chlorohydra* was so receptive that it was possible to establish a free-living *Oocystis* in it.

Of first importance is the ability of the alga to resist digestion by the animal it inhabits. Intracellular symbionts lie freely in the cytoplasm of their carrier cell, and are therefore not exposed to as high a concentration of enzymes as they would be were they within a food vacuole. But this is merely to pose the question, how do they get into that position, for presumably they are ingested by way of a food vacuole? Ability to destroy and prevent the re-formation of vacuole membranes would seem to be prerequisite for life within a phagocyte; but this takes us beyond the scope of our discussion (see, however, Kitching, 1956; Elberg, 1960; Karnovski, 1962). Another view expressed by van Trigt (1919) is that the sponge is capable of slow extra-vacuolar digestion of its symbiotic algae.

No doubt the cellulose wall possessed in some degree by most endozoic algae affords them a modicum of protection; indeed, feeding experiments with lamellibranch larvae and calanoid copepods have proved *Chlorella* a tough proposition even for habitual herbivores (Cole, 1936; Marshall & Orr, 1955).

Origins

Apart from *Peliaina*, *Cyanophora*, and those remarkable associations with other algae, *Glaucocystis* and *Glaucochaete*, whose origins are obscure, the original opportunity for endo-symbiosis is the animal habit of phagotrophy. Having regard, therefore, to the feeding habits of animals, Yonge (1935, 1944) has emphasized that symbiosis must have originated rather differently in carnivores than in animals accustomed to feed on algae. With the latter, the main consideration was resistance to digestion; one sees in the sponge how precarious life with a herbivore can be. The hypothesis, as we have seen, does not account for the fact that most herbivorous animals are not inhabited by those algae which are admittedly resistant: there must also be some weakening of the animal. With carnivores, on the other hand, in so far as they may neither ingest nor digest plant cells, entry of the alga may have been more difficult, but once access had been gained there would be fewer hazards.

Assuming that the history of the individual can recapitulate the history of the race, Goetsch (1924) visualized the initial stages of symbiosis in carnivores as a parasitic infection, a wholly one-sided affair, only later to be followed by adjustment on the part of the animal. In the work referred to experimental infection of *Hydra attenuata* was only achieved after the animals were weakened by starvation and exposure to high temperature, and this was accompanied by pathological symptoms, but the animals recovered and permanent associations resulted.

Goetsch's postulate reflects the ideas of Elenkine (1906). It could also be applied to herbivorous animals, which after all, should experience some discomfort on finding indigestible foreign protoplasts within their tissues. A possible instance of this can be seen in *Paramoecium*. Siegel (1960) observed the initial spread of an infection throughout the population to be non-random, for after some 15 doublings of the protozoan a relatively large number of chlorellae (10–100 per cell) was confined to a very small percentage of the population. A depression of growth rate in animals containing symbionts most simply explains the observation (although other inferences were drawn by Siegel). Since Siegel measured no significant difference in average growth rate between established white and green clones of *P. bursaria*, any initial handicap due to the presence of algae must eventually have been overcome, as it was by *Hydra attenuata*.

Adaptations

Once established a symbiosis could lead in the long run to modification of either partner; We may consider (1) functional and (2) structural modifications rendering the symbiosis more efficient, and (3) modifications tending to make it more necessary.

(1) *Response to light.* Nearly all animals associating with algae show photopositive behaviour at some point in their life history. The response of marine coelenterates may vary with the number of algal cells (Kawaguti, 1941, 1944*b*), and ceases with elimination of the symbiont (Zahl & McLaughlin, 1959).

The response is probably secondary, since it depends on the presence of algae. In *Paramoecium*, at least, the alga can be considered the photoreceptor, for the action spectrum of the response has been observed by Engelmann (1882) to correspond to that of chlorophyll. Furthermore, since the effect of light was noticeable only under conditions of limiting oxygen tension, Stanier & Cohen-Bazire (1957) conclude that the intermediary to which the animal responds is the internal oxygen gradient. A correlative observation is that green *P. bursaria* are less active swimmers than colourless ones (Siegel, 1960).

The mating reaction of *Paramoecium bursaria*, which also depends on light, is, on the contrary, independent of the presence or absence of chlorellae (Ehret, 1953). But the best known instance of a direct response is provided by the two species of *Convoluta* which have different and complex reactions clearly related to their respective inter-tidal niches. The positive photokinesis in, for example, *C. roscoffensis* is operated by green light, presumably absorbed by the orange ocelli of the worm (Gamble & Keeble, 1904).

Whether an animals' ability to react to light determines the symbiosis, or whether the response arose as a result of it, one cannot say. Irritability experiments with artificial associations could decide whether or not the reaction to an internal oxygen gradient is inherent.

(2) *Structural adaptations.* The prevalence of the coccoid habit among symbiotic algae is better regarded as a matter of selection than of

adaptation, for it occurs commonly among free-living forms. Moreover, the life history of *Symbiodinium* described by Freudenthal (1962) also has its parallel in free-living algae. The life history of the alga of *Convoluta roscoffensis*, on the other hand, is unique and can only be accounted for in terms of a response to an endozoic existence. One may recapitulate this familiar history: a free-living autonomous life cycle with typical chlamydomonad features; ingestion; multiplication within the worm, at first normally then, as the cell wall disappears, by budding, so as to form an irregular mass of tissue with degenerating nuclei with seemingly no function left to it other than photosynthesis; and finally, complete degeneration and digestion by the animal. Symbiosis for this alga is a fatal, though luxurious, episode, from which there is no return. Therefore, although the response belongs to the plant, it is clear that the evolutionary pressures leading to its invocation have acted entirely on the animal.

The best examples of structural adaptation are found in Tridacnidae, which 'conserve', rather than exploit, their symbionts. Now, however, it is the animal which is specialized, and the structural changes result from a deep-seated evolutionary response to the symbiosis (Yonge, 1936). These changes involve a re-orientation of the shell and mantle of the bivalve, and great enlargement of the latter, so that its edges now lie exposed over the uppermost margins of the shell. Here the zooxanthellae are literally farmed in vast numbers; they are concentrated around certain 'hyaline organs', which allow light to penetrate deep into the mantle.

(3) *Loss of function*. Whereas the symbiotic relationship in ciliates and sponges is casual and, indeed often rather precarious for the alga, in *Dalyellia*, *Chlorohydra* and the majority of marine coelenterates a more balanced state of affairs is achieved, and although the symbiosis is not obligatory, the animals (*Dalyellia* excepted) are probably never devoid of symbionts in nature.

Marine coelenterates are more or less easily depleted of their algal cells by exposure to continuous darkness and suffer no great harm provided they are fed properly, but absolute elimination of the symbionts has never been attempted (Yonge & Nicholls, 1931b; Smith, 1939; Zahl & McLaughlin, 1959). Whitney (1907) and Goetsch (1924), however, obtained alga-free *Chlorohydra*, while the production of white clones of *Paramoecium bursaria* is now standard practice in genetics laboratories (Siegel, 1960).

The effects of habitual association begin to be visible with *Chlorohydra*, in which white individuals have not the vigour of the green ones.

In other cases the loss of function is more advanced; in *Convoluta* dependence on its algae is probably connected with an impaired power of excretion (Keeble, 1910). *Convoluta* is manifestly unable to excrete its nitrogen wastes unaided, since deposits of uric acid accumulate in animals kept free of algae and disappear again when they become infected. On the other hand, an absence of excretory organs (flame cells) cannot be attributed to the symbiotic habit, since the majority of acoelous Turbellaria are not symbiotic and none have flame cells. In other animals the feeding habits may be interfered with. Some Alcyonaria, according to Gohar (1940), are unable to capture prey, while in others (Pratt, 1906) the digestive mesenteries are degenerate. Such animals become absolutely dependent on the photosynthesis of their algae for nourishment.

Except for the one instance of *Convoluta roscoffensis*, we have no evidence of loss of function in the algal partner, for, as we shall see, no symbiont yet cultivated *in vitro* has proved more nutritionally exacting than its free-living counterpart. This, however, could be a matter of selection, as few have been cultivated. Fritsch (1952) pointed out that the absence of zooxanthellae in plankton hauls was not proof of their inability to maintain a free-living existence.

Hereditary endo-symbiosis

The term 'plasmid' was coined by Lederberg (1952) to cover any extra-chromosomal intracellular hereditary factor, whether organelle, plasmo-gene or symbiont. The interest of symbiotic algae in this context is that they are clearly of foreign origin and, in so far as the concept is relevant, may illustrate incorporation of foreign material into the genetic apparatus of the cell.

Ideally three properties are associated with plasmids: (1) hereditary transmission, (2) no independent existence, and (3) interference in cell metabolism. How far have symbiotic algae progressed along the path towards this ideal? In the first place, there are Turbellaria with obligate symbionts, which are, however, no more hereditary than say, the shell of a caddis worm; then, there are protozoa with facultative symbionts usually transmitted hereditarily, with the symbiont also capable of a free-living existence; thirdly, there are coelenterates with symbiosis an incidental affair for individuals (but probably a factor in the evolution of their communities) the algae being transmitted in a hereditary fashion, even though they may be capable of independent life; and lastly, and most significantly, we have protozoa and algae in which the symbiont may be transmitted as precisely as a cell organelle, is invariably present, and is unknown outside the host. The origin of such infection is buried deep in the past. *Peliaina* is usually cited as example of the third property of plasmids, for whereas the combination of alga and flagellate forms starch as a storage product, neither does so in isolation. The phototactic response of symbiont-containing animals is also quoted in this context.

On a humbler plane, there are Siegel's experiments with *Paramoecium bursaria*, which show that there is an incipient host specificity, in spite of the casual nature of the symbiosis in the ciliate (Siegel, 1960). Of marine coelenterates with their stricter laws of transmission nothing is known, but McLaughlin & Zahl's zooxanthellae await implantation into completely depleted polyps should they become available.

NUTRITION

As long ago as 1882 Geddes proposed the following for the role of algal cells in the animal's economy: (1) production of oxygen, (2) removal of carbonic acid and nitrogenous wastes, (3) supply of carbohydrate material during life, and (4) the supply of other nutrients by their death. These elementary headings will also serve to introduce the present discussion.

Oxygen

Photosynthesizing algae excrete oxygen which the animal partner can obviously use. But how necessary has this oxygen become to the animal, either as an individual or as a member of a community?

The early experiments on anemones by Brandt (1883b), Trendelenberg (1909) and Pütter (1911) established that oxygen evolved in the light by symbiotic algae usually exceeds the amount used by both host and alga in respiration. In corals there may be a fivefold excess, though a lower figure is more usual (Verwey, 1931; Yonge, Yonge & Nicholls, 1932). On the other hand, when the whole of a 24-hour period is taken into account, almost invariably rather more oxygen is found to have been consumed than produced (Marshall, 1932; Yonge *et al.* 1932; Welsh, 1936). Extremely high respiratory rates occasionally measured in corals (Verwey, 1931) at one time seemed to indicate an essential role for the oxygen produced by zooxanthellae, but it was later shown that most of this apparent demand was due to oxidation of mucus secreted by experimental animals (Yonge, 1937b).

All these experiments, therefore, have simply proved that the symbiont is capable of photosynthesizing at a greater rate than the combined respiratory rate of animal and alga; none establishes that photosynthesis is necessary to the animal. Indeed, as Yonge (1940) pointed out, the oxygen tension of reef water could just as efficiently be maintained by phytoplankton free in the water as by zooxanthellae (that is, if there were any phytoplankton). On the other hand, reef corals only thrive in nature in shallow water where the light is sufficient for a net excess of photosynthesis over respiration in plants.

Carbon dioxide

Zooxanthellae of all investigated coelenterates have been shown to take up carbon dioxide during photosynthesis, as have the zoochlorellae of *Paramoecium bursaria* (Parker, 1926; Yonge & Nicholls, 1931*a*; Muscatine & Hand, 1958)—it is likely that most symbiotic algae do the same. One may note in parentheses, however, that some marine Volvocales photosynthesize anaerobically with the aid of acetic acid in place of carbon dioxide (Pringsheim & Wiessner, 1960; Droop, 1961). It has, as a matter of fact, never been necessary to supply acetic acid to the few symbiotic algae that have been cultivated free of bacteria (Limberger, 1918; Genevois, 1924; Loefer, 1936*a*; McLaughlin & Zahl, 1959).

The value of respiratory carbon dioxide to the symbiotic algae is difficult to assess. On the one hand, there is normally abundant carbon dioxide both in the sea and fresh waters. On the other, the concentration of photosynthesizing cells (locally up to 30,000 per mm.^{3} in the tissues of a contracted polyp) would probably be sufficient to cause a severe local deficiency were it not for a source so near at hand as that respired by the animal.

When photosynthesis is prevented respiratory carbon dioxide has of course to be excreted into the surrounding medium. In this there would be no great disadvantage to the animal under natural conditions, particularly in a well-buffered marine environment. Mayer (1924) showed that the respiration of corals was unimpaired when the carbon dioxide content of the water was raised 200-fold by bubbling the gas into it until pH (usually 8·2) had fallen to 5·85. There is, however, one animal function affected specifically by the carbon dioxide balance; this is the process of calcification, for which reef corals are noted.

Skeleton building in corals. Some effect upon calcification is probable on purely chemical grounds; an accumulation of carbon dioxide might tend to dissolve the skeleton (Verwey, 1930). Moreover algae, by raising pH, should favour the precipitation of calcium carbonate. The participation of algae in the process was demonstrated by Kawaguti & Sakumoto (1948) and by Goreau & Goreau (1959), who found that light was necessary for the uptake of calcium by several zooxanthella-bearing species. Using a tracer technique, Goreau & Goreau obtained an average 10:1 ratio between calcium deposition in the light and that in the dark, while corals previously depleted of their algae showed a uniformly low rate of calcification independent of light (Goreau, 1959).

The coral skeleton is an external structure, secreted by ectodermal cells, and there is strong evidence that the calcium is inorganic in origin (Goreau, 1961). Nevertheless, calcification is, in part at any rate, controlled by the enzyme carbonic anhydrase in the tissues which

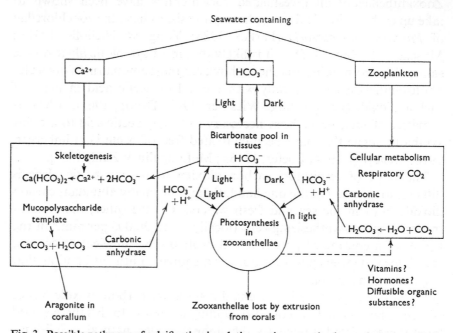

Fig. 3. Possible pathways of calcification in relation to tissue respiration and photosynthesis by zooxanthellae in reef corals. From Goreau (1961).

catalyses the reaction $H_2CO_3 \rightleftharpoons H_2O + CO_2$. In the dark, or in the absence of symbionts, specific inhibitors of this enzyme reduce calcium deposition to extremely low levels, whereas the reduction is far less severe in the light when algae are present (Goreau & Goreau, 1959). The deposition of calcium carbonate (as aragonite) from calcium and bicarbonate ions results in the production of hydrogen ions ($Ca^{2+} + HCO_3^- \rightleftharpoons CaCO_3 + H^+$), whose removal is presumably effected by carbonic anhydrase and excretion of carbon dioxide. It is not difficult to imagine, therefore, that photosynthesizing algae assist calcification merely by maintaining the animal's tissue pH.

Possible pathways of calcification are shown in Fig. 3.

Nitrogen and phosphorus

The nitrogen and phosphorus excreted by animals are frequently in forms that can be utilized directly by plant cells; such as orthophosphate, nucleotide phosphate, the ammonium ion (provided pH is not too high), guanine, uric acid and urea. In addition, very many lower algae are able to obtain their nitrogen from a variety of amino acids (Algeus, 1946; Reinhardt, 1950; Droop, 1955, 1961; Gibor, 1956; McLaughlin, 1958; Miller & Fogg, 1958). Limberger (1918) cultivated the alga from *Euspongilla* on a medium containing ammonium chloride. Organic nitrogen in the form of gelatine or peptone was not tolerated in high concentration by this alga, though in *Paramoecium* Loefer (1936a) remarked that the tolerance of the alga in this respect was higher than that of the host. McLaughlin & Zahl's recent studies on the nutrition of *Symbiodinium microadriaticum* from various coelenterates confirm that the zooxanthella is well equipped to make use of animal excretory products. In addition to the inorganic salts, this alga can utilize as a sole nitrogen source urea, uric acid, guanine, adenine or any of 12 amino acids, and as sole phosphorus source any of the following phosphoric acids: glycero-phosphoric, cytidylic, adenylic, or guanylic (McLaughlin & Zahl, 1959).

It has been shown on several occasions that ammonia and phosphate produced by the metabolism of coelenterates are completely taken up by their contained algae. Furthermore, it would appear that the host is unable to satisfy the whole demand of its symbionts for these ions, because although they are excreted normally by animals without their symbionts, animals with algae actually absorb them from the medium (Pütter, 1911; Yonge & Nicholls, 1931a; Smith, 1939; Kawaguti, 1953).

Nitrogen and phosphorus are the two nutrients most frequently scarce in natural waters, so that in respect of these two nutrients at least, the alga unquestionably benefits from the association. While some benefit to the animal is also to be presumed, since excretory products are being efficiently removed, this is not so obvious, except in isolated instances such as *Convoluta*.

Organic material

Some animals can thrive without taking in solid food, and are presumed to depend entirely on their contained algae for the bulk of their nourishment. We may instance adult Radiolaria (although their young, which contain few algae, do feed), *Paulinella* with its two cyanellae which is never known to feed (Brandt, 1883a; Pascher, 1929a), *Convoluta roscoffensis* and the Alcyonaria previously referred to. Both

Euspongilla (Brandt, 1882) and *Paramoecium bursaria* (Pringsheim, 1928) can be made to survive considerable periods without food in a mineral medium in the light.

Protozoa and smaller metazoa living in a eutrophic environment, on the other hand, may well be able to obtain dissolved nourishment from their surroundings in the manner of *Tetrahymena, Paramoecium aurelia* or the brine shrimp *Artemia* (Kidder & Dewey, 1951; Wagtendonk, 1955; Provasoli, Shiraishi & Lance, 1959). The very marked stimulatory effect of glucose on illuminated pure cultures of *P. bursaria*, observed by Loefer (1936*b*), indicates that the net status of the *P. bursaria/ Chlorella* combination is that of a chemotroph. However, Pütter's ideas on the general significance of dissolved organic matter in zooplankton nutrition have not yet been substantiated (Pütter, 1908).

By no means all the animals which harbour symbiotic algae are able to live entirely at their expense: the madreporarian coral, as repeatedly shown by Yonge and his colleagues, immediately begins to expel its zooxanthellae when it is starved of its normal food of zooplankton (Yonge & Nicholls, 1931*b*). But the important fact is that some animals can and do rely on their algae; that the majority may not is another matter. The products of algal metabolism may nevertheless be of some benefit to the animal, either during the life of the alga or after its death.

Dissolved substrates. In culture algae commonly excrete a great deal of organic matter, possibly the equivalent of as much as 50 % of the carbon fixed in photosynthesis (Allen, 1956). Much of this material is of a mucilaginous nature (Lewin, 1956; McLaughlin, Zahl, Nowak & Marchisotto, 1962) and if excreted by symbionts might not be meta-bolized—coelenterates, for example, appear to lack carbohydrases (Yonge & Nicholls, 1930; Smith, 1936). But peptides and organic acids have also been identified, and glycollic acid in particular is found in supernatants from cultures of *Chlorella* and other green algae (Fogg, 1952; Allen, 1956; Tolbert & Zill, 1956; Lewin, 1957). Some of the substances found in culture fluids occur also in natural waters harbouring phytoplankton (Vallentyne, 1957).

In view of this, it would be remarkable if no organic matter passed from symbiotic algae to the tissues of their host, although it cannot be assumed that the animal necessarily makes any use of the material. Coral algae photosynthesize at a high rate. Gardiner (1931) calculated that if all the products of photosynthesis were directed to algal growth the animal would quickly be ruptured by the mass of plant material accruing. The small size of the pyrenoids in the chlorellae of *Chloro-*

hydra and *Dalyellia* (Haffner, 1925) suggest that the products are not being stored by these algae either. Actual proof of the passage of assimilatory material from algae to their host has been furnished by Muscatine & Hand (1958), who allowed sea anemones to photosynthesize in sea water containing $^{14}CO_2$. Autoradiographs taken from sections of animals killed at intervals revealed the isotope fixed by the algae in the endoderm. At first the isotope was confined to that tissue, but later became widely distributed in the ectoderm which is virtually free of algae. Some recent autoradiographic measurements on two West Indian corals (Goreau & Goreau, 1960) suggest, however, that the amount of material passing from alga to host would not be enough to nourish the animal.

Algal chemotrophy. Organic substrates pass in the opposite direction, that is from the animal to the plant, whenever the latter lives chemotrophically at the animal's expense, as when, for instance, *Chlorohydra* or *Paramoecium bursaria* are kept in the dark and the chlorellae continue to multiply, albeit at a reduced rate (Haffner, 1925; Siegel, 1960). According to Limberger (1918) the chlorella from *Euspongilla* utilizes glucose *in vitro*, as in fact do many Chlorococcales. Indeed, the threat of parasitism inherent here may well prove a feature of zoochlorellae in general. On the other hand, there is every indication that zooxanthellae are not able to live chemotrophically; they do not thrive in animals kept in darkness. Furthermore, no photosynthetic dinoflagellate yet studied in culture has shown any propensity to chemotrophy, and McLaughlin & Zahl's zooxanthella, *Symbiodinium microadriaticum*, is apparently no exception (McLaughlin & Zahl, 1959).

Even chemotrophic algae are in general rather limited in the substrates they can utilize for growth in the dark, the most commonly available to them being acetic acid, ethanol and glucose. A poor permeability—one of the factors responsible for obligate phototrophy—could be an advantage for life in a medium as rich in possibly toxic substrates as the interior of an animal cell.

Micronutrients. Organic micronutrients and other physiologically active substances may pass between animal and plant, but little can be said of the significance of such exchanges for want of precise knowledge of nutritional requirements, particularly of the animal. Animals, like *Convoluta roscoffensis*, which do not take in food, may be entirely dependent upon their algae for their supply of vitamins; or the carnivores among those that do feed, may fail to get all they want from an exclusively animal diet. *Paramoecium aurelia*, though not strictly carnivorous, needs steroids of specifically vegetable origin (Wagtendonk,

1955). The requirements of *P. bursaria* are not known. It can be cultivated free of bacteria if it contains algae, and then all its nutritional
needs are met by a peptone medium (Loefer, 1936*b*), whereas if it is
depleted of its algae living bacteria must be supplied in its diet. This
suggests strongly that this ciliate also has need of a lipid. However, it
is doubtful whether the symbiotic relation as such is instrumental in
meeting this requirement, for *P. bursaria* continuously releases chlorellae
to the medium and browses on their progeny (Siegel, 1960).

In view of the general pattern of vitamin requirements found in algae,
one would expect to find symbionts, in so far as they may be auxotrophic,
in need of vitamin B_{12} or thiamine, and possibly biotin (Droop, 1962).
As it happens, all zoochlorellae cultivated axenically have proved not
to be auxotrophic (Limberger, 1918; Genevois, 1924; Loefer, 1936*a*;
Siegel, 1960), though there may be some significance in the fact that no
one has yet managed to cultivate the chlorella from *Chlorohydra* in
isolation. Zooxanthellae probably need vitamin B_{12}, for free-living
photosynthetic dinoflagellates have always been found to do so.
However, McLaughlin & Zahl (1959) were unable to establish an absolute
requirement in *Symbiodinium microadriaticum* for thiamine or vitamin
B_{12}, but they note that these vitamins have a stimulatory effect. Provasoli & Pintner (1953) reported a vitamin B_{12} deficiency in the flagellate
Cyanophora paradoxa. Since this was an overall requirement on the
part of the symbiotic combination, either both partners must be
deficient or the cyanella must be unable to meet the flagellate's needs on
top of its own; one suspects the latter, because the requirement had the
high specificity unknown in the Monera (Droop, McLaughlin, Pintner
& Provasoli, 1959).

Antibiotics. An interesting side-light on the interrelationship of zooxanthellae and their hosts is provided by some recent work of Ciereszko
(1962) on antimicrobial compounds from Alcyonaria. In an earlier
paper Ciereszko, Sifford & Weinheimer (1960) reported several terpinoid
compounds with bacteriocidal properties from a number of species, and
it was suggested that their function was to keep the colonies clean. Now
it appears that in at least one species an antimicrobial terpene can be
harvested from the zooxanthella.

Symbiotic algae as a source of food

Endozoic algae constitute a considerable source of potential nourishment for the animal, although few animals, it appears, draw regularly
upon this source. Apart from *Convoluta roscoffensis*, which devours its
symbionts as a matter of course towards the end of its life in the weeks

preceding egg-laying (Keeble & Gamble, 1907), those animals which do regularly digest their algae are typical phytoplankton-feeders such as *Euspongilla* and *Tridacna* previously mentioned. In other instances (e.g. *Paramoecium bursaria*, *Convoluta paradoxa* and *Dalyellia*) the algae are digested only when either the animal or the alga has been subjected to some stress, such as temperature or starvation (Keeble, 1908; Haffner, 1925; Pringsheim, 1928).

Coelenterates, and corals in particular, are controversial. It is not universally admitted that they are strict carnivores, although most zoologists would hold that view. The arguments in favour of their ability to resort to a diet of algae when necessary rest on the observation of disintegrating algal cells in the tissues of the digestive region of the mesentery in starved polyps (Boschma, 1925). The arguments against this are twofold: in the first place, the part of the mesentery in question has been shown also to be the region of excretion (Mouchet, 1930; Yonge, 1931), and corals do eject their algae wholesale when stressed; and in the second place, no cellulolytic enzyme has yet been found in a coelenterate, and such other carbohydrases as have been recognized have been traced to the algal cells (Yonge & Nicholls, 1930; Smith, 1936). Neither Hadzi (1906) nor Beutler (1924) found an amylase in *Chlorohydra*.

With one or two notable exceptions, animals do not 'eat' their symbionts, although some may do so when occasion arises. One does not in general kill the goose that lays the golden egg.

Ecological significance

Only in few instances can the association between algae and animals be regarded as truly reciprocal or even obligatory for either partner. For the alga there are certain obvious benefits to be had, notably the convenience of being confined to a situation abundant in nutrients and in this fact may lie the primary cause of algal/invertebrate symbiosis. As for the animal, except in the few cases of obligatory association, it also cannot be said to gain more than incidental advantage from the abundance of oxygen and other assimilatory products and from the removal of its own wastes. However, this does not preclude an association, however casual, from having great ecological value, for an advantage, seemingly of marginal physiological consequence when viewed in isolation, may assume an overriding significance in the face of nature. Removal of waste products is a case in point. A tropical animal frequently harbours symbionts while its relatives from temperate and colder waters do not; more advantage may be gained by symbiosis in a warm than a cold climate. Although the reason for this may be complex, a tropical animal has a higher rate of metabolism than its temperate counterpart and consequently may also have a greater need for the complementary service of an alga.

Productivity of coral reefs

We discussed the role played by carbon dioxide in the building of the coral skeleton. Goreau (1961) considers that the synergistic effect of algae on calcification rate is one of the decisive factors in the evolution of coral reefs for, as he says 'the development of such enormous communities in the face of constant battering by heavy seas was possible only when processes of limestone deposition became fast enough for the rate of accumulation to exceed the rate of loss by organic and inorganic attrition'. The capacity for growth is relatively enormous—as much as 41 cm. per annum has been recorded, though 2·5 cm. would appear to be a more usual figure (Yonge, 1940). Even so, the lower figure represents a very high rate of productivity when taken over a reef as a whole.

It is possible to obtain an estimate of the gross primary organic productivity from measurements of oxygen production and utilization on a reef by a method due to Sargent & Austin (1949). There are now data from two atolls in the Marshall Islands (Sargent & Austin, 1954; Odum & Odum, 1955) and a fringing reef in Hawaii (Kohn & Helfrich, 1957). All three reefs reveal gross productivity values of the same high order (1500–3500 g. carbon fixed per m.^{2} per year), which is from 10 to 100 times as high as corresponding values obtained in open waters of the tropical Pacific, and from 4 to 8 times as high as the most productive regions of temperate seas (Kohn & Helfrich, 1957). Attached plant communities elsewhere, however, show comparable high rates.

The reef as a whole is 'autotrophic', producing more organic matter than it consumes. In the words of Sargent & Austin (1954), 'neither the amount of material (in the current drawn across the reef) caught in the plankton net nor the amount available for oxidation by bacteria...is sufficient to satisfy the requirements of the reef community'. A reef is also a closed community in the sense that it does not enrich the surrounding seas with the products of its metabolism, for the latter are intercepted by the associated algae. Free phytoplankton and consequently zooplankton will be fewer and the re-cycling of nutrients that is such a feature of the plankton economy of temperate seas will also be less evident. High productivity of the benthic reef communities may thus in the long run contribute to the poverty of ocean water in the tropics. The situation is summed up in Yonge's colourful phrase 'imprisoned phytoplankton'.

Coral tissues may contain as many as 30,000 zooxanthella cells per mm.[3] (Marshall, 1932); nevertheless, according to Odum & Odum (1955), this represents as little as 5 % of the total biomass of algae associated with the coral, the remainder consisting of algae living in and on the dead skeleton. Although these authors believe that the non-zooxanthella algae contribute as directly to the polyp's well being as do the actual zooxanthellae, in so far as most of the reef's primary production is external to the coral tissues, it follows that the greater part of the cycle of organic matter is also external (though within the reef as a whole). Part of the cycle will, of course, involve small animals and the carnivorous habit of the coral polyp. Whether the carnivorous diet, which is admittedly essential to the coral, forms the major source of nourishment, or whether, on the other hand, its function is to supply specific growth factors of animal origin is an unanswered question. So also must remain unanswered the question of the fundamental role of the zooxanthellae in the coral economy, although we do know that organic matter passes from the algae to the coral and that the algae assist significantly in the calcifying process.

POSTSCRIPT

Very many problems in symbiosis involve nutrition either directly or indirectly. Other problems, notably those concerned with taxonomy of symbiotic algae, evolution and genetics of symbiotic associations, and above all the physiology of the host/symbiont relationship at all levels can be elucidated only by the study of pure cultures. Given sufficient application modern culture techniques are probably adequate for the vast majority of symbiotic algae. The algae need to be cultivated in sufficiently wide variety for comparative purposes and detailed nutritional comparisons are needed, for instance, between Loefer's chlorella from *Paramoecium bursaria* and free-living strains, as well as later isolates from *P. bursaria*. McLaughlin & Zahl's isolation of zooxanthellae is a good beginning in the marine field. No cyanellae have been cultivated.

Axenic culture of the animal is also needed, and progress will be less sure. Axenic *Paramoecium bursaria* would be a triumph, which indeed may not be too distant, in view of achievements with some other ciliates. Loomis' work with *Hydra* and *Chlorohydra* points the way to axenic coelenterates, while axenic planarians have already been obtained (Loomis, 1959; Miller & Johnson, 1959). Why not *Convoluta* also, once the feeding stimulus is found? Meanwhile, a great deal could be learned from *Cyanophora paradoxa*, which shares with *P. bursaria* the honour of being in bacteria-free culture.

ACKNOWLEDGEMENTS

I am indebted to my colleagues Drs Shiena M. Marshall and A. P. Orr for the benefit of their experience of the Great Barrier Reef Expedition and to Professor C. M. Yonge for reading the typescript of this article.

Fig. 2 was redrawn from the *Journal of Protozoology* and Fig. 3 reproduced from *Endeavour*. I am grateful to the authors and authorities concerned.

REFERENCES

ALGEUS, S. (1946). Untersuchungen über die Ernährungsphysiologie der Chloro-phyceen. *Bot. Notiser*, **21**, 129.

ALLEN, M. B. (1956). Excretion of organic compounds by *Chlamydomonas*. *Arch. Mikrobiol.* **24**, 163.

DE BARY, A. (1879). *Die Erscheinung der Symbiose.* Strassburg: Trübner.

BERNARD, N. (1909). L'Évolution dans la symbiose. Les orchidées et leurs cham-pignons commensaux. *Ann. Sci. nat. (Bot. Sér. 9)*, **9**, 1.

BEUTLER, R. (1924). Experimentelle Untersuchungen über die Verdauung bei *Hydra*. *Z. vergl. Physiol.* **1**, 1.

BEYERINCK, M. W. (1890). Kulturversuche mit Zoochlorellen Lichengonidien und anderen niederen Algen. *Bot. Ztg.* **48**, 725.

BOSCHMA, H. (1924). On the food of Madreporaria. *Proc. Acad. Sci. Amst.* **27**, 13.

BOSCHMA, H. (1925). On the symbiosis of certain Bermuda coelenterates and zoo-xanthellae. *Proc. Amer. Acad. Arts Sci.* **60**, 451.

BRANDT, K. (1882). Ueber die morphologische und physiologische Bedeutung des Chlorophylls bei Thieren. *Arch. Anat. Physiol., Lpz.*, 125.

BRANDT, K. (1883*a*). Über die morphologische und physiologische Bedeutung des Chlorophylls bei Thieren. *Mitt. zool. Sta. Neapel*, **4**, 191.

BRANDT, K. (1883*b*). Über Symbiose von Algen und Thieren. *Arch. Anat. Physiol., Lpz.*, 445.

BRANDT, K. (1885). Die kolonienbildenden Radiolarien des Golfes von Neapel. *Fauna Flora Neapel.* **13**, 65.

BUCHNER, P. (1930). *Tier und Pflanze in Symbiose.* Berlin: Gebrüder Borntraeger.

CHODAT, R. (1924). Sur les organismes verts qui vivent en symbiose avec les Tur-bellariées rhabdocèles. *C.R. Soc. Phys. Hist. nat. Genève*, **41**, 130.

CIERESZKO, L. S. (1962). Chemistry of coelenterates. III. Occurrence of anti-microbial terpinoid compounds in the zooxanthellae of Alcyonarians. *Trans. N.Y. Acad. Sci.* Ser. II, **24**, 502.

CIERESZKO, L. S., SIFFORD, D. H. & WEINHEIMER, A. J. (1960). Occurrence of terpenoid compounds in gorgonians. *Ann. N.Y. Acad. Sci.* **90**, 917.

COLE, H. A. (1936). Experiments in the breeding of oysters (*Ostrea edulis*) in tanks, with special reference to the food of larva and spat. *Fish. Invest., Lond.*, Ser. 2, **15** (4), 1.

DANGEARD, P. A. (1900). Les zoochlorelles du *Paramoecium bursaria*. *Botaniste*, **7**, 161.

DOYLE, W. L. & DOYLE, M. M. (1940). The structure of zooxanthellae. *Pap. Tortugos Lab.* **32**, 127.

DROOP, M. R. (1955). Some new supra-littoral protista. *J. mar. biol. Ass. U.K.* **34**, 229.

DROOP, M. R. (1961). *Haematococcus pluvialis* and its allies. III. Organic nutrition. *Rev. Alg.*, N.S. **4**, 247.

DROOP, M. R. (1962). Organic micronutrients. In *Physiology and Biochemistry of Algae*, p. 141. Ed. R. A. Lewin. New York and London: Academic Press. (In the Press.)

DROOP, M. R., MCLAUGHLIN, J. J. A., PINTNER, I. J. & PROVASOLI, L. (1959). Specificity of some protophytes towards vitamin B_{12}-like compounds. *Preprints 1st Int. Congr. Oceanogr., New York*, p. 916. Ed. M. Sears. Washington: Amer. Ass. Advanc. Sci.

EHRET, C. F. (1953). An analysis of electro-magnetic radiations in the mating reaction of *P. bursaria*. *Physiol. Zoöl.* **26**, 274.

ELBERG, S. S. (1960). Cellular immunity. *Bact. Rev.* **24**, 67.

ELENKINE, A. (1906). La symbiose comme équilibre instable des organismes cohabitants. *Trav. Soc. Nat. St-Pétersb.* (*Léningr*), no. 37. (In Russian; French abstract, *Année biol*, 1906, p. 336.)

ENGELMANN, T. W. (1882). Über Licht- und Farbenperception niederster Organismen. *Pflüg. Arch. ges. Physiol.* **29**, 237.

FOGG, G. E. (1952). The production of extracellular nitrogenous substances by a blue-green alga. *Proc. Roy. Soc.* B, **139**, 372.

FRASER, E. A. (1931). Observations on the life history and development of the hydroid *Myrionema amboinense*. *Sci. Rep. Gr. Barrier Reef Exped.* **3**, 135.

FREUDENTHAL, H. D. (1962). *Symbiodinium* gen. nov. and *Symbiodinium microadriaticum* sp.nov., a zooxanthella: taxonomy, life cycle, and morphology. *J. Protozool.* **9**, 45.

FRITSCH, F. E. (1952). Algae in association with heterotrophic and holozoic organisms. *Proc. Roy. Soc.* B, **139**, 185.

GAMBLE, F. W. & KEEBLE, F. (1904). The bionomics of *Convoluta roscoffensis*, with special reference to its green cells. *Quart. J. micr. Sci.* **47**, 363.

GARDINER, J. S. (1931). *Coral Reefs and Atolls*. London: MacMillan.

GEDDES, P. (1882). On the nature and function of the 'yellow cells' of radiolarians and coelenterates. *Proc. Roy. Soc. Edinb.* B, **11**, 377.

GEITLER, L. (1959). Syncyanosen. In *Handbuch der Pflanzenphysiologie*, **11**, 530. Ed. W. Ruhland. Heidelberg: Springer-Verlag.

GENEVOIS, L. (1924). Contribution à l'étude de la symbiose entre zoochlorelles et Turbellariées rhabdocèles. *Ann. Sci. nat.* (*Bot.* 10), **6**, 53.

GIBOR, A. (1956). Culture of brine algae. *Biol. Bull., Woods Hole*, **111**, 223.

GOETSCH, W. (1924). Die Symbiose der Süsswasser-Hydroiden und ihre künstliche Beeinflussung. *Z. Morph. Ökol. Tiere*, **1**, 660.

GOETSCH, W. & SCHEURING, L. (1926). Parasitismus und Symbiose der Algengattung *Chlorella*. *Z. Morph. Ökol. Tiere*, **7**, 220.

GOHAR, H. A. F. (1940). Studies on the Xeniidae of the Red Sea. *Publ. Mar. biol. Sta. Ghardaa*, **2**, 25.

GOREAU, T. F. (1959). The physiology of skeleton formation in corals. I. A method for measuring the calcium deposition by corals under different conditions. *Biol. Bull., Woods Hole*, **116**, 59.

GOREAU, T. F. (1961). Problems of growth and calcium deposition in reef corals. *Endeavour*, **20**, 32.

GOREAU, T. F. & GOREAU, N. I. (1959). The physiology of skeleton formation in corals. II. Calcium deposition by hermotypic corals under various conditions in the reef. *Biol. Bull., Woods Hole*, **117**, 239.

GOREAU, T. F. & GOREAU, N. I. (1960). Distribution of labelled carbon in reef building corals with and without zooxanthellae. *Science*, **131**, 668.

HADZI, J. (1906). Vorversuche zur Biologie von *Hydra*. *Arch. EntwMech. Org.* **22**, 38.

HADZI, J. (1911). Über die Symbiose von Xanthellen und *Halecium ophiodes*. *Biol. Zbl.* **31**, 85.

HAFFNER, K. VON (1925). Untersuchungen über die Symbiose von *Dalyellia viridis* und *Chlorohydra viridissima* mit Chlorellen. *Z. wiss. Zool.* **126**, 1.

HOOGENRAAD, H. R. (1927). Zur Kenntnis der Fortpflanzung von *Paulinella chromatophora* Lauterb. *Zool. Anz.* **72**, 140.

HOVASSE, R. (1937). Les zooxanthelles sont les dinoflagellés. *C.R. Acad. Sci., Paris*, **205**, 1015.

HOVASSE, R. & TEISSIER, G. (1923). Peridiniens et zooxanthelles. *C.R. Acad. Sci., Paris*, **176**, 716.

KARNOVSKI, M. L. (1962). Metabolic basis of phagocytic activity. *Physiol. Rev.* **42**, 143.

KAWAGUTI, S. (1941). Study on the invertebrates associating with unicellular algae. *Palao trop. biol. Stud.* **2**, 307.

KAWAGUTI, S. (1944a). On the physiology of reef corals. VII. Zooxanthella of the reef corals is *Gymnodinium* sp., Dinoflagellata; its culture *in vitro*. *Palao trop. biol. Stud.* **2**, 675.

KAWAGUTI, S. (1944b). Zooxanthellae as a factor of positive phototropism in those animals containing them. *Palao trop. biol. Stud.* **2**, 681.

KAWAGUTI, S. (1950). Observations on the heart shell *Corculum cardissa* (L.), and its associated zooxanthellae. *Pacif. Sci.* **4**, 43.

KAWAGUTI, S. (1953). Ammonium metabolism of the reef corals. *Biol. J. Okayama Univ.* **1**, 171.

KAWAGUTI, S. & SAKUMOTO, D. (1948). The effect of light on the calcium deposition of corals. *Bull. oceanogr. Inst. Taiwan*, **4**, 65.

KEEBLE, F. (1908). The yellow-brown cells of *Convoluta paradoxa*. *Quart. J. micr. Sci.* **52**, 431.

KEEBLE, F. (1910). *Plant-Animals: A Study in Symbiosis*. Cambridge University Press.

KEEBLE, F. & GAMBLE, F. W. (1907). The origin and nature of the green cells of *Convoluta roscoffensis*. *Quart. J. micr. Sci.* **51**, 167.

KIDDER, G. W. & DEWEY, V. C. (1951). The biochemistry of ciliates in pure culture. In *Biochemistry and Physiology of Protozoa*, **1**, 324. Ed. A. Lwoff. New York and London: Academic Press.

KITCHING, J. A. (1956). Food vacuoles. In *Protoplasmatologia*, vol. 3, Art. D 3b. Ed. L. V. Heilbrun and F. Weber. Vienna: Springer-Verlag.

KOHN, A. J. & HELFRICH, P. (1957). Primary organic productivity of a Hawaiian coral reef. *Limnol. Oceanogr.* **2**, 241.

KORSCHIKOFF, A. A. (1924). Protistologische Beobachtung. I. *Cyanophora paradoxa*. *Arch. russes Protist.* **3**, 57.

LEDERBERG, J. (1952). Cell genetics and hereditary symbiosis. *Physiol. Rev.* **32**, 403.

LEWIN, R. A. (1956). Extracellular polysaccharides of green algae. *Canad. J. Microbiol.* **2**, 665.

LEWIN, R. A. (1957). Excretion of glycolic acid by *Chlamydomonas*. *Bull. Jap. Soc. Bot.* **5**, 74. (In Japanese; English summary.)

LIMBERGER, A. (1918). Über die Reinkultur der Zoochlorella aus *Euspongilla lacustris* und *Castrada viridis* Volz. *S.B. Acad. Wiss. Wien*, **127**, 395.

LOEFER, J. B. (1936a). Isolation and growth characteristics of the 'zoochlorella' of *Paramoecium bursaria*. *Amer. Nat.* **70**, 184.

LOEFER, J. B. (1936b). Bacteria-free culture of *Paramoecium bursaria* and concentration of the medium as a factor in growth. *J. exp. Zool.* **72**, 387.

LOOMIS, W. F. (1959). Control of sexual differentiation in *Hydra* by pCO_2. *Ann. N.Y. Acad. Sci.* **77**, 73.

McLAUGHLIN, J. J. A. (1958). Euryhaline chrysomonads: nutrition and toxigenisis in *Prymnesium parvum* with notes on *Isochrysis galbana* and *Monochrysis lutheri*. *J. Protozool.* **5**, 75.

McLAUGHLIN, J. J. A. & ZAHL, P. A. (1959). Axenic zooxanthellae from various invertebrate hosts. *Ann. N.Y. Acad. Sci.* **77**, 55.

McLAUGHLIN, J. J. A., ZAHL, P. A., NOWAK, A. & MARCHISOTTO, J. (1962). Some constituents of zooxanthellae grown in axenic culture. *Proc. 1st Int. Congr. Protozool. Prague* 1961 (in the Press).

MANGAN, J. (1909). The entry of zooxanthellae into the ovary of *Millepora* and some particulars concerning the medusae. *Quart. J. micr. Sci.* **53**, 697.

MARSHALL, S. M. (1932). Notes on oxygen production in coral planulae. *Sci. Rep. Gr. Barrier Reef Exped.* **1**, 253.

MARSHALL, S. M. & ORR, A. P. (1955). On the biology of *Calanus finmarchicus*. VIII. Food uptake, assimilation and excretion in adult and stage V *Calanus*. *J. mar. biol. Ass. U.K.* **34**, 495.

MAYER, A. G. (1924). Structure and ecology of Samoan reefs. *Publ. Carneg. Inst.* no. 340, 1.

MILLER, C. A. & JOHNSON, W. H. (1959). Preliminary studies on the axenic cultivation of a planarian (*Dugesia*). *Ann. N.Y. Acad. Sci.* **77**, 87.

MILLER, J. D. A. & FOGG, G. E. (1958). Studies on the growth of Xanthophyceae in pure culture. II. The relation of *Monodus subterraneus* to organic substances. *Arch. Mikrobiol.* **30**, 1.

MOUCHET, S. (1930). L'excrétion chez les actinies. *Notes Sta. océanogr. Salammbô*, **15**, 1.

MUSCATINE, L. & HAND, C. (1958). Direct evidence for the transfer of materials from symbiotic algae to the tissues of a coelenterate. *Proc. nat. Acad. Sci. Wash.* **44**, 1259.

NAVILLE, A. (1926). Notes sur les eolidies. Un eolidien d'eau saumâtre. Origine des nématocystes. Zooxanthelles et homochromie. *Rev. Suisse Zool.* **33**, 251.

ODUM, H. T. & ODUM, E. P. (1955). Trophic structure and productivity of a windward coral reef community on Eniwetok Atoll. *Ecol. Monogr.* **25**, 291.

OEHLER, R. (1922). Die Zellverbindung von *Paramoecium bursaria* mit *Chlorella vulgaris* und anderen Algen. *Arb. staatsinst. exp. Ther. Frankfurt.* **15**, 5.

PARKER, R. C. (1926). Symbiosis in *Paramoecium bursaria*. *J. exp. Zool.* **46**, 1.

PASCHER, A. (1929a). Über einige Endosymbiosen von Blaualgen in Einzellern. *Jb. wiss. Bot.* **71**, 386.

PASCHER, A. (1929b). Über die Natur der blaugrünen Chromatophoren des Rhizopoden *Paulinella chromatophora*. *Zool. Anz.* **81**, 189.

PASCHER, A. (1930). Eine neue, stigmatisierte und phototaktische Amöbe. *Biol. Zbl.* **50**, 1.

PRATT, E. M. (1906). The digestive organs of the Alcyonaria and their relation to the mesogloeal cell plexus. *Quart. J. micr. Sci.* **49**, 327.

PRINGSHEIM, E. G. (1928). Physiologische Untersuchungen an *Paramoecium bursaria*. *Arch. Protistenk.* **64**, 289.

PRINGSHEIM, E. G. & WIESSNER, W. (1960). Photo-assimilation of acetic acid by green organisms. *Nature, Lond.* **188**, 919.

PROVASOLI, L. & PINTNER, I. J. (1953). Ecological implications of *in vitro* nutritional requirements of algal flagellates. *Ann. N.Y. Acad. Sci.* **56**, 839.

PROVASOLI, L., SHIRAISHI, K. & LANCE, K. (1959). Nutritional idiosyncrasies of *Artemia* and *Tigriopus* in monoxenic culture. *Ann. N.Y. Acad. Sci.* **77**, 250.

PÜTTER, A. (1908). Der Stoffhaushalt des Meeres. *Z. allg. Physiol.* **7**, 321.

PÜTTER, A. (1911). Der Stoffwechsel der Aktinien. *Z. allg. Physiol.* **12**, 297.

REINHARDT, K. (1950). Der Stoffwechsel heterotropher Flagellaten. VII. Experimentelle Untersuchungen über die Stickstoffernährung. *Arch. Mikrobiol.* **15**, 278.

SARGENT, M. C. & AUSTIN, T. S. (1949). Organic productivity of an atoll. *Trans. Amer. geophys. Un.* **30**, 245.

SARGENT, M. C. & AUSTIN, T. S. (1954). Biologic economy of coral reefs. *Prof. Pap. U.S. geol. Surv.* **260** E, 293.

SIEGEL, R. W. (1960). Hereditary endo-symbiosis in *Paramoecium bursaria. Exp. Cell. Res.* **19**, 239.

SMITH, H. G. (1935). On the presence of algae in certain ascidians. *Ann. Mag. nat. Hist.* **15**, 615.

SMITH, H. G. (1936). Contribution to the anatomy and physiology of *Cassiopeia frondosa. Pap. Tortugas Lab.* **31**, 17.

SMITH, H. G. (1939). The significance of the relationship between actinians and zooxanthellae. *J. exp. Biol.* **16**, 334.

STANIER, R. Y. & COHEN-BAZIRE, G. (1957). The role of light in the microbial world: some facts and speculations. In *Microbial Ecology. Symp. Soc. gen. Microbiol.* **7**, 56.

TOLBERT, N. E. & ZILL, L. P. (1956). Excretion of glycolic acid by algae during photosynthesis. *J. biol. Chem.* **222**, 895.

TRENDELENBERG, W. (1909). Versuche über den Gaswechsel bei Symbiose zwischen Alga und Tier. *Arch. Anat. Physiol., Lpz.*, 42.

TRIGT, H. VAN (1919). A contribution to the physiology of the freshwater sponges (Spongillidae). *Tijdschr. ned. dierk. Ver.* 2 Ser., **17**, 1.

VALLENTYNE, J. R. (1957). The molecular nature of organic matter in lakes and oceans, with lesser reference to sewage and terrestrial soils. *J. Fish. Res. Bd Can.* **14**, 33.

VERWEY, J. (1930). Depth of coral reefs and penetration of light. With notes on oxygen consumption of corals. *Proc. Pacif. Sci. Congr.* **4**, II A, 277.

VERWEY, J. (1931). Coral reef studies. II. The depth of coral reefs in relation to their oxygen consumption and the penetration of light in the water. *Treubia*, **13**, 169.

WAGTENDONK, W. J. VAN (1955). The nutrition of ciliates. In *Biochemistry and Physiology of Protozoa*, **2**, 57. Ed. S. H. Hutner & A. Lwoff. New York and London: Academic Press.

WELSH, M. (1936). Oxygen production by zooxanthellae in a Bermudan Turbellarian. *Biol. Bull., Woods Hole*, **70**, 282.

WESENBERG-LUND, C. (1909). Beiträge zur Kenntnis des Lebenszyklus der Zoochlorellen. *Int. Rev. Hydrobiol.* **2**, 153.

WHITNEY, D. D. (1907). Artificial removal of green bodies of *Hydra viridis. Biol. Bull., Woods Hole*, **13**, 291.

WICHTERMAN, R. (1942). A new ciliate from a coral of Tortugas and its symbiotic algae. *Pap. Tortugas Lab.* **33**, 107.

WINTER, F. W. (1907). Zur Kenntnis der Thalamophoren. *Arch. Protistenk.* **10**, 1.

YONGE, C. M. (1931). Studies on the physiology of corals. III. Assimilation and excretion. *Sci. Rep. Gr. Barrier Reef Exped.* **1**, 83.

YONGE, C. M. (1935). Symbiosis between invertebrates and unicellular algae. *Proc. Linn. Soc. Lond.* **147**, 90.

YONGE, C. M. (1936). Mode of life, feeding, digestion and symbiosis with zoo-xanthellae in the Tridacnidae. *Sci. Rep. Gr. Barrier Reef Exped.* **1**, 283.

YONGE, C. M. (1937a). Evolution and adaptation in the digestive system of the Metazoa. *Biol. Rev.* **12**, 87.

YONGE, C. M. (1937b). Studies on the biology of Tortugas corals. III. The effect of mucus on oxygen consumption. *Pap. Tortugas Lab.* **31**, 207.

YONGE, C. M. (1940). The biology of reef building corals. *Sci. Rep. Gr. Barrier Reef Exped.* **1**, 353.

YONGE, C. M. (1944). Experimental analysis of the association between inverte-brates and unicellular algae. *Biol. Rev.* **19**, 68.

YONGE, C. M. (1958). Ecology and physiology of reef-building corals. In *Per-spectives in Marine Biology*, p. 117. Ed. A. A. Buzzati-Traverso. Berkeley and Los Angeles: University of California Press.

YONGE, C. M. & NICHOLAS, H. M. (1940). Structure and function of the gut and symbiosis with zooxanthellae in *Tridachia crispata* (Oerst.) Bgh. *Pap. Tortugas Lab.* **32**, 287.

YONGE, C. M. & NICHOLLS, A. G. (1930). Studies on the physiology of corals. II. Digestive enzymes, with notes on the speed of digestion. *Sci. Rep. Gr. Barrier Reef Exped.* **1**, 59.

YONGE, C. M. & NICHOLLS, A. G. (1931a). Studies on the physiology of corals. IV. The structure, distribution and physiology of the zooxanthellae. *Sci. Rep. Gr. Barrier Reef Exped.* **1**, 135.

YONGE, C. M. & NICHOLLS, A. G. (1931b). Studies on the physiology of corals. V. The effect of starvation in light and darkness on the relationship between corals and zooxanthellae. *Sci. Rep. Gr. Barrier Reef Exped.* **1**, 177.

YONGE, C. M., YONGE, M. J. & NICHOLLS, A. G. (1932). Studies on the physiology of corals. VI. The relationship between respiration in corals and the pro-duction of oxygen by their zooxanthellae. *Sci. Rep. Gr. Barrier Reef Exped.* **1**, 213.

ZAHL, P. A. & MCLAUGHLIN, J. J. A. (1959). Studies in marine biology. IV. On the role of algal cells in the tissues of marine invertebrates. *J. Protozool.* **6**, 344.

SYMBIOSIS AND APOSYMBIOSIS IN ARTHROPODS*

MARION A. BROOKS

Department of Entomology and Economic Zoology,
University of Minnesota

PREVALENCE OF SYMBIOTIC MICRO-ORGANISMS

Insects, mites and ticks, constituting as they do the vast majority of species of multicellular animals, offer tremendous variations in both anatomy and physiology which have been exploited by numerous symbiotic micro-organisms. The insects have successfully colonized all terrestrial and freshwater habitats of the world, largely as a consequence of their evolutionary adaptability in feeding habits. There seems to be no plant or animal which is not attacked by insects in at least some parts of the world. Perhaps only the bacteria excel the insects in their ubiquity. But whereas the range of substrates suitable for bacteria is determined by such properties of the bacteria as membrane permeability or the production of exogenous enzymes, the substrate range of the insects is determined principally by their means of locomotion and their mouthparts. Modifications of the mouthparts permit them to chew, pierce, suck, lap or siphon. Thus in classifying insects, while the first division is according to the presence or absence of wings, and their characteristics when present, the next major divisions are made according to the types of mouthparts. Correlated with the mouthparts are modifications of the intestine. For example, insects which chew or gnaw woody or other rough materials have a series of chitinized grinding plates in the anterior part of the gut. On the other hand, bugs which pierce plant tissues and imbibe large quantities of aqueous plant sap have long, very much coiled intestines and long, convoluted excretory tubules which filter the sap. Weevils and beetles which infest stored grains have variable numbers of longitudinal rows of pouches or blind sacs attached externally to the epithelium of the gut, and communicating with the gut lumen. Larvae of beetles or flies which live in humus or wood possess enlargements of the intestine or pouches forming fermenting chambers. Or, there may be a ring of enlargements en-

* Paper no. 1117 Miscellaneous Journal Series, Minnesota Agricultural Experiment Station, St Paul 1, Minnesota. It is a pleasure to acknowledge the financial assistance for the original parts of this work from the National Institutes of Health, U.S. Public Health Service.

circling the anterior midgut and opening into it. Certain leaf-eating beetles and a number of families of plant-sucking bugs (the so-called 'higher' families) have similar arrangements of two or four longitudinal rows of caeca or crypts along the posterior section of the midgut.

Anatomical arrangements

It is notable that micro-organisms of several types—but a constant type in each species of insect—are found within these specialized organs or regions. The micro-organisms may be extracellular, that is, in the lumen of the gut or its attachments; or they may be intracellular in the cytoplasm of the cells forming these structures.

There has been a tendency for the intestinal crypts or girdling swellings to detach from the gut. Four successive stages in this transition can be seen in contemporary beetle larvae of the Anobiid family (Gräbner, 1954). Whereas in one genus, the swellings are circumferentially sessile, in two other genera they protrude on short connecting necks. In another genus the necks are elongated and narrowed. Finally in a fifth genus, the swellings have become enlarged, paired, blind pouches which have migrated forward in the body cavity but still retain their connexions with the midgut through a pair of slender ducts with extremely narrow lumens (less than 1 μ diameter). No symbionts are seen in these ducts. It is easy to imagine how the extension of this process in other insects could have resulted in ultimate severance of the pouches from the gut so that they have become discrete organs in the haemocoele, as they are in aphids and some bugs and beetles. Finally, individual cells or syncytia filled with micro-organisms are found embedded in the visceral fat body, as in cicadas, scale insects and cockroaches.

Micro-organisms were first observed in insects about a century ago, when the compound microscope became available as a biological tool, by embryologists and histologists examining insects. One of the earliest to report such a discovery was Leydig (1850), who saw the large, egg-like organ in aphids which contains the symbiotic yeasts and which he called the 'pseudovitellus', not knowing its true function. Leydig (1854) also first observed budding, yeast-like forms in the blood of a scale insect. In neither case did he seem to recognize that these objects were living micro-organisms.

In 1879 deBary coined the term 'symbiosis' familiar to all biologists. It is not so well known that he also named cells and organs containing symbiotic yeasts 'mycetocytes' and 'mycetomes' respectively. These terms have persisted with respect to all such structures even though the micro-organisms within them may not be yeasts. It was Blochmann (1886) who first considered that the objects seen in the eggs and follicular membranes of ants and wasps were bacteria. Upon examining cockroaches, Blochmann (1888) was convinced that the organisms in the mycetocytes of the fat body, in the eggs, and in the embryos, were bacteria, and he attempted to culture them.

Many investigators have subsequently verified that insects contain micro-organisms, that the anatomical provisions for housing them are numerous, and that the mechanisms for transmitting them are sometimes elaborate. For a complete and well-documented account, the reader is referred to Buchner (1953).

Distribution among arthropods

With our present knowledge of the host range of symbiotic micro-organisms, an attempt may be made to trace the evidence for a phylogenetic relationship between arthropods and the type of symbiont they possess. Without regard to the cellular location of the organisms, their presence in the hosts is shown in Table 1. (Clearly we are excluding from this summary arthropod vectors of disease organisms.)

It is difficult to see any clear-cut phylogenetic trend in the table. The most one could venture to say is that certain of the lower orders possess symbiotic protozoa while the rickettsiae are found only in the more advanced or specialized orders. The yeasts and bacteria are scattered about with some overlapping, not only within orders but also within species, since some species contain both types. Furthermore, the identity of the yeasts and bacteria is uncertain in many instances because the organisms have not been cultured. Some bugs, Fulgorids for instance, have several bizarre forms which defy identification. The apparent distribution of micro-organisms among the orders of insects may actually be an artifact due to our incomplete knowledge. It is noteworthy that most of the orders in which no symbionts are known are of no economic importance, while many of the species with symbionts are in orders of economic interest to man and therefore subject to more thorough scrutiny.

Relation to diet

A significant fact not brought to light in Table 1 is the irregular distribution of symbionts *within* orders. In the following orders, all examined species possess symbionts: cockroaches, termites, chewing and sucking lice, and ticks. In the others, only certain genera are known to possess symbionts, and this may be related to the biology of the arthropods. It is well-documented that there is, in general, a relationship between the

Table 1. *Distribution of symbionts in orders of Arthropods*

Order of Arthropods	Flagellates	Yeasts	Bacteria	Rickettsiae
Thysanura	.	.	.	.
Collembola	.	.	.	.
Ephemeroptera	.	.	.	.
Odonata	.	.	.	.
Blattariae	+	.	+	.
Mantodea	.	.	.	.
Isoptera	+	.	+	.
Plecoptera	.	.	.	.
Embiidina	.	.	.	.
Orthoptera	.	.	.	.
Phasmida	.	.	.	.
Dermaptera	.	.	.	.
Psocoptera	.	.	.	.
Zoraptera	.	.	.	.
Mallophaga	.	.	+	.
Anoplura	.	.	+	.
Thysanoptera	.	.	.	.
Hemiptera	.	.	+	.
Homoptera	.	+	+	.
Coleoptera	.	+	+	.
Strepsiptera	.	.	.	.
Neuroptera	.	.	.	.
Mecoptera	.	.	.	.
Hymenoptera	.	.	+	+
Diptera	.	.	+	+
Siphonaptera	.	.	.	.
Trichoptera	.	.	.	.
Lepidoptera	.	.	.	.
Acarina	.	.	.	+

arthropod's diet and its possession of symbionts. In the introductory paragraphs, it was stated that micro-organisms are frequently housed in the anatomically modified intestine. Insects with such modifications are mostly eaters of materials such as wood, stored dry cereal grains, forest humus, feathers, hair or wool, which are rich in cellulose, poor in nitrogen, and deficient in vitamins. Insects which feed on plant sap (low in nitrogen) and insects and ticks feeding on vertebrate blood (low in vitamins) constitute the remainder of the symbiotic groups. Furthermore, in any order of insects, such as the Hemiptera, only the plant-sap feeders possess symbionts while the carnivorous forms do not. Similarly among fly larvae, the humus dwellers possess symbionts while larvae breeding in less decomposed organic matter do not. (This is true only if we define a symbiont as a constant companion. The larvae of flies which breed in excreta harbour various species of coliform and enteric bacteria which are beneficial but the flora changes with a change in habitat.) If we select the blood feeders, we find still another dichotomy.

Only lice, ticks and 'kissing bugs', which eat blood from birth possess symbionts, while horseflies, mosquitoes and fleas, which eat a varied diet of organic detritus in the larval forms and blood only in the adult forms, are without them. None of the carnivorous insects (dragonflies, mantids, predatory bugs, carrion beetles, strepsiptera, robberflies, blowflies) are known to possess symbionts. With possibly a few exceptions, insects which eat leaves, fruits or vegetables do not have symbionts.

The general observation that symbionts are found in insects which eat a restricted diet deficient in certain essential nutrients has suggested to workers in the field of arthropod symbiosis that symbiotic micro-organisms are involved in synthesizing these deficiencies for the benefit of their hosts. There is very little evidence that any reciprocal benefit accrues to the micro-organisms.

There are some exceptions to the above generalization, notably in the cockroaches and some of the ants. One does not need training as a professional entomologist to know that cockroaches are omnivorous and ants are scavengers. Yet all cockroaches examined (about twenty genera) possess intracellular bacteria in the fat body and consistently transmit them intraovarially (Plates 1 and 2), and in spite of the general diet of the hosts, these organisms are nutritional necessities. Those genera of cockroaches which are wood-feeders (e.g. *Parcoblatta*, *Cryptocercus*), living under the bark of fallen trees and in rotting tree stumps, are equipped with two sets of symbionts; namely the intracellular organisms ('bacteroids') of the omnivorous feeders, and the intestinal flagellates similar to those of the wood-eating termites. Among the termites, all of the wood-eaters (but not the mound-builders, which eat leaves and fungi) possess intestinal flagellates. An interesting Australian relict termite, *Mastotermes darwiniensis* (Froggatt), also possesses two sets of symbionts: the intestinal flagellates of wood-eaters, and the intra-cellular bacteria morphologically identical with those of the cock-roaches. The fossil record shows that cockroaches and termites had probably split off from a common ancestor by the beginning of the Carboniferous Period, and that the cockroaches had completed their development by the end of this era. The presence of similar forms of intracellular symbiosis in both the cockroaches and the relict termite means that this symbiotic relationship is at least 300 million years old (Buchner, 1953).

The loss of the intracellular symbionts by the higher families of termites with restricted diets, that now have only intestinal flagellates, presents a challenging question. One would expect that as the termites evolved more restricted dietary habits, it would be an advantage to have retained both sets of symbionts. But in actuality, the possession of both bacteroids and flagellates by *Mastotermes* apparently does not confer an advantage to it. Dr L. R. Cleveland (personal communication) has noted that following ecdysis, when intestinal defaunation occurs, *Mastotermes* survives a shorter time if not re-faunated than do any of the higher termites which do not have bacteroids.

Transmission to offspring

In some insects elaborate systems exist for the transmission of symbiotic micro-organisms to the offspring (see Koch, 1962). The modes of trans- mission are related to the definitive location of the symbionts in the adult insect, and to the trophic structure of the ovary. In the louse *Pediculus vestimenti* De Geer, the symbionts migrate in the blood from the stomach disc, attached to the wall of the midgut, to the ovaries, from whence the organisms enter the eggs. The stomach disc can be seen, even after it has been emptied, through the translucent body wall of the insect. In cockroaches, there is a migration of mycetocytes to the im- mature ovaries. The vitelline membranes form microvilli, discernible only with the electron microscope, which seem to surround the bacteria and thus draw them into the egg cytoplasm (Bush & Chapman, 1961). This results in a uniform, or apolar, infection of the entire periphery of the egg, from whence the organisms are carried inward during the subsequent processes of cleavage and blastoderm formation. That the fat-body mycetocytes in the developing embryo are specialized cells (not arising because of chance infection) is indicated by their formation in the embryo in anticipation of subsequent infection. If experimental manipulations prevent the bacteria from entering the mycetocytes, these cells nevertheless remain distinct from the other fat body cells and migrate to the ovaries (Brooks & Richards, 1955a). Similarly, empty mycetomes have been produced in the beetle, *Oryzaephilus surinamensis* (Linnaeus) (Koch, 1962).

The cockroach type of infection process is consistent with a panoistic type of ovary, in which there are no nurse cells and all the surrounding follicular epithelium nourishes the eggs. The interspersion of myceto- cytes around the ovarioles and between the follicles is quite comparable to the arrangement of nurse cells in more advanced insects with poly- trophic ovaries. It is tempting to speculate whether the nurse cells which we now regard as insect cells may long ago have been mycetocytes, or aggregates of micro-organisms. During evolutionary advancement, the microbial inclusions may have become an integral part of the insect's cytoplasm. Certainly many of the trophic functions performed by the nurse cells and by the mycetocytes are similar.

Comparable egg infections occur in *Mastotermes*, the ants *Camponotus* and *Formica*, the powder-post beetle *Lyctus*, and related wood-boring beetles.

In the majority of egg infections, however, the symbionts are restricted to entrance at either the anterior or posterior pole. This type occurs in aphids, scale insects and other bugs which possess telotrophic ovaries, that is, ovaries in which the nurse cells are all congregated at the anterior tip of each ovariole. The nurse cells nourish the chain of successively ripening eggs by means of long trophic cords passing backward to each egg. In contrast to the cockroach, in which the follicular epithelial cells apparently play no active role in the passage of the symbionts, in polar types of infection the follicular cells at the pole move apart for a short period of time, producing lacunae through which the symbionts pass. The exact means by which this is accomplished are unknown since no one has studied the ultrastructure of these stages. The openings are eventually closed and the chorion is secreted. In those eggs which receive a more or less spherical mass of symbionts, this mass is known as a transitory embryonic mycetome. Ultimately the symbionts are dispersed to the definitive mycetocytes of the developing embryo. Various complications may intervene at this step. There may be more than one morphological type of micro-organism in the symbiont ball—up to six kinds have been described—and all the organisms may remain mixed together. But in some bugs they are actually sorted out, so that one or more kinds are isolated in separate compartments.

The remarkable thing about these movements is that they all seem to be accomplished by the activity of the host cells. Motility is unknown among symbiotic micro-organisms. Their translocations are passive, resulting from waves of contractions, or extensions of pseudopodia-like structures or microvilli, of the adjacent cells of the insect host or embryo. In other instances, the symbionts may be carried by the haemolymph.

Ovarial infections of the types mentioned result in the intracytoplasmic location of symbionts in internal organs or cells. The early embryology of insects in this category is complicated by these involvements. In other insects the symbionts are restricted to the lumen of the gut or its attachments, or to the epithelial cytoplasm. Here infection may not be ovarial; if not, transmission to the offspring is nevertheless assured by one of two general means. Either the female has accessory organs or ducts which permit her to smear the surface of each egg as it is laid, or the behaviour of the hatchlings is stereotyped to guarantee their eating the symbionts. In the former instance, certain species simply eliminate some faecal material containing symbionts as each egg is extruded, or pouches connecting the gut and the vagina or ovipositor provide a coating of symbionts to each egg. The organisms may enter the egg shell via the micropyle, and be eaten when the insect hatches; but usually they simply remain on the egg shell and some are consumed when the hatchling eats part or all of the shell as it forces its way out. Organisms transmitted in this way must be capable of resisting desiccation for several days.

At least three examples are known of insects in which both the maternal and filial behaviour are stereotyped to assure infection of the offspring. One of these, a plant sap feeding bug which lays eggs in loose sand, remains near the eggs until they hatch. The hatchlings crawl over the abdomen of the mother and avidly consume droplets of a bacterial suspension which she exudes from time to time from her rectum. Similarly, the wood-eating termites which live in well-organized social colonies infect their young and newly moulted juveniles by proctodaeal feeding. Another bug (*Coptosoma scutellatum* Geoffr.) encloses packets of symbiotic bacteria in a secretion from the gut which hardens to form little cocoons. One of these packets is deposited between each pair of eggs as they are laid. When the nymphs emerge from their egg shells, they promptly insert their beaks into the packets and suck up a dose of bacteria before departing.

It is the exception for symbiotic insects to leave the infection of their offspring to chance. Insects which live congregated in crevasses among their own excreta, such as wood roaches and *Rhodnius prolixus* Stål. (the blood-sucking bug), become infected naturally by eating faeces or egg shells contaminated with faecal material.

It remains to be learned how pure cultures result in the guts of insects which must be originally infected with a variety of organisms. It is apparently only chance infections which occur in lamellicorn beetles and crane-flies whose larvae hatch and live in humus or decaying vegetable matter, which of course has a rich microbial flora. Several older studies (see Buchner, 1953) report that pouch-like extensions of the hindgut of these larvae contain bacteria capable of decomposing cellulose. These bacteria may be identical with free-living forms in the soil.

EFFECTS OF ELIMINATING
EXTRACELLULAR SYMBIONTS

The brief descriptions given above serve only as an indication of the astonishing array of symbiotic combinations now known, and may serve as an introduction to further inquiry. The first line of investigation has been directed toward an understanding of the contributions of symbiotic micro-organisms to nutrition, including digestion or enzymic breakdown, since it was assumed that organisms present in the gut must act catabolically on the foodstuffs present. Such determinations are experimentally possible only if the symbionts can be separated and maintained in a satisfactory physiological condition.

Cellulose decomposition

Termites have been subjects of physiological studies for years because of their peculiar habit of eating wood and their fantastic flagellate populations. Four termite families, Mastotermitidae, Kalotermitidae, Hodotermitidae, and Rhinotermitidae, harbour vast numbers of Hypermastigina and Polymastigina. The large family of highly specialized termites, the Termitidae, mostly do not eat wood; the few wood-eating species in this family contain bacteria which are believed to serve the same function as the protozoa in the other families (Hungate, 1955). In fact, the protozoa themselves harbour numerous bacteria, some of them adhering in oriented groups which may be mistaken for cilia; others are intracellular (especially common in the food vacuoles) and intranuclear. Cleveland (1925) succeeded in killing the flagellates in the gut of several species of termites by subjecting the animals to oxygen tensions greater than one atmosphere. The flagellates, being anaerobic, cannot tolerate this much oxygen, although it is not clear how the oxygen gains access to the intestine at an effective level. At any rate, the insects are unharmed and this method has led to a long series of studies on the physiology of wood-eating insects. Defaunated insects cannot survive for more than a month on a diet of wood or cellulose. Cook (1943) found that defaunated insects, fed glucose, lived longer than faunated controls which were starved. This substantiated the theory of enzymic breakdown of cellulose by cellulase produced by the flagellates, and also suggested the possibility that atmospheric nitrogen may be fixed by the flagellates. Problems of maintaining the flagellates *in vitro* for sufficiently long make analytical determinations difficult. Trager (1932) demonstrated cellulase activity in an extract of protozoa and in *in vitro* survivors of *Trichomonas termopsidis*. Succeeding later in obtaining limited cell division and growth of this species, Trager (1934) demonstrated the conversion of cellulose. Since, however, one strain of bacteria was never eliminated, bacterial metabolism was still suspect. Hungate (1943) made quantitative manometric measurements of cellulolytic activity in concentrated, washed suspensions of surviving flagellates. He determined that although cellulose is converted to glucose, glucose as such is not released to the surrounding media but is anaerobically broken down to hydrogen, carbon dioxide, acetic and other unidentified acids. The insects then utilize acetic acid as a carbon source. Hungate also presented evidence, based on gut ligations, that the intestinal epithelium is permeable to acetic acid, which is probably taken up by the haemolymph.

There is no evidence for nitrogen fixation by flagellates. In the gut there is a bulbous, thin-walled enlargement of the hindgut in which the majority of the flagellates live; this is separated by sphincters at both ends. The mass of flagellates is so compacted that individual bodies are plastically deformed, and represent one-third to one-seventh of the insect's total weight. As the organisms die, they are undoubtedly digested, contributing to the utilizable nitrogen. The habit, common to both termites and woodroaches, of scavenging on the dead and incapacitated bodies of their fellows, and of proctodaeal feeding, recycles the nitrogen of the undigested wood as well as of the dead flagellates. The death rate of flagellates in the intestine is very low and the amounts of nitrogen made available by this means will also be small. Probably the most effective contributors to the nitrogen economy are saprophytic fungi which may concentrate the nitrogen of the wood.

Since very little success has been achieved with maintaining a self-reproducing culture of flagellates, it is apparent that the gut milieu furnishes conditions not easily duplicated *in vitro*. Oxygen, of course, is highly detrimental to them. With anaerobic culture media containing reduced glutathione or thioglycollate, most species survive only a few hours.

Vitamin synthesis

That symbiotic micro-organisms perform synthetic functions of value to the host was nicely demonstrated in two species of Anobiid beetles (*Lasioderma serricorne* F. and *Stegobium paniceum* L.). Larvae of these insects possess mycetomes at the junction of the fore- and midgut, in which symbiotic yeasts abound, both in the lumen and in the cytoplasm of some cells. Koch (1933) prevented transmission of the yeasts, normally accomplished by contamination of the egg shell, by sterilizing the surface of eggs with 5 % chloramine in 70 % alcohol, and found that the 'sterilized' insects grew poorly unless their diet was supplemented with yeast.

Blewett & Fraenkel (1944) found that the symbionts could be replaced by feeding five B vitamins and yeast sterols. Pant & Fraenkel (1950) succeeded in culturing the yeasts on a simple solution (Hansen's solution). The identity of the cultured organisms was proved simply by smearing sterile eggs with them and thus obtaining normally symbiotic insects again. Even though the mode of symbiosis in these two closely related beetles is similar and each insect can be infected reciprocally by the yeast of the other, yet the yeasts are not identical. Pant & Fraenkel (1954) found that the one from *Lasioderma* grows more prolifically in culture and is a more efficient substitute in the alternate host. The yeasts did not change their morphology or growth characteristics when they were grown in culture or in a foreign host. These authors suggested that the yeast cells are digested in the intestine to release the vitamins. However, Parkin (1952) has described the anatomy of *Ptilinus pectinicornis* L., a closely related Anobiid, in which the mycetomes are far forward of the fore- and midgut junction, and are connected to the gut by slender ducts. The lumen of the ducts is so narrow that no yeast cells are found in them. This is evidence that in some insects secretions from the symbionts pass into the gut while the symbionts themselves remain intact.

Because external transmission of symbionts can easily be interrupted, the branched bacterium, *Nocardia rhodnii* Erikson, symbiotic in the lumen of the midgut of the blood-sucking bug, *Rhodnius*, has been studied in some detail. Wigglesworth (1936) had shown that the cultured organisms are a source of B vitamins, when fed to sterile *Lucilia* larvae. Brecher & Wigglesworth (1944) found that *Rhodnius* could be rendered aposymbiotic by surface-sterilizing the eggs with crystal violet; but Baines (1956) found that the procedure could be even more simple. He merely collected fresh eggs from clean surroundings and kept the insects free from subsequent contamination. Baines fed his sterile bugs on rabbits or mice whose blood had been enriched with injections of B vitamins. The bugs were judged to be dependent on the bacteria for pyridoxin, calcium pantothenate, nicotinamide, and thiamine, but not folic acid or biotin. Unfortunately, technical difficulties were encountered in that the mammals could not tolerate high enough concentrations of riboflavin, so that the requirement for this vitamin could not be determined.

Recently Harington (1960) devised a technique for feeding aposymbiotic bugs a concentrated dose of vitamins through a membrane. Although the number of bugs used was too small to give decisive results, those obtained indicated that thiamine is required. Contrary to Baines's findings, aposymbiotic bugs also had an apparent requirement for folic acid. Both vitamins were synthesized by cultures of *Nocardia rhodnii*, probably in excess of the organism's own requirements.

Certainly the symbiotic relationships in *Rhodnius* are still far from clear. This brings into focus several of the obstacles that must be overcome in determining nutritional needs. Although it is easy to prevent infection of the bugs, and to culture the organisms *in vitro*, one can still not be sure that the synthetic activity *in vitro* is the same as *in vivo*. Many organisms are adaptable enough to survive under conditions which may not be optimal for synthetic activities. One can only estimate the growth requirements of the organisms themselves comparing them with related non-symbiotic forms, as was done by Bewig & Schwartz (1955). Obviously this method cannot give very dependable figures for the synthesis of excess vitamins. Finally, as emphasized by Geigy, Halff & Kocher (1954), unless the proper ratios of vitamins are used in feeding a synthetic diet to insects, the animals may not respond properly and the entire picture may become confused.

The importance of determining the proper vitamin dose has been a deterrent in assaying the contributions of the symbionts of the blood-sucking louse, *Pediculus*. Puchta (1955), by elaborating the work of earlier investigators, who used centrifugation to prevent infection, obtained offspring free from symbionts. But overdoses of either yeast extract or B vitamins, or unbalanced proportions of vitamins, were toxic to both normal and aposymbiotic lice. It was concluded that the symbiotic bacteria function in supplementing blood with the proper dosages of particular B vitamins.

One of the nicest ways to prevent infection of offspring is possible with the bug, *Coptosoma*, described above (Müller, 1956). By manually removing the packets of symbiotic bacteria from the egg clusters, the hatchling bugs are rendered aposymbiotic. Müller found increased mortality, prolonged development time, and lack of reproduction in these bugs. Diet was again implicated, inasmuch as germinating seeds of the host plant, rather than leaves, partially compensated for the lack of symbionts.

MEANS OF ELIMINATING
INTRACELLULAR SYMBIONTS

Very little has been accomplished towards eliminating symbionts which are transmitted in the cytoplasm of the egg. Since mechanical techniques such as centrifugation and surface-sterilization will not work with these eggs, techniques which influence the interior milieu have been employed.

Examples of these are ultra-violet irradiation; incubation of the eggs at extreme temperature; or chemotherapeutic treatment of the developing egg by feeding antibiotics and bactericidal drugs to the female. Most attempts of this sort produce deleterious effects on the treated insects, resulting in high mortality or atrophied gonads. The fumigant, methyl bromide, apparently eliminates the symbionts in weevils infesting stored grain; but the successive generations survive aposymbiotically (Musgrave, Monro & Upitis, 1961). Another exception is *Oryzaephilus*, the saw-toothed grain beetle, which produces sterile eggs when incubated at 36° without itself being harmed. The high temperature causes degeneration of only the short infective stage of the symbionts in the ovaries, in eggs, and in embryos, but not of the other forms in the fat body. This means that only the later *offspring* of heat-treated individuals are aposymbiotic. It was rather startling to learn that these aposymbiotic beetles, retaining apparently empty fat-body mycetomes, were able to survive and reproduce successive generations (Huger, 1954). With further study it was, however, found that there is usually a residue of heat-resistant micro-organisms in the mycetomes. Huger (1956) demonstrated that in the closely related beetle *Rhyzopertha*, this residue can re-populate the insect.

Antibiotic-resistant symbionts

The same phenomenon has been observed in the cockroach, *Blattella germanica* (Linnaeus). Heating this insect for a long enough time (17 days at 39°) to destroy the bacteria in the eggs kills about two-thirds of the insects, which is unsatisfactory for obtaining a colony of aposymbiotic offspring. The cockroach is more amenable to antibiotic treatment. Brooks & Richards (1955a) reared the insects—from hatching throughout life—on chlortetracycline added at the rate of 0·1 % to the regular diet. The offspring appeared aposymbiotic. Their cuticles were an unnatural light colour, an effect which has been reported for other species of aposymbiotic insects. The nymphs were weak, many of them falling down and dying soon after hatching. In examining complete series of histological sections of a large sample of the nymphs, we found that the bacteria were practically annihilated; only rarely could one or two organisms be identified in an occasional mycetocyte (Pl. 3, fig. 7). Siblings of these specimens were fed the stock diet of ground dog biscuit and although some of them remained alive for months, they failed to grow. A little yeast added to the food was of no consequence. But one part of yeast to three parts of dog biscuit enabled them to grow slowly; eventually some of them matured and reproduced the aposymbiotic F_2 generation, which again was weak and light-coloured. However, we have found, after maintaining such a colony for many months and having produced six or eight successive generations, that inevitably some dark, strong individuals are produced; and the proportion of them in the colony increases with each generation. This phenomenon can be explained only on the basis of a resistant residue of bacteria remaining, which re-populates the insect and its offspring gradually during the course of several generations. The effect seems to be due to antibiotic resistance, since these forms develop under a constant regimen of chlortetracycline fed to all generations. The dark insects have not become re-infected through some external portal of entry, as it is impossible to infect aposymbiotic nymphs by feeding or injecting bacteria into them. Our only method of maintaining aposymbiotic insects is by periodically initiating a new colony from the stock culture.

The paleontological evidence, coupled with the persistence of empty mycetomes and the resistant properties of symbionts, has impressed many authors with the phylogenetic significance of symbiotic relationships. Some interesting exceptions occur in tropical insects which, on the basis of their relation to other forms, would be expected to have symbionts but lack them as far as we know. The interested reader is referred to Buchner (1953) or Koch (1962) for a detailed treatment of this subject.

There seems little doubt that a normally symbiotic insect deprived completely of its symbiotic micro-organisms is a handicapped insect. The severity of the handicap may vary with different species; alternatively, degrees of viability may reflect degrees of success in the complete removal of the symbionts. For obvious reasons, it is illogical to assume that one can demonstrate complete absence of micro-organisms in the body of any animal.

The form of handicap always seems to be a nutritional deficiency. In its simplest form, a few B vitamins and sterols can replace the symbionts. In a more subtle form, the vitamins must be carefully proportioned. Perhaps some large molecules such as polypeptides, or intermediates such as coenzyme A or diphosphopyridine nucleotide, are synthesized by the symbionts. The aposymbiotic German cockroach cannot grow on a diet in which balanced B vitamins replace yeast, nor on a diet in which amino acids replace casein (Brooks, 1958). Growth is not the only impairment in aposymbiotic insects, because adults which grow up on a yeast-fortified diet always have difficulty in reproducing. Females may lay few viable eggs, a few unviable eggs, or no eggs. This is not a matter of lack of copulation, because observers report that the insects are sexually active. Undoubtedly the same nutritional factors resulting in deficient eggs affect sperm also. It was shown in the German cockroach, by mating aposymbiotic females to either aposymbiotic or symbiotic males, that the latter matings resulted in more viable eggs (Brooks & Richards, 1955 a). In this species, the symbionts do not penetrate the testes and thus are never transmitted with the sperm; so the effect on male fertility must be mediated through soluble nutritional factors.

Only rarely are symbionts physically transmitted with the sperm, the Bostrychid beetles being the only known instances (Mansour, 1934).

SEX DETERMINATION

It is completely within the realm of experience in other fields to learn that the nutritional effects of micro-organisms in insects affect insect fertility. But it is surprising to find that the physical presence of micro-organisms within the fertilized egg can determine the sex of the embryo. This reputedly occurs in two species of scale insects studied by Buchner (1954, 1955). The symbionts, in one instance a yeast, in the other a bacterium, are not uniformly distributed between the eggs after they once gain entrance into the ovaries. They pass through openings in the follicular epithelium of the eggs nearest them, and subsequently these eggs develop into females while the remainder become males. Since there is no detectable difference between the two kinds of eggs prior to the entrance of the micro-organisms, it is believed that the presence of the latter in some way prevents the loss of a chromosome (the cytological mechanism by which males are produced) or conversely, causes the retention of the chromosome to produce females (Hughes-Schrader, 1948). In males, the large cells, homologous with those cells in females which receive the symbionts, persist possibly as relics of the time when in phylogeny males also possessed symbionts. If this hypothesis is tenable, then bisexuality must be more recent than unisexuality in these insects.

The unique condition of unisexual progenies in certain strains of *Drosophila* is thought to be due to the presence of a small spirochete in the blood of females (Poulson & Sakaguchi, 1961). This organism, if it is the responsible agent, causes mortality of male zygotes.

Among the aphids, there is one (*Pemphigus spirothecae* Pass.) that makes galls on poplars, in which the course of phylogeny is said to be traceable in the contemporary embryos. Aphids are unique insects for several reasons. They are polymorphic, being winged or wingless; they alternate generations, sometimes reproducing sexually and sometimes parthenogenetically; and they are ovoviviparous. But their development is telescoped to the extent that three generations are visible, one within the other, in virgins on the chief host plant. In sections of female *Pemphigus* containing sexupara daughter embryos, one may distinguish in these embryos the germinal tissue of embryos of the third generation (Lampel, 1959). There are two male embryos in the anterior part of the abdomen and six female embryos in the posterior part. During subsequent development of these embryos, before invagination of the germ band, the females become infected with symbionts while the males remain sterile. During the blastula stage, gigantic mycetome nuclei can be seen inside of the cleavage cavity, in both sexes; but only the mycetomes of the female contain symbionts; and the sterile mycetomes of the male soon degenerate. After birth the sterile mycetomes are no longer identifiable. Koch (1962) and Lampel cite these phenomena as demonstrating that in males symbiosis occurred at one time but proved to be superfluous and was abandoned during the course of phylogeny.

I find this hypothesis difficult to reconcile with the facts of cockroach development. In these insects symbiosis is not superfluous in the male, since both sexes require the symbiotic bacteria for growth and reproduction. The fat-body mycetocytes and the bacteria are indistinguishable in sections or smears taken from the two sexes. There are, however, two differences in the symbiosis in males: the bacteria do not enter the testes, as mentioned above, and the mycetocytes degenerate earlier in adult males than they do in adult females. But, at any rate, the life span of adult males is abbreviated compared to that of females. The evidence which we have at present favours the hypothesis that in these insects the germinal tissue controls the entrance of the symbionts, rather than that the distribution of symbionts determines the sex of the organism. Specifically, there is the work of Bush & Chapman (1961) showing that microvilli developing from the vitelline membrane of the egg of *Periplaneta americana* (Linnaeus) appear to enclose and draw inward the bacteria which surround the egg. Of course, there may be an alternative explanation of the origin of these microvilli, as proposed at the same time by Gresson & Threadgold (1960), who believe that the microvilli originate from the bacteria. The evidence of published electron micrographs favours the suggestion of the former authors. It is pertinent at this point to call attention to the electron micrographs of Meyer & Frank (1957), in which there is no indication of microvilli surrounding bacteria in the fat-body mycetocytes.

Furthermore, there is evidence from the embryology of *Blattella germanica*, in which the bacteria enter ovaries *after* they are distinguishable from testes. The male of the German cockroach is an anatomical hermaphrodite—probably this is true of all cockroaches although they have not been examined in this respect. Heymons (1890), in describing the development of the gonads of the male embryo of this species, showed that the early genital ridges possess 'Anlagen' (initials) of both ovaries and testes. These initials are pulled apart, drawn in opposite directions by contractions of their suspending filaments and attached ducts. Halfway through embryonic development, small but apparently perfect ovaries have formed in the identical position taken by ovaries in females. While the testes continue to develop, the ovaries remain rudimentary, and can be identified as such even in adult males. In some individuals one or two follicles enlarge as if they were forming yolk. Relevant to the present discussion is the fact that in our laboratory we have histological preparations showing that the bacterial symbionts infect these rudimentary ovaries just as they do functional ovaries in females. When the insects are 14–21 days old, female ovaries become infected. The microscopically identical process occurs at the same age in the males, whereas the testes never become infected (unpublished data). In my opinion, this clearly supports the theory that the sex of these cells was determined prior to their infection by the symbionts. Admittedly, this is but one insect; and since insects exhibit so many variations in development, it is unsafe to generalize from a single example.

REGULATION OF SYMBIONTS BY HOST CELLS

There are additional instances in insect development which favour the idea that the host cells regulate the symbiotic relationship. For example, Schomann (1937) presents a figure of the intestinal epithelium of the newly hatched larva of the Cerambycid beetle, *Oxymirus cursor* L. The young larva had just infected its gut with the symbiotic yeast while eating the egg shell which its mother had smeared with the yeasts. When the yeast cells reach a particular zone in the intestine, large epithelial cells with correspondingly large nuclei lose their brush border and send cytoplasmic processes out into the lumen to engulf the yeasts.

Foeckler (1961) attempted to re-infect aposymbiotic *Stegobium* larvae with various non-symbiotic strains of yeasts and also with symbiotic strains from other insects. In only one series of experiments did the intestinal cells become infected; but then the infection was widespread throughout the length of the midgut, and was not restricted to the cells of the blind sacs which ordinarily contain yeast cells. When the larvae pupated, the foreign yeast was expelled. Koch (1960) points out that in the normal symbiotic relationship the 'host remains sovereign over its guests' while in the experimental case the host loses control over them. He likens the latter to parasitism; undoubtedly much can be learned about immunological processes from the behaviour of insects in controlling their symbionts.

The literature is full of descriptions of mycetocytes and mycetomes, histologically different from neighbouring cells, which develop in insect embryos in anticipation of infection. Buchner (1961) portrays many distinctive large cells in the intestinal epithelium and Malpighian tubules of *Ips* beetles which are the only cells to become invaded by symbiotic bacteria. And in those insects with fat-body mycetocytes, such as the cockroach, if embryonic infection of the mycetocytes is prevented, the 'empty mycetocytes' (Pl. 3, figs. 8 and 9) nevertheless persist, remaining histologically recognizable (Brooks & Richards, 1955a). In cytoplasmic infections of this sort, the symbionts are at the mercy of the host cells for distribution. If the mitotic multiplication of the mycetocytes is delayed by a deficiency of symbionts, the symbionts have no other way to disperse throughout the body (Brooks & Richards, 1955b, 1956).

We are led by such observations to further consideration of the things the host does for its symbiotic micro-organisms.

Of all the possible advantages that micro-organisms might realize from a symbiotic life, the ones most commonly suggested are protection from desiccation, provision of suitable metabolic substrates, geographic dispersion, and various ill-defined factors lumped in the category of a favourable 'abode'. (It is unclear to me why geographic dispersion should be an advantage to an organism which is always confined to an internal body habitat.) The foregoing suggestions are supported in general by the facts that symbiotic micro-organisms are never found free-living and with very few exceptions are refractory to *in vitro* cultivation. Even such forms as the yeasts from *Lasioderma* and *Stegobium*, which are easily cultured, have not been identified beyond their genus (*Saccharomyces*). The actinomycete from *Rhodnius* is found nowhere else. Furthermore, seemingly all the bacterial symbionts are non-motile and do not form endospores, two properties which would be superfluous in an internal abode.

Practically no work has been done on the fine structure of symbionts; only the studies of Roshdy (1961) on tick rickettsiae and of Meyer & Frank (1957) and Bush & Chapman (1961) on the bacteroids of cockroaches come to mind. The latter report that the cell wall, 50–90 Å. thick, is the thinnest bacterial cell wall they have ever seen. They suggest that the protective function of the cell wall has been assumed by the host and that this thinness may contribute to the difficulties encountered in attempts to culture the organism. Beneath the cell wall is an extremely thin cytoplasmic membrane. The cytoplasm possesses a diffuse nuclear region, a fine, dense thread-like structure within the nuclear area, a granular ground cytoplasm, peripheral bodies associated with transverse cell-wall formation and membranous structures associated with septum formation. There are no flagella. The diffuse nature of the nuclear region resembles that in blue-green algae.

Two relatively new techniques have been employed to study cockroach symbionts. Henry & Block (1960) fed and injected radioactive inorganic sulphate into cockroaches, and recovered labelled sulphur amino acids. Autoradiograms showed that the mycetocytes were radioactive, although less so than other tissues. Manunta & Bernardini (1958) isolated mycetocytes by differential centrifugation and analysed them chromatographically for vitamins.

While lacking spores or any means of locomotion, bacterial and yeast symbionts are highly pleomorphic. In some insects this is quite confusing. Certain Fulgorid bugs show many forms of symbionts, assumed to be different strains, each undergoing a cycle of development and increasing in size and complexity as the host develops from the egg (Buchner, 1953). Usually there is a smaller form of symbiont in the ovaries, known as the infective stage. Koch (1962) places considerable emphasis on pleomorphism in different parts of the insect's body, and at different times in the insect's life, comparing these with pleomorphism evoked in cultured organisms by different environmental conditions, such as salt composition of the medium.

Dietary factors affecting transmission

In a long series of experiments on effects of nutrition on transmission of cockroach symbionts, we have found that there are several factors in the insect's diet which stimulate or inhibit the survival of the bacteria in the ovarial eggs. This was quite unexpected in the face of all the evidence for the persistence of this age-old relationship. The first indication that the bacteria in the eggs were more susceptible to interfering factors than were the bacteria in the fat body came from the results of feeding antibiotics and sulfa drugs. Earlier workers had injected or force-fed antibiotics for short periods of time and obtained some adverse effects on the bacteria, but these were accompanied by high mortality and degeneration of the ovaries (Brues & Dunn, 1945). By constantly feeding antibiotics from the day the insects hatched we were able to obtain offspring without bacteria (Brooks & Richards, 1955*a*), and noticed that the results were different with the same antibiotic in different carrier diets. If the carrier was a commercial dog biscuit, reproduction was satisfactory for our purposes. But if the carrier was an artificial diet, containing casein, vitamins, a salt mixture, etc., reproduction practically ceased. In fact, controls on an artificial diet without antibiotics produced only a few weak nymphs which proved upon examination to be aposymbiotic. This led to a study of factors that might be deleterious to the bacteria in a diet which was supposedly wholesome, since the parental generation grew very rapidly on it and survival was excellent. Ultimately the salt mixture was found to be the cause (Brooks, 1960). The minerals manganese, zinc and calcium all affect the transmission, but in different ways. Manganese is the most critical, since it is required first for growth of the insect. If manganese is minimal, just a few bacteria are found in the resulting offspring. By increasing the manganese, more and more bacteria appeared in the eggs and the offspring were laden with them. Zinc is not needed for growth of the insect, but it acts as a synergist to manganese, or expressed differently, has a sparing action if manganese is low. Finally, calcium is relatively neutral in its effect on growth, and is itself without effect on bacterial transmission, but it antagonizes manganese. Thus the molar ratio of calcium: manganese becomes highly significant. Not only that, but calcium in the mother's diet promotes plump embryos, that is, embryos with well-developed fat body. Therefore, if the mother's diet contains adequate manganese, a small amount of zinc, and a favourable ratio of calcium, the healthy offspring hatch with well-filled mycetocytes spaced nicely throughout abundant fat body in the haemocoele. It was finally determined that satisfactory levels of these minerals in the diet are as follows: zinc, 1·0 mM/kg.; manganese, 1·5 mM/kg.; and calcium, 21·0 mM/kg. While the insects tolerate calcium over a very wide range, the optimum for healthy offspring is a calcium: manganese ratio of approximately 14.

As far as we were able to determine, other essential minerals do not affect ovarial transmission independently of growth or survival.

Beginning life with a plump fat body is important to the young insect in some subtle way which has not as yet been clarified. Nymphs which hatch with big mycetocytes compressed together in a haemocoele practically devoid of fat-body tissue seem to be at a disadvantage if their diet in turn places a stress on them (unpublished observations).

Cell-wall formation, or rather lack of it, seems to be implicated in the results obtained from the use of antibiotics or deficient minerals. In culturable organisms, penicillin inhibits cell-wall formation, and zinc is thought to act catalytically in cell-wall formation. The tetracyclines probably act as chelating agents in preventing the use of enzyme catalysts such as manganese (Burkholder, 1959). The unique part of the symbiont story is the especially sensitive stage of the bacteria in the ovarial eggs. While in the fat-body mycetocytes, these organisms are well-nigh impervious to any treatment. It is only the short infective forms in the ovary which are sensitive to heat, antibiotics, mineral deficiencies, and lipid deficiencies (discussed below). Obviously the intracellular milieu of the mycetocytes affords protection to the bacteria. Since the ovary is the site where the symbiont is most easily destroyed the harmful agents probably act only for the brief period while the bacteria are extracellular—between leaving the mycetocytes and before being drawn into the egg cytoplasm by the microvilli. Any effect on the egg membranes has not been investigated; one would expect that calcium might be involved.

One must also consider the effect of reserves of substances passed on to the offspring through the egg which prevent the detection of dietary deficiencies in the first generation started from eggs of the stock culture. There is enough for the growth of the insect and the growth of the mycetocytes. We have shown that the mycetocytes multiply intermittently, once during each intermoult cycle, so that the number of mycetocytes relative to the body size remains constant (Brooks & Richards, 1955b). This implies a generation time of 10 days for the bacteria, since the insect moults once every 10 days. But in the mature female, the developing eggs enlarge tremendously, increasing from a follicle of about 25 μ diameter to a ripe egg measuring 0·5 mm. × 1 mm. within 7 days. This implies a very much speeded-up generation time for the bacteria to keep pace with this growth. This is why the antibiotics may be effective in the egg, the greater rate of turnover in effect increasing the dose. Similarly, the mineral deficiencies are expressed during a period of rapid growth. We do not know whether the deficiency is exerted directly on bacterial growth or indirectly by making the egg cytoplasm an inferior substrate, or whether the bacteria are killed outright or simply prevented from reproducing themselves. That they get into the small follicles can be demonstrated microscopically, and supravital staining with neotetrazolium chloride shows that they are still viable in small eggs that have not begun to enlarge. But the course of events beyond this is uncertain.

I must emphasize that the loss of the symbionts is not demonstrable in the fat body of the treated generation. It is true that there are some adverse effects; for example, the bacteria give a lessened reaction with tetrazolium after prolonged feeding of chlortetracycline. But in our experience they are never completely killed unless the dose is high enough to kill the host. Cessation of treatment is usually followed by recovery of the bacteria. Thus the only way to obtain aposymbiotic insects is by preventing the original infection in the offspring.

In addition to minerals, we have found that there are two other harmful factors in food which one might otherwise consider good enough for cockroaches (Brooks, 1962). These are oxidized linoleic acid and urea. Although urea is not normally a constituent of diets, it is frequently used as a source of nitrogen in experimental diets and we used it in the form of urea-complexed fatty acids to stabilize the fats in the diet and to prevent their going rancid. We thus confirmed the observation that the parental generation grows very well without any fatty acid in its diet and is not harmed by up to 4 % of urea, but both the lack of linoleic acid and the presence of 8 % or more urea in the diet prevent transmission of bacteria through the eggs. The urea-complexed linoleic acid could not be used without further investigation into the effects of each component studied separately.

Linoleic acid in the diet did not affect the growth response of the parental generation, but it had an effect on egg production. Without linoleic acid, as many eggs were laid as on an adequate diet, but all the eggs shrivelled and died. Diets containing linoleic acid at 0·02, 0·05, 0·10, 0·20, 0·40, 0·60, 0·80, 1·00 and 1·50 % caused no real difference in the number of eggs laid, but the hatchability increased up to 0·20 % linoleic acid and then remained constant. (Hatchability of 100 % has never been reached on anything but crude natural diets such as dog biscuit.) The effect of linoleic acid deficiency is still under investigation.

Turning next to the effect of urea we found first that there was no significant difference in either growth rate or survival with urea at zero, 2 or 4 %. At 8 % the growth rate declined considerably and survival to the adult stage dropped to half. At 12 % survival dropped to about one-quarter. Regardless of how poorly the individuals of the parent generation may have grown, if they reached maturity they laid eggs on urea levels up to 8 %. Samples of the hatchling nymphs were taken from every diet, and every nymph contained symbiotic bacteria. But at 12 % urea, no eggs were laid. At the termination of the experiment, several surviving females on each diet were sacrificed for examination of their ovaries. The ovaries were crushed, smeared on slides, and stained with Gram's stain. The bacteria within the ovaries gave a Gram-positive reaction, which is the normal reaction, on the diets up to 8 % urea. But on 12 % urea the bacteria gave a Gram-negative reaction. The effect on egg viability was an all-or-none reaction; that is, as long as the bacteria were Gram-positive, viable eggs were laid. When the bacteria were Gram-negative, no eggs were laid. This reaction differs in kind from the reactions to antibiotics or mineral deficiencies, which permit the production of viable eggs without bacteria.

Interpretation of the urea effect remains a matter of speculation. There is a direct relationship between absence of bacteria and early accumulation of urate concretions in the fat body. These concretions, thought to be a form of storage excretion, accumulate around the mycetocytes and become more numerous with advancing age (Pl. 4, fig. 10). Senile insects appear to have the mycetocytes literally crushed by the urates. Insects fed a high protein diet (above 65 %) accumulate the urates at an earlier age and the fat body soon takes on a chalky texture (Haydak, 1953). An aposymbiotic insect begins to accumulate urates while still very young; when it is mature its fat body is like that of a normal insect fed a high protein diet. We interpret this as indicating that the bacteria are involved in mobilization of nitrogenous wastes. Perhaps a high intake of urea taxes this process. The inhibition is more specific than merely high nitrogen intake, however, since the nitrogen in a diet containing 8 % urea and 30 % casein does not reach the high levels of nitrogen intake studied by Haydak.

The reverse of the situation in aposymbiotic insects has been observed in symbiotic cockroaches from which two of the endocrine glands, the corpora allata and the corpora cardiaca, have been removed (Bodenstein, 1953). Removal of these glands in adults of *Periplaneta americana* caused the disappearance of urates, while reimplantation of the glands restored the deposition of urates.

Hormonal regulation of life cycles of symbionts

Inevitably, observations such as those above lead us to consider the possibility of symbionts functioning as endocrines. Their effect on growth, colour, metabolism and reproduction all mimic hormonal action. But there is no supporting evidence. In fact, Cleveland and his associates have recently shown that the hormones of the wood-eating roach, *Cryptocercus punctulatus*, control its symbionts!

Cryptocercus can be maintained for months in the laboratory after capture in the spring. Following natural winter conditioning, which involves a period of 3 months of near-freezing temperatures, the nymphs moult almost in synchrony during the summer, and the hatching of the eggs coincides with the peak of the moulting season. The newly hatched nymphs gain their initial infections partially by feeding on faecal pellets containing encysted flagellates which have been eliminated by the older, moulting nymphs (Cleveland & Nutting, 1955; Nutting, 1956).

Cryptocercus harbours fourteen genera and over thirty species of hypermastigote and polymastigote flagellates closely related to the flagellates of termites (Cleveland, Burke & Karlson, 1960; Cleveland, Hall, Sanders & Collier, 1934). However, the roach flagellates differ from the termite flagellates in that many of the former encyst during the moulting period of the host, while cysts have not been observed in termites. The encystment process is correlated with the behaviour of the insects. Some, but not all, of the encysted flagellates are expelled with faecal pellets a few hours after ecdysis, and in the encysted form the flagellates withstand desiccation. When the pellets are eaten by young nymphs, the flagellates excyst and resume development. Termites do not pass dry faecal pellets, but rather a viscous fluid which contains active protozoa. When termites moult, the flagellates do not encyst and all die. Since termites do not moult in synchrony, newly moulted and defaunated termites regain their infection by soliciting proctodaeal food—the material containing the active protozoa—from non-moulting members of the colony.

Cleveland reported in 1947 that during the moulting period of *Cryptocercus*, the protozoa change from asexual to sexual methods of reproduction; and in an extensive series of papers which followed, he described in detail the morphology of these forms (for complete references see Cleveland *et al.* 1960).

In brief, the asexual forms of the protozoa round off and encyst, and within the cyst nuclear and cytoplasmic divisions give rise to one male and one female gamete, which can be distinguished from one another by staining reactions and morphology (Cleveland, 1956, 1957). In some species, all the asexual cells are haploid, in others, all diploid. In the haploid genera, gametogenesis and meiosis are separate processes, but in diploids they occur at the same time. Thus meiosis may precede or follow fertilization. In some species, the gametes escape from the cyst and then cross-fertilize, but in others, the gametes immediately fuse within the gametocyst. The zygote stage may be completed by one or two meiotic divisions before ecdysis, in which case pseudoencystation occurs during ecdysis. Commonly, the gametocyst stage persists until after ecdysis; some species never encyst. In some species, escape from the cyst, fertilization, and meiosis are all completed only after entering a new host. The length of the entire process also varies with the species, the longest starting 44 days, at 20°, before ecdysis.

Although there are numerous differences in the timing of major events and in the details of cell organelles in meiotic and mitotic divisions, the overall picture in all the genera is strikingly similar. They undergo profound morphological differentiation involving cytoplasmic, nuclear and chromosomal changes. After much careful comparison with events in non-moulting roaches, Cleveland has interpreted these processes as sexual stages which have been evoked by the hormonal condition of the roach during the moulting period. While the interpretation of these cytological changes as *sexual* stages has been criticized (Grassé, cited by Hungate, 1955), at least there is no doubt that they are processes fundamentally different from those which occur in the non-moulting period of the insect.

Cleveland & Nutting (1955) demonstrated conclusively that the environment provided by the moulting host is responsible for initiating (and seemingly maintaining) the sexual cycles in the protozoa. This was accomplished by transfaunations. Moulting and intermoult nymphs, and adults, were defaunated by oxygen pressure; then refaunated by injecting into the rectum suspensions of protozoa removed from the hindgut of donors. The intermoult age of all recipient nymphs was determined precisely. The course of events was traced for three major genera of flagellates. In all instances, only nymphs which had begun their moulting process were able to initiate sexual cycles in the protozoa, and the cycles could be maintained only in moulting nymphs. Transferring sexual protozoa to non-moulting nymphs or adults resulted in the death of the protozoa. Asexual stage protozoa were unaffected by a loss of moulting hormone; but transferring asexual protozoa from non-moulting insects to nymphs about to begin moulting initiated sexual development in the protozoa. Nearly 100 % of the individuals of some species undergo sexual reproduction in a moulting host. Each genus of protozoa has its own time-schedule in the moulting process for beginning sexual stages; this is a reflexion of different thresholds of susceptibility to the host's hormone titre.

Aiming to elucidate the nature of the host's environment which is responsible for the sexual cycles, Nutting & Cleveland (1958) found that surgical removal or cauterization of the pars intercerebralis of the forepart of the brain, containing numerous neurosecretory cells, resulted in the loss of control. The role of this part of the brain in controlling moulting was indecisive, since some insects moulted after extirpation. But the importance of the brain in regulating the protozoa was shown by the fact that if extirpation was done after moulting had started, still the sexual stage protozoa died, and the stage was not initiated in the other species. Species which ordinarily would not be changing at the time of extirpation were unaffected.

Moulting (that is, the growth of a new cuticle) is a consequence of the presence of a growth and differentiation hormone, called ecdysone, which is produced by the prothoracic glands under the control of the brain. The exact mechanism of this control is still unknown. Experiments with transfaunations and brain extirpations have left unanswered the question of whether the moulting hormone affected the protozoa directly or whether it acted indirectly through changes in the host tissue. Further light was shed on this question when ecdysone was made available in crystalline form and could be injected into the roaches. Cleveland (1959) found that when ecdysone was injected into non-moulting nymphs or adults, the protozoa were more sensitive to the hormone than was the host! Quantities of hormone insufficient to initiate moulting did induce sexual cycles. It is only gametogenesis—the first step—which is stimulated in haploids; fertilization and meiosis can be completed without ecdysone. This can be demonstrated by removing protozoa from injected nymphs after gametogenesis is completed and implanting them into an adult, where the sexual cycle is completed. But in diploids, in which gametogenesis and meiosis occur concomitantly, both are stimulated by ecdysone and only fertilization is completed without the hormone. This represents a more precise observation than those of the transfaunation experiments, which had indicated that once the cycle was initiated, it could not be maintained without the moulting hormone.

Continued studies on the effect of ecdysone indicate that, if it is maintained at an abnormally high titre, as opposed to its normal decline after ecdysis, aberrations develop in some steps of the sexual cycle (Cleveland *et al.* 1960). Internal re-organizations such as autogamy or endomitosis may occur in species which do not do this in nature. Or the final step, meiosis, may not begin, being retarded as long as the ecdysone titre is kept high. Furthermore, manipulating the ecdysone content of the gut, that is, causing it to fluctuate by squeezing out the gut contents, causes one species of flagellate to interrupt its sexual cycle and return to the asexual stage. It is evident that the responses of the protozoa are not as rigid as was once thought.

In transferring protozoa to hosts with various titres of hormones, Cleveland & Burke (1960) have been able to modify the sexual cycles by lengthening or shortening them according to the ecdysone titre. Such experiments demonstrate that the flagellates must adapt themselves to the conditions in which they find themselves. Ordinarily adaptation is gradual; abrupt experimental changes make adaptation impossible.

All of these findings on the effects of the host on its symbionts are striking additions to the idea of merely providing food and shelter. Cleveland is impressed with the evolutionary advantages or potentials which sexuality has bestowed on the flagellates, considering this to be more important than the contribution of the flagellates to the insect. As Hungate (1955) points out, there are two aspects to sexuality in the flagellates which are still confusing. First, since there is intensive inbreeding of both the protozoa and the roach, which leads to homozygosity, sexuality can do little to increase variation with so little heterozygosity present. Secondly, the successful survival of the asexual termite protozoa, which respond to their host's moulting influence only by dying, shows that sexuality has not been essential for their evolution. Perhaps we should not always look on the evolution of these forms as having reached its highest potential; perhaps the potential from sexuality is to be realized in the future.

IN VITRO CULTURE OF SYMBIONTS

The last aspect which merits discussion is culture of symbionts *in vitro*. Mention has been made throughout this paper of ways to separate host and symbiont, with emphasis on the fact that the host may be harmed by the treatment. In the case of intracytoplasmic symbiosis, the microorganisms are destroyed in the process. There have been several conclusive demonstrations of the nutritional impairment of growth and reproduction in aposymbiotic insects. There are also some derangements in metabolism and colour suggestive of an endocrine lack in aposymbiotic insects. But successful cases of culturing the isolated symbionts, determining their needs and functions, and identifying them are rare indeed. This is not because of lack of enthusiasm for the subject, for persons working in this field have engaged in lively arguments from time to time as to the nature of the internal objects called symbionts. Since culture of the objects is so rarely achieved, there are many people who believe they are mitochondria, cell organelles, or waste products. Pure culture may answer these questions, but not necessarily so, since there is no reason to believe that cell fragments cannot be cultured as well as whole cells.

This paper does not pretend to be a complete review of the literature, and so no effort will be made to list or evaluate published results of culture attempts. Only a few generalizations will be considered.

First, it appears unlikely that intracytoplasmic micro-organisms can be cultured on simple nutrient broths but instead will require complex media like those used for tissue culture, perhaps enriched with insect growth and differentiation hormones. Secondly, it appears unlikely, even if suitable media for such organisms were found, that growth would be rapid, because in the insect the organisms grow slowly. Thirdly, the naturally occurring pleomorphism of symbionts within their hosts leads to much confusion with cultured contaminants. Investigators have fallen back on the excuse of pleomorphism to justify some bizarre forms in their cultures. Fourthly, Koch's postulates must be fulfilled, but this requires an aposymbiotic host in which to test the cultured organisms. As we know, this is not only difficult to achieve with intracytoplasmic organisms but also capricious to maintain. The cockroach, for example, as shown by several investigators, will not lose all of its symbionts by feeding or injecting with antibiotics. If the cultured organisms are injected into 'aposymbiotic' insects which have a residual population of symbionts, subsequent finding of symbionts in these insects does not constitute proof that the cultured organisms were the symbionts.

CONCLUSION

Summarizing briefly, arthropods are veritable cultures of micro-organisms, harbouring rickettsiae, bacteria, yeasts and protozoa. The micro-organisms are located in modifications of the intestine, in the Malpighian tubules, in the gonads, and in special cells of the fat body or organs in the haemocoele. The micro-organisms have assumed the appearance of cytoplasmic elements in many cells. There are as many ways of transmitting them as there are ways of harbouring them. Although numerous technical problems make analyses difficult, there is general agreement that the relationships are mutually beneficial, although not always of equal benefit to each partner. The hosts derive nutritional factors of benefit to growth and reproduction. The micro-organisms are dependent on the hosts for specific nutritional factors, translocation, and morphological differentiation.

REFERENCES

BAINES, S. (1956). The role of the symbiotic bacteria in the nutrition of *Rhodnius prolixus* (Hemiptera). *J. exp. Biol.* 33, 533.
DEBARY, A. (1879). *Die Erscheinung der Symbiose*. Strassburg: Trübner.
BEWIG, F. & SCHWARTZ, W. (1955). Über die Symbiose von Triatomiden mit *Nocardia rhodnii* (Erikson) Waksman und Henrici. *Naturwissenschaften*, 42, 423.
BLEWETT, M. & FRAENKEL, G. (1944). Intracellular symbiosis and vitamin requirements of two insects, *Lasioderma serricorne* and *Sitodrepa panicea*. *Proc. Roy. Soc.* B, 132, 212.

BLOCHMANN, F. (1886). Über eine Metamorphose der Kerne in den Ovarialeiern und über den Beginn der Blastodermbildung bei den Ameisen. *Verhandl. Naturh.-Med. Ver.* **3**, 243.

BLOCHMANN, F. (1888). Über das regelmässige Vorkommen von bakterienähnlichen Gebilden in den Geweben und Eiern verschiedener Insekten. *Z. Biol.* **24**, 1.

BODENSTEIN, D. (1953). Studies on the humoral mechanisms in growth and metamorphosis of the cockroach, *Periplaneta americana*. III. Humoral effects on metabolism. *J. exp. Zool.* **124**, 105.

BRECHER, G. & WIGGLESWORTH, V. B. (1944). The transmission of *Actinomyces rhodnii* Erikson in *Rhodnius prolixus* Stål (Hemiptera) and its influence on the growth of the host. *Parasitology*, **35**, 220.

BROOKS, M. A. (1956, 1958). Nature and significance of intracellular bacteroids in cockroaches. *Proc. Xth Int. Congr. Ent., Montreal*, **2**, 311.

BROOKS, M. A. (1960). Some dietary factors that affect ovarial transmission of symbiotes. *Proc. helminth. Soc., Wash.* **27**, 212.

BROOKS, M. A. (1962). The relationship between intracellular symbiotes and host metabolism. *Proc. simposio intern. biol. sper. celebrazione Spallanzaniana* (Reggio & Pavia, Italy, 1959).

BROOKS, M. A. & RICHARDS, A. G. (1955 a). Intracellular symbiosis in cockroaches. I. Production of aposymbiotic cockroaches. *Biol. Bull., Woods Hole*, **109**, 22.

BROOKS, M. A. & RICHARDS, A. G. (1955 b). Intracellular symbiosis in cockroaches. II. Mitotic division of mycetocytes. *Science*, **122**, 242.

BROOKS, M. A. & RICHARDS, A. G. (1956). Intracellular symbiosis in cockroaches. III. Re-infection of aposymbiotic cockroaches with symbiotes. *J. exp. Zool.* **132**, 447.

BRUES, C. T. & DUNN, R. C. (1945). The effect of penicillin and certain sulfa drugs on the intracellular bacteroids of the cockroach. *Science*, **101**, 336.

BUCHNER, P. (1953). *Endosymbiose der Tiere mit pflanzlichen Mikroorganismen.* Basel: Verlag Birkhäuser.

BUCHNER, P. (1954). Endosymbiosestudien an Schildläusen. I. *Stictococcus sjoestedti. Z. Morph. Ökol. Tiere*, **43**, 262.

BUCHNER, P. (1955). Endosymbiosestudien an Schildläusen. II. *Stictococcus diversiseta. Z. Morph. Ökol. Tiere*, **43**, 397.

BUCHNER, P. (1961). Endosymbiosestudien an Ipiden. I. Die Gattung *Coccotrypes. Z. Morph. Ökol. Tiere*, **50**, 1.

BURKHOLDER, P. R. (1959). Antibiotics. *Science*, **129**, 1457.

BUSH, G. L. & CHAPMAN, G. B. (1961). Electron microscopy of symbiotic bacteria in developing oocytes of the American cockroach, *Periplaneta americana. J. Bact.* **81**, 267.

CLEVELAND, L. R. (1925). Toxicity of oxygen for protozoa *in vivo* and *in vitro*: animals defaunated without injury. *Biol. Bull., Woods Hole*, **48**, 455.

CLEVELAND, L. R. (1947). Sex produced in the protozoa of *Cryptocercus* by molting. *Science*, **105**, 16.

CLEVELAND, L. R. (1956). Brief accounts of the sexual cycles of the flagellates of *Cryptocercus. J. Protozool.* **3**, 161.

CLEVELAND, L. R. (1957). Correlation between the molting period of *Cryptocercus* and sexuality in its protozoa. *J. Protozool.*, **4**, 168.

CLEVELAND, L. R. (1959). Sex induced with ecdysone. *Proc. nat. Acad. Sci., Wash.* **45**, 747.

CLEVELAND, L. R. & BURKE, A. W., Jr. (1960). Modifications induced in the sexual cycles of the protozoa of *Cryptocercus* by change of host. *J. Protozool.* **7**, 240.

CLEVELAND, L. R., BURKE, A. W., Jr. & KARLSON, P. (1960). Ecdysone induced modifications in the sexual cycles of the protozoa of *Cryptocercus*. *J. Protozool.* **7**, 229.

CLEVELAND, L. R., HALL, S. R., SANDERS, E. P. & COLLIER, J. (1934). The wood-feeding roach *Cryptocercus*, its protozoa, and the symbiosis between protozoa and roach. *Mem. Amer. Acad. Arts & Sci.* **17**, 185.

CLEVELAND, L. R. & NUTTING, W. L. (1955). Suppression of sexual cycles and death of the protozoa of *Cryptocercus* resulting from change of hosts during molting period. *J. exp. Zool.* **130**, 485.

COOK, S. F. (1943). Non-symbiotic utilization of carbohydrates by the termite, *Zootermopsis angusticollis*. *Physiol. Zoöl.* **16**, 123.

FOECKLER, F. (1961). Reinfektionsversuche steriler Larven von *Stegobium paniceum* L. mit Fremdhefen und die Beziehungen zwischen der Entwicklungsdauer der Larven und dem B-Vitamingehalt des Futters und der Hefen. *Z. Morph. Ökol. Tiere*, **50**, 119.

GEIGY, R., HALFF, L. A. & KOCHER, V. (1954). L'acide folique comme élément important dans la symbiose intestinale de *Triatoma infestans*. *Acta Trop.* **11**, 163.

GRÄBNER, K.-E. (1954). Vergleichende morphologische und physiologische Studien an Anobiiden- und Cerambyciden-Symbionten. *Z. Morph. Ökol. Tiere*, **41**, 471.

GRESSON, R. A. R. & THREADGOLD, L. T. (1960). An electron microscope study of bacteria in the oocytes and follicle cells of *Blatta orientalis*. *Quart. J. micr. Sci.* **101**, 295.

HARINGTON, J. S. (1960). Synthesis of thiamine and folic acid by *Nocardia rhodnii*, the micro-symbiont of *Rhodnius prolixus*. *Nature, Lond.* **188**, 1027.

HAYDAK, M. H. (1953). Influence of the protein level of the diet on the longevity of cockroaches. *Ann. ent. Soc. Amer.* **46**, 547.

HENRY, S. M. & BLOCK, R. J. (1960). The sulfur metabolism of insects. IV. The conversion of inorganic sulfate to organic sulfur compounds in cockroaches. The role of intracellular symbionts. *Contr. Boyce Thompson Inst.* **20**, 317.

HEYMONS, R. L. (1890). Über die hermaphroditische Anlage der Sexualdrüsen beim Männchen von *Phyllodromia* (*Blatta* L.) *germanica*. *Zool. Anz.* **13**, 451.

HUGER, A. (1954). Experimentelle Eliminierung der Symbionten aus den Myzetomen des Getreidekapuziners, *Rhizopertha dominica* F. *Naturwissenschaften*, **41**, 170.

HUGER, A. (1956). Experimentelle Untersuchungen über die künstliche Symbiontenelimination bei Vorratsschädlingen: *Rhizopertha dominica* F. (Bostrychidae) und *Oryzaephilus surinamensis* L. (Cucujidae). *Z. Morph. Ökol. Tiere*, **44**, 626.

HUGHES-SCHRADER, S. (1948). Cytology of coccids (Coccoidea–Homoptera). *Advanc. Genet.* **2**, 127.

HUNGATE, R. E. (1943). Quantitative analyses on the cellulose fermentation by termite protozoa. *Ann. ent. Soc. Amer.* **36**, 730.

HUNGATE, R. E. (1955). Mutualistic intestinal protozoa. In *Biochemistry and Physiology of Protozoa*, Vol. II. Ed. S. H. Hutner & A. Lwoff. New York: Academic Press, Inc.

KOCH, A. (1933). Über das Verhalten symbiontenfreier *Sitodrepa* Larven. *Biol. Zbl.* **53**, 199.

KOCH, A. (1960). Intracellular symbiosis in insects. *Annu. Rev. Microbiol.* **14**, 121.

KOCH, A. (1962). Grundlagen und Probleme der Symbioseforschung. *Med. Grundlagenforschung*, **4**, 63.

LAMPEL, G. (1959). Geschlecht und Symbiose bei den Pemphiginen. *Z. Morph. Ökol. Tiere*, **48**, 320.

LEYDIG, F. (1850). Einige Bemerkungen über die Entwicklung der Blattläuse. *Z. wiss. Zool.* **2**, 62.

230 MARION A. BROOKS

LEYDIG, F. (1854). Zur Anatomie von *Coccus hesperidum*. *Z. wiss. Zool.* 5, 1.
MANSOUR, K. (1934). On the intracellular microorganisms of some Bostrychid beetles. *Quart. J. micr. Sci.* 77, 243.
MANUNTA, C. & BERNARDINI, P. (1958). Saggi cromatografici preliminari su batteriociti di Blatta (*Blaberus cranifer*). *Ist. Lombardo (Rend. Sci.)* B, 92, 507.
MEYER, G. F. & FRANK, W. (1957). Elektronenmikroskopische Studien zur intracellulären Symbiose verschiedener Insekten. I. Untersuchungen des Fettkörpers und der symbiontischen Bakterien der Küchenschabe (*Blatta orientalis* L.). *Z. Zellforsch.* 47, 29.
MÜLLER, H. J. (1956). Experimentelle Studien an der Symbiose von *Coptosoma scutellatum* Geoffr. (Hem. Heteropt.). *Z. Morph. Ökol. Tiere,* 44, 459.
MUSGRAVE, A. J., MONRO, H. A. U. & UPITIS, E. (1961). Apparent effect on the mycetomal micro-organisms of repeated exposure of the host insect, *Sitophilus granarius* (L.) (Coleoptera), to methyl bromide fumigation. *Canad. J. Microbiol.* 7, 280.
NUTTING, W. L. (1956). Reciprocal protozoan transfaunations between the roach, *Cryptocercus*, and the termite, *Zootermopsis*. *Biol. Bull., Woods Hole,* 110, 83.
NUTTING, W. L. & CLEVELAND, L. R. (1958). Effects of glandular extirpations on *Cryptocercus* and the sexual cycles of its protozoa. *J. exp. Zool.* 137, 13.
PANT, N. C. & FRAENKEL, G. (1950). The function of the symbiotic yeasts of two insect species, *Lasioderma serricorne* F. and *Stegobium (Sitodrepa) paniceum* L. *Science,* 112, 498.
PANT, N. C. & FRAENKEL, G. (1954). Studies on the symbiotic yeasts of two insect species, *Lasioderma serricorne* F. and *Stegobium paniceum* L. *Biol. Bull., Woods Hole,* 107, 420.
PARKIN, E. A. (1952). Symbiosis in *Ptilinus pectinicornis* L. *Nature, Lond.* 170, 847.
POULSON, D. F. & SAKAGUCHI, B. (1961). Nature of 'sex-ratio' agent in *Drosophila*. *Science,* 133, 1489.
PUCHTA, O. (1955). Experimentelle Untersuchungen über die Bedeutung der Symbiose der Kleiderlaus *Pediculus vestimenti* Burm. *Z. Parasitenk.* 17, 1.
ROSHDY, M. A. (1961). Observations by electron microscopy and other methods on the intracellular rickettsia-like microorganisms in *Argas persicus* Oken (Ixodoidea, Argasidae). *J. Insect Path.* 3, 148.
SCHOMANN, H. (1937). Die Symbiose der Bockkäfer. *Z. Morph. Ökol. Tiere,* 32, 542.
TRAGER, W. (1932). A cellulase from the symbiotic intestinal flagellates of termites and of the roach, *Cryptocercus punctulatus*. *Biochem. J.* 26, 1763.
TRAGER, W. (1934). The cultivation of a cellulose-digesting flagellate, *Trichomonas termopsidis*, and of certain other termite protozoa. *Biol. Bull., Woods Hole,* 66, 182.
WIGGLESWORTH, V. B. (1936). Symbiotic bacteria in the blood sucking insect, *Rhodnius prolixus* Stål. (Hemiptera, Triatomidae). *Parasitology,* 28, 284.

EXPLANATION OF PLATES

PLATES 1 AND 2

Photomicrographs of material taken from normal (i.e. symbiotic) *Blattella germanica*; unless otherwise noted, fixed in Carnoy's solution and stained with Delafield's haematoxylin; cut at 6 μ.

Fig. 1. Cross-section through mid-abdominal area of young nymph. Ventral nerve cord visible in lower part of photograph, gut visible in upper part. Arrow (*m*) points to one of many mycetocytes in fat body.

Fig. 2. Section through several mycetocytes in fat body.

PLATE 1

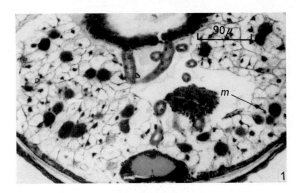

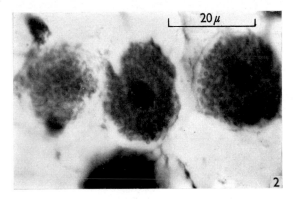

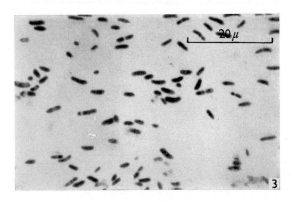

PLATE 2

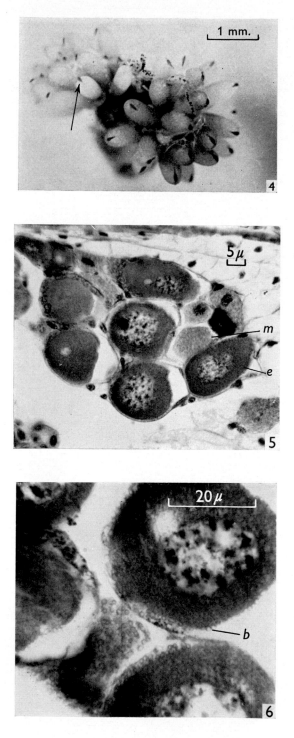

PLATE 3

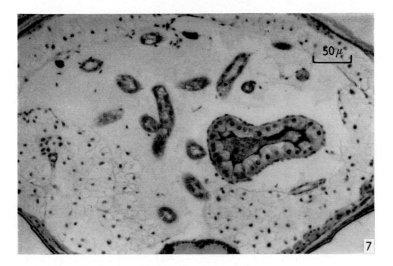

50μ

7

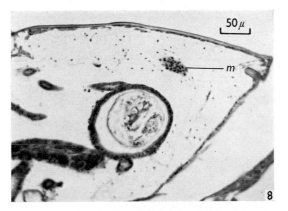

50μ

m

8

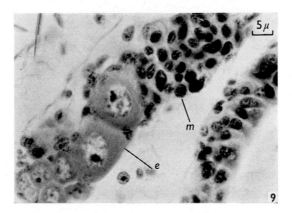

5μ

m

e

9

PLATE 4

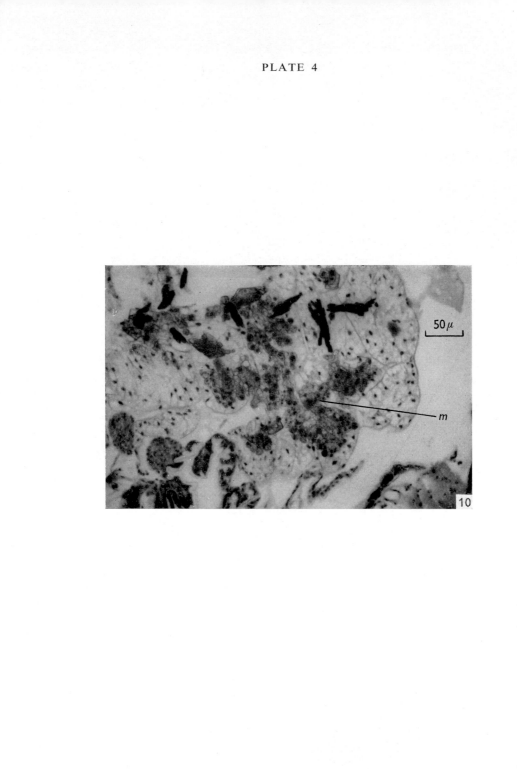

Fig. 3. Smear of fat body, Gram stained, showing individual bacteroids.

Fig. 4. Excised ovaries of mature ♀ supravitally stained with neotetrazolium chloride. At proximal end of each ripe egg may be seen the successively younger eggs (arrow), which stain intensely because of the relatively thick peripheral layer of bacteroids.

Fig. 5. Section through ovary of very young ♀ *in situ*, showing mycetocytes (*m*) insinuated between egg follicles (*e*).

Fig. 6. Section through ♀ slightly older than the one in fig. 5 showing bacteroids (*b*) beneath vitelline membranes of egg follicles.

PLATES 3 AND 4

Photomicrographs of material taken from aposymbiotic B. germanica
prepared as in figs. 1–6.

Fig. 7. Cross-section through mid-abdomen of nymph lacking bacteroids. Compare with fig. 1.

Fig. 8. Cross-section through one of the clusters of 'empty mycetocytes'.

Fig. 9. Cross-section showing a cluster of 'empty mycetocytes' (recognizable by large, dark nuclei) apposed to ovary of young ♀. (*m* = mycetocytes; *e* = eggs.)

Fig. 10. Cross-section through fat body of cockroach in which many urates have accumulated; mycetocytes (*m*) scarcely recognizable.

AMBROSIA BEETLES AND THEIR FUNGI, WITH PARTICULAR REFERENCE TO *PLATYPUS CYLINDRUS* FAB.

J. M. BAKER

Forest Products Research Laboratory,
Princes Risborough, Buckinghamshire

Insects which feed upon woody tissue utilize a particularly unpromising food source. The material consists largely of cellulose and lignin, both of which are resistant to degradation by enzymes normally found in the animal kingdom, and it is deficient in essential B vitamins and sterols which the insect cannot synthesize for itself. As Schedl (1958) has pointed out, the ancestral wood-eating Coleoptera probably entered their host only after decay had commenced, relying on the activity of fungi and bacteria to reduce the wood to a more suitable physical and nutritional state. Later came xylophagous beetles which could utilize dead but undecayed wood by virtue of their endosymbiotic flora of yeasts or bacteria. The vast body of work on endosymbiosis and its physiological significance is reviewed in Buchner's monograph (1953) and by Koch (1960).

Among the most successful wood-inhabiting beetles are the Scolytidae and Platypodidae, well known as forest insects causing damage of great economic significance to trees and timber. The two families are frequently grouped into the superfamily Scolytoidea, and though Browne (1961 a) has recently denied the validity of this taxon it is useful to retain it for descriptive convenience. The Scolytoidea are best known in the temperate regions of the world from the Scolytid bark beetles which attack and frequently kill living trees. The bark beetles are phloeophagous, that is they feed on the living tissues under the bark: phloem, cambium and outer sapwood. Phylogenetically this mode of nutrition was an advance on simple xylophagy which among present-day Scolytidae occurs only in the primitive group Hylesininae. The bark beetles were able to colonize living and recently felled trees, and as their food material is highly nutritious the specialized endosymbiotic flora so evident in xylophagous forms appears to be of less importance. However, the bark beetles are regularly associated with a wide range of fungi and yeasts which they transmit from tree to tree but which, in many instances, may not be nutritionally significant to the insect. A specialization of the wood-boring habit was achieved by those Scolytoids which re-entered the deep wood tissue to form the group of wood-boring insects known as the ambrosia beetles. These beetles although wood-borers are not wood-feeders: the adults bore into wood and introduce into their tunnels 'ambrosia fungi' which grow on the tunnel walls and serve as the chief source of food for all stages of the insect. The fungi are constantly associated with the insect and are transmitted by the adult, frequently by means of specialized organs. Eggs are laid in the tunnel either free or in niches, and the developing larvae browse the ambrosia fungus. The group has long interested entomologists; the name 'ambrosia' was first applied to the glistening lining of the tunnels by Schmidberger, as long ago as 1836, before its fungal nature was known.

Schedl (1958) proposed the new term 'xylomycetophagy' to describe this mode of nutrition, since the word 'ambrosia' is also applied to the fungus gardens of certain termites and to the food fungi of some gall-inhabiting Cecidomyid midges. However, for the purposes of this account the more euphonious term 'ambrosial' is used as synonymous with xylomycetophagous in the sense of Schedl. While the xylophagous beetles rely on their endosymbiotic micro-organisms to aid in the digestion of wood, the ambrosial forms allow the fungi to grow independently and to act as intermediaries producing a concentrated food source from the wood through which the tunnels run. Buchner (1953) regards this relationship as an ectosymbiosis.

In this paper the distribution and general biology of the ambrosial habit in the Coleoptera will be reviewed, and this will be followed by a detailed study of the ambrosial relationship in the oak pinhole borer *Platypus cylindrus* (briefly reported by Baker, 1960).

DISTRIBUTION OF THE AMBROSIAL HABIT IN THE COLEOPTERA

Scolytoidea

Browne (1961 *a*) sets out a classification which adopts Schedl's concepts and which is followed in outline here.

Family Scolytidae

(1) *Hylesininae*. This subfamily contains the more primitive xylophagous members of the Scolytidae but is predominantly composed of phloeophagous species among which are many of great economic significance in the temperate regions.

(2) *Scolytinae*. Distributed throughout the holarctic and neotropical regions, this group consists entirely of phloeophagous members.

(3) *Ipinae*. This subfamily of world-wide distribution exhibits great variation in habits and feeding patterns. Within it occur almost all the ambrosial Scolytids which, however, are not placed in a single taxon but in five distinct tribes.

(4) *Scolytoplatypinae*. This family consists of the single genus *Scolytoplatypus* with several species, all ambrosial, in Africa and Oriental regions.

The phylogenetic status of the five ambrosial tribes of the Scolytidae is uncertain but it is generally agreed (Browne, 1961 *a*; Wood, 1957) that there have been two main evolutionary lines—Trypodendrini and Corthylini in one line and Xyleborini, Eccoptopterini and Webbini in the other. The similarities both in morphology and social organization between an individual ambrosial tribe and its apparently most closely related phloeophagous cousin are closer than the similarities between one ambrosial group and another. This leads to the speculation that the ambrosial habit is of polyphyletic origin within the Scolytidae.

Family Platypodidae

The exclusively ambrosial family Platypodidae is far more homogeneous than the Scolytidae both in morphology and habit. Although the adults of the Platypodidae differ greatly in their appearance from those of the Scolytidae, their larval forms (Crowson, 1960) and their internal anatomy are similar (Chapman, 1961). The phylogenetic origin of the Platypodids is quite unknown—whereas fossil remains of phloeophagous Scolytids allied to the present ambrosial Scolytids are found in profusion in the Baltic amber, no authentic fossil Platypodids have been discovered. Either the Platypodids are remains of an older Rhyncophoran stock which have independently adopted the ambrosial habit, or they have evolved relatively rapidly from an already specialized ambrosial Scolytid line.

Lymexilidae

This family of beetles is unrelated to the Scolytoidea and arrived independently at the ambrosial habit. Only certain species within the group are truly ambrosial, but in these the form of fungal growth in the tunnels shows remarkable convergence with that in the Scolytoidea (Batra & Francke-Grosmann, 1961). The Lymexilids are not mentioned further in this account which is concerned only with the Scolytoid ambrosia beetles.

If we regard as ambrosial only those insects which regularly transmit their food fungi, and which culture them strictly in association with their tunnels, we must exclude those 'space parasitic' species of the families Curculionidae and Brenthidae which inhabit the tunnels of Scolytoids and feed on the ambrosia fungi of their hosts. Also excluded is the Lymexilid *Melittomma insulare* which introduces into coconut wood a bacteria-yeast complex which grows ahead of the insect rendering the tissues suitable for food (Brown, 1954*a*). A similar association is seen in the woodwasps where the female while egg-laying introduces a wood-rotting Basidiomycete into the wood. The fungus once established grows indiscriminately. Although the wasp possesses specialized mechanisms for transmission of the fungus, the lack of close association of fungus with insect inside the wood, and the fact that the larva is feeding on softened wood rather than on the fungus itself, preclude the recognition of this type of relationship as ambrosial.

GENERAL BIOLOGY OF AMBROSIA BEETLES

In Northern temperate forests the Scolytoidea are best known from the numerous species of bark beetles, the ambrosia beetles being comparatively uncommon. In tropical forests, however, the position is reversed and throughout the world as a whole the species of ambrosia beetles far outnumber those of the bark beetles. As may be supposed from a consideration of their heterogeneous nature and the widely varying climates in which they occur, the habits and life histories of the ambrosia beetles are exceedingly diverse and no generalized account applicable to all types can be given. Fisher, Thompson & Webb (1953–54) review the general biology of the group and give a comprehensive bibliography, and more recent reviews are those of Bletchly (1961) on the factors which predispose trees and logs to attack, and of Rudinsky (1962) on the ecology of the Scolytidae. The habits of the different species within a geographical region have been the subject of several detailed studies among which must be mentioned the monographs of Beeson (1941), Browne (1961*a*), Chamberlin (1939, 1958) and Kalshoven (1958–59, 1960) on the Indian, Malayan, North American and Indonesian species respectively.

AMBROSIA FUNGI OF SCOLYTOID AMBROSIA BEETLES

Ambrosia fungi in natural growth and in culture

The 'ambrosia' which Schmidberger had thought to be a product of plant sap was first recognized as a fungus by Hartig (1844) who gave the name *Monilia candida* to the ambrosia of *Anisandrus dispar*. Considerable confusion has arisen over the name *Monilia* because it has been used *sensu* Persoon, 1801, for a genus in the Moniliales and also for a number of yeasts of medical importance, including '*Monilia candida*' Bonorden, which have subsequently been transferred to the genus *Candida* Berkhout. A fuller account of the nomenclature is given by Leach, Hodson, Chilton & Christensen (1940) and Lodder & Kreger-van Rij (1952). In descriptions of ambrosia fungi, the genus *Monilia* has been employed *sensu* Persoon, and although Webb (1945) did not consider that these fungi agreed with Persoon's characterization, the genus is still used for certain ambrosia fungi (Mathiesen-Käärik, 1953, 1960 *a*).

After Hartig's description there was little work on the relations of ambrosia beetles and fungi until Hubbard (1897) described in general terms the fungi and feeding habits of species of *Platypus*, *Corthylus*, *Xyleborus*, *Anisandrus* and *Trypodendron* in the U.S.A. He recognized two distinct forms of ambrosial growth: erect hyphae with terminal conidia where larvae lived free in the galleries, e.g. *Platypus*; and tangled monilioid chains of spores where larvae were reared in separate niches as in *Trypodendron* galleries. Hubbard noticed small wood fragments among the ambrosia and believed that the beetles cultivated the fungi on prepared beds of wood chips. A fuller picture of the 'ambrosial pattern' in tunnels emerged from the work of Neger (1908 *a*, *b*, 1909) and Beauverie (1910) who by examining microscopical sections of wood, showed that the fungus penetrates a few millimetres into the wood adjacent to the tunnel filling the cells, particularly of parenchyma, with a tangled mass of dark hyphae. It also forms, around the tunnel wall, a stroma from which a palisade of hyaline, spore-bearing hyphae projects into the lumen of the tunnel. The spores are often arranged in monilioid chains. Yeast-like or oidial forms are frequently present.

Neger (1908 *a*, *b*) speculated on the nature of the fungi concerned. At first he regarded the ambrosia fungi of the Scolytids *Anisandrus dispar* and *Trypodendron domesticum* as phases of *Ceratocystis piceae*, but later abandoned this hypothesis and suggested that they were specialized species, perhaps related to the staining fungi but having different characteristics in culture. In particular the ambrosia fungi had a pleasant ester-smell and the medium became intensely stained. Beauverie (1910), although confirming these peculiarities, looked on the fungi as possible stages of *Macrophoma*. His *Macrophoma* would now be regarded as *Botryodiplodia*, a widely distributed sap-staining fungus (Findlay, 1959).

Since then there have been many investigations on ambrosia beetles from different parts of the world with descriptions of the fungi in the tunnels and in culture, often with identification of the species which each worker considered to be the chief component of the ambrosia. The findings are outlined in Table 1.

The picture of the ambrosial habit built up by Neger (1908 a, b) and Beauverie (1910) was in general confirmed by later workers, but the nature of the fungi concerned remained controversial. Common moulds such as *Penicillium* were frequently cultured from tunnels and in several instances were regarded as true ambrosia fungi (Smith, 1935; Fischer, 1954). In most instances, however, workers adopted one of the two interpretations which Neger had offered, and either regarded the ambrosial habit as a particular growth phase of fungi commonly causing blue-stain in timber, or considered the ambrosia fungi as highly specialized species not found elsewhere.

Further evidence of the specialized nature of ambrosia fungi was provided by Schneider-Orelli (1913). The fact that the ambrosia fungus of *Anisandrus dispar* survives within the hibernating insect and is transmitted to a new site was clearly demonstrated (though the mode of transmission was not), and its growth in the tunnels and in culture was described. In culture, the monilioid ambrosia growth was lost and the fungus took on a cottony appearance; dark staining of the medium and ester formation were noted. Hadorn (1933), in a detailed account of *Trypodendron lineatum* and its fungus, agreed with Schneider-Orelli's views and though he failed to find the fungus in the overwintering beetle, argued that it must be carried internally and sown by the action of the female. Leach *et al.* (1940) and Verrall (1943) described and named several ambrosia fungi which agreed with the concept of specialized symbiotic species.

By contrast, Webb (1945) dealing with the Australian Platypodid, *Platypus subgranosus*, described the blue-stain fungus *Leptographium lundbergii* Lagerb. & Melin as the principal ambrosia fungus, and in a review of work on other ambrosia fungi sought to show that all species described up to that time fell within the genus *Leptographium*. Bakshi (1950) isolated several species of *Ceratocystis* and its conidial forms from tunnels of *Trypodendron domesticum* and *T. lineatum* and regarded them as true ambrosia fungi. It should be mentioned at this point that the genus *Ceratocystis* Ellis & Halst. was revived by Bakshi (1950), and following Hunt (1956) is used for staining fungi previously assigned to several genera, especially *Ceratostomella* Sacc. and *Ophiostoma* Syd. The staining fungus *Botryodiplodia theobromae* has been cultured from the brood chambers of *Xylosandrus* (*Xyleborus*) *morstatti* (Gregory, 1954) and likewise accorded ambrosial status.

Table 1. *Fungi reported in association with ambrosia beetles*

Beetle	Fungus reported as chief ambrosia	Other associated fungi	Reference
Scolytidae			
Trypodendrini			
Trypodendron lineatum	Unnamed specialized monilioid fungus	—	Hadorn (1933)
T. lineatum	*Ceratocystis piceae, Leptographium lundbergii*	*Fomes annosus*	Bakshi (1950, 1952)
T. lineatum	A specialized monilioid fungus, *Monilia ferruginea* sp.nov.	*Ceratocystis* spp., yeasts	Mathiesen-Käärik (1953)
T. domesticum	*Ceratocystis ambrosia* sp.nov.	—	Bakshi (1950)
T. domesticum	Specialized species close to *Monilia ferruginea*	*Ceratocystis* spp., yeasts	Mathiesen-Käärik (1953), Francke-Grosmann (1956*a*, 1958)
T. signatum	Specialized species close to *Monilia ferruginea*	—	Francke-Grosmann (1956*a*, 1958)
T. retusum, T. betulae	Unnamed specialized monilioid fungus similar to that of *A. dispar*	—	Leach *et al.* (1940)
Corthylini			
Monarthrum scutellare, M. detigerum	*Monilia,* unnamed species	—	Doane & Gilliland (1929)
Pterocyclon mali, P. fasciatum	A specialized monilioid fungus, *Monilia brunnea* sp.nov.	General staining fungi and yeasts	Verrall (1943)
P. fasciatum	—	*Aspergillus* sp. (status unknown)	Beal & Massey (1945)
Corthylus columbianum	*Ceratocystis piceae; Pichia* sp.	*Fusarium*	Wilson (1959)
Xyleborini			
Gnathotrichus sulcatus	*Ceratocystis* sp.	—	Doane & Gilliland (1929)
Anisandrus dispar	Specialized, ester producing, monilioid fungus (= *Monilia candida* Hartig 1844)	Yeasts	Schneider-Orelli (1913), Francke-Grosmann (1956*a*, 1958)
Xylosandrus germanus		*Ceratocystis ulmi*	Buchanan (1940)
X. germanus	Unnamed specialized monilioid fungus		Francke-Grosmann (1956*a*, 1958)
X. (Xyleborus) morstatti, haberkorni, morigerus, bicornis	*Monilia* sp.	*Fusarium* spp.	Muller (1933)

Beetle	Fungus reported as chief ambrosia	Other associated fungi	Reference
Xyleborini (cont.)			
Xylosandrus (Xyleborus) morstatti	Botryodiplodia theobromae	—	Gregory (1954)
X. (Xyleborus) morstatti	Cladosporium cladosporioides, Penicillium pallidum	—	Brown, (1954b)
Xyleborus saxesini	Specialized fungus, physiologically like Monilia sp., but without monilioid growth	—	Schneider-Orelli (1913), Francke-Grosmann (1956a, 1958) Fischer (1954)
X. saxesini	Penicillium sp.	—	Francke-Grosmann (1956a, 1958)
X. monographus	As for Xyleborus saxesini	—	Verrall (1943)
X. affinis	Cephalosporium pallidum sp.nov.	—	Verrall (1943)
X. peccanis	Cephalosporium pallidum sp.nov.	—	Gadd & Loos (1947)
X. fornicatus	Specialized fungus Monacrosporium ambrosium sp.nov.	—	
X. mascarensis	—	Unidentified blue-stain	Schneider (1959)
Xyleborus sp.	Ambrosiamyces zeylandicus sp.nov. (incertae sedis)		Trotter (1934)
Ambrosiodmus lecontei, A. linderae	'Sporotrichium' (no details)	—	Beal & Massey (1945)
Platypodidae			
Crossotarsus grevillae	Penicillium sp.	—	Smith (1935)
Platypus subgranosus	Leptographium lundbergii, Endomycopsis 2 spp. (unnamed)	—	Webb (1945), Hogan (1948)
P. compositus	'Endomyces bispora sp.nov.' (unaware of earlier use of name by Beck, 1922)	—	Verrall (1943)
P. difficilis	—		Vital (1951)
Platyscapulus auricomus	Monilioid fungus	Stysanus microsporus, Ceratocystis, Diplodia and moulds	Cachan (1957)
Doliopygus dubius	Cephalosporium	—	Browne (1961b)

The researches of Mathiesen-Käärik and of Francke-Grosmann have in the last decade provided strong evidence for the hypothesis that the ambrosia fungi are specialized and cannot be regarded as normal blue-stain organisms. Mathiesen-Käärik (1960a) summarized the work on the association of blue-staining *Ceratocystis* spp. with Scolytid bark beetles. She showed that there was a gradation in the closeness of the association from species such as *C. pluriannulatum* which could be wind-dispersed or carried by many different insects, to those which were regularly and constantly associated with one or a few species, for instance *C. canum* with *Blastophagus minor*. Biochemical investigations showed that those species of *Ceratocystis* most closely associated with insects required more complex culture media than less closely associated species. Wright (1935) described the blue-stain fungus *Trichosporium ambrosium* as symbiotically associated with the bark beetle *Scolytus ventralis*, and Mathiesen-Käärik (1950, 1960a) pointed out that the similar species *T. tingens* differed from *Ceratocystis* in being unable to utilize certain inorganic nitrogen sources and in requiring unknown growth factors in addition to B vitamins. It also differed in assimilating fats as a source of carbon. Rennerfelt (1950) and Francke-Grosmann (1952) demonstrated that the association between this fungus and the bark beetles *Blastophagus minor* and *Ips acuminatus* is closer than the usual bark beetle-*Ceratocystis* association. *Trichosporium tingens* lines the larval chambers of the insects, producing abundant conidia which are eaten by the larvae, but it does not form a stroma and palisade structure. Francke-Grosmann (1952) suggested that this species is intermediate between normal blue-stain and ambrosia fungi.

The ambrosia beetle *Trypodendron lineatum* was found (Mathiesen-Käärik, 1953) to transmit not only regularly associated *Ceratocystis*, as Bakshi (1950) had indicated, but also another fungus morphologically distinct in culture from all species of *Ceratocystis*. Unlike *Ceratocystis* it required protein, could utilize fat as the sole source of carbon, and produced an oily stain and sweet-smelling esters. Mathiesen-Käärik regarded it as a true ambrosia fungus belonging to a group not closely related to *Ceratocystis* and named it *Monilia ferruginea*. Francke-Grosmann's findings (1956a), discussed in the next section, that this fungus and others of similar characteristics are transmitted by the ambrosia beetles in specialized organs and are dependent upon secretions of the insects for maintenance of their growth, was of fundamental significance. She found that the ambrosia fungus of *Trypodendron lineatum*, which she confirmed as Mathiesen-Käärik's *Monilia ferru-*

ginea, and that of *T. domesticum* did not readily germinate on malt agar, but would do so on casein or peptone agar or in pure olive oil to give an ambrosial form of growth. When these fungi were grown on malt agar they produced only a sterile aerial mycelium, but if olive oil was poured over this aerial growth the formation of ambrosial sporodochia with monilioid chains of spores was induced. Staining of the medium was caused by dark hyphae but more by an oily exudate. Later, Francke-Grosmann (1958) compared the ambrosia fungi of several species of beetles in the Scolytid tribes Trypodendrini and Xyleborini (Table 1), finding great similarities in the fungi within each group.

The experimental approach of rearing ambrosia beetles on their own ambrosia fungi in artificial culture might at first sight appear promising, but such work has in fact yielded conflicting results. Only two successful attempts have been reported. Gadd (1947) reared *Xyleborus fornicatus* from egg to adult on its associated *Monacrosporium ambrosium*, a species which fits in with Francke-Grosmann's concept of a specialized ambrosia fungus. The other successful result was that of Buchanan (1941) who raised *Xylosandrus germanus* through to adults not on the specialized ambrosia fungus but on cultures of *Ceratocystis ulmi*, *C. piceae* and species of *Pestalotia*. This unexpected result might indicate that despite the undoubtedly complex character of the ambrosial relationship in nature, the insect is not rigidly adapted to utilize only its specific ambrosia for food but can in fact assimilate a wide range of fungi. Mycetophagous insects in general are able to feed on sundry species of fungi and ambrosia beetles may perhaps behave as unspecialized mycetophagous insects under *in vitro* conditions.

The experiments so far reported, however, are open to the criticism that the larvae were placed on fungus growing on an artificial medium which may itself be nutritious to the larvae.

Transmission of ambrosia fungi

Scolytid ambrosia beetles. The demonstration by Francke-Grosmann (1956a) that those Scolytid ambrosia beetles which hibernate maintain their ambrosia in yeast-like form throughout the overwintering period in specialized integumental organs where oily secretions accumulate, ended the speculation which had surrounded the problem since the group had first been studied (Neger, 1911; Schneider-Orelli, 1913; Hadorn, 1933; Fisher *et al.* 1953–54). The localization of the organs, which are found only in the female, varies from species to species, but essentially they consist of depressions or of flask-like invaginations of the body surface in which secretions of oil accumulate. The oil is produced either from specialized hypodermal cells which line the invaginated integument, or by glandular hairs. Francke-Grosmann (1956 a, b) suggests that in ancestral species the glands had the primary function of secreting oil for lubrication of the insect during boring, and as a water repellent protecting it against excessive sap flow. The presence of these organs then allowed the development of a symbiosis with fungi of the sap-stain type in wood. The fungi which were favoured by natural selection were those which utilized fats and which were able to colonize the oil glands and grow there in oidial or yeast-like form. After a long association with the ambrosia beetles the fungi have become highly specialized symbiotic organisms adapted to their environment, losing their resemblance to the normal blue-staining fungi.

In order to bring fungi into culture from the transmission organs, Francke-Grosmann (1956a) used an elegant, simple technique which she called 'fractional sterilization'. Hibernating or flying beetles were caged on moist filter-paper for 48 hr., then on oven-dried paper for a further 48 hr.; this cycle was repeated three times. Chance moulds, and even the regularly associated blue-stain organisms, germinated in the moist phase and died in the dry phase, but the symbiotic ambrosia fungi, protected within the oil, were unaffected. Treated beetles were broken up and implanted on gelatin slides where the germination of the overwintering oidia and subsequent growth could be observed.

Francke-Grosmann believes the sowing of ambrosia fungus in new tunnels is achieved during active boring, when the output of secretion from the glands is increased, and slowly washes out the oidial cells which then germinate on the tunnel wall. Their growth is assisted by the oil smeared on to the wall during this process. She demonstrated this with *Trypodendron lineatum*. Normally the fungi remain in the transmission flasks throughout the winter, but if boring was induced in the laboratory during this period, the flasks were slowly emptied during a period of 10 days; other individuals which were allowed to remain quiescent, retained their fungi within the flasks. Farris & Chapman (1957) observed independently that the fungus of this species was deposited at a number of scattered points in the tunnel during its construction.

Francke-Grosmann (1958), after investigating a further series of Scolytid transmission organs, showed that they could be arranged in order of increasing complexity. In three European species of *Xyleborus* the spores are carried between tufts of glandular hairs on the abdominal tergites under the elytra. In the series *Xylosandrus germanus*, *Anisandrus dispar* and *Eccoptopterus sexspinosus*, progressive modification of the intersegmental membrane between pro- and mesonotum occurs, forming an invaginated glandular sac under the pronotum. In *Eccoptop-terus* the invagination is rolled into a scroll. The three species of *Trypo-dendron* have paired ∩-shaped tubes lying deep within the body and opening at the bases of the prothoracic coxae. The tubes of *Trypodendron* are unlike the transmission organs of other Scolytids, and Francke-Grosmann (1956a) believes they have been derived from ventral lubricating glands associated with a specific pattern of tunnelling behaviour.

The comparative morphology of the transmission organs may, Francke-Grossmann (1958) suggests, form a basis for reconsideration of the taxonomy of the group. However, later investigations have revealed complex organs in several tropical species of *Xyleborus*. Fernando (1959) describes paired sacs in the head of the female of *X. fornicatus* which lie alongside the brain and open into the inside of the mouth. *X. mascarensis* has similar paired pockets behind the labrum and mandibles (Francke-Grosmann & Schedl, 1960). The coffee shoot borer, *Xylosandrus* (*Xyleborus*) *morstatti*, has a single invaginated flask-like organ which lies under the pronotum and strongly resembles the arrangement in *Anisandrus dispar* (Lhoste & Roche, 1959). It seems likely that further investigation will reveal such organs in many more species and it is too soon yet to assess their significance as taxonomic criteria. The wide variation in their site and structure seems to point to the polyphyletic origin of the Scolytid ambrosia beetles. The discovery of complex transmission organs in tropical Scolytids which do not over-winter must lead to a revision of Francke-Grosmann's (1956a) state-ment that only species which overwinter possess these organs—it may be that even with species which spend a short time out of the log, the fungi require a phase of growth in the beetle's secretion.

Blastophagus minor and *Ips acuminatus*, transitional between the phloeophagous and the ambrosial bark beetles, carry their associated fungi in a groove running the length of the inner edge of the left elytron and on the metathorax under the elytra. The spores survive the fractional sterilization technique and Francke-Grosmann (1956a) is inclined to regard these insects as 'ambrosia beetles *sensu lato*'.

Platypodid ambrosia beetles. Whereas the mode of transmission of the fungi in the Scolytidae was unknown before the work of Francke-Grosmann (1956a) there were several instances where the external carriage of ambrosial spore masses by the Platypodidae had been demonstrated. Strohmeyer (1911) found them on *Crossotarsus* at the base of hairs surrounding a depression on the front of the head, and in other species they were packed around the maxillary bristles. Similar spore masses have recently been found among frontal hairs of *Doliopygus dubius*, in both sexes, but especially in the female (Browne, 1961 *b*). Beeson (1917) reported that *Diapus furtivus* had large prothoracic pores which contained fat to which spores of ambrosia adhered.

Francke-Grosmann (1956 *a*) did not find specialized transmission organs in *Platypus cylindrus* or *Doliopygus serratus*—in both species clumps of spores were carried on various parts of the surface. The spores were, however, as in the Scolytids, always associated with fatty or oily secretions either from glandular hairs or minute tubules which open on to the surface. Both sexes carry spores but the glandular secretions are more evident in the female. Francke-Grosmann suggests that the significance of the oil is the same as in the Scolytids: it serves for nutrition of the fungus and its maintenance in pure culture, not merely, as Beeson (1917) had supposed, for mechanical adhesion of spores.

Recent communications (Roche & Lhoste, 1960; Lhoste & Roche, 1961) have drawn attention to specialized transmission organs in two species of *Periomatus* and in *Platypus hintzi* but do not state whether they are present in both sexes. The organs in *Periomatus* are simple elongated pits in the surface of the prothorax, containing large chlamydospores of the ambrosia fungus. In *Platypus hintzi* there are paired 'utricules'—spherical pits, 0·3 mm. in diameter, opening to the exterior by small pores on the exposed surface of the pronotum, in a region of thick chitin. Specialized hypodermal glands drain into the pits. These organs differ from the pockets of the Scolytidae which are derived from invaginated tegumental folds of thin chitin. No differentiated transmission organs could be detected in three species of *Doliopygus* which Lhoste & Roche (1961) examined.

It appears therefore that the transmission organs of the Platypodidae are as variable as those of the Scolytidae, but are in general less highly organized.

Blue-stain fungi and yeasts associated with ambrosia beetles

The regularity with which blue-stain species have been found in ambrosia beetle tunnels leaves no doubt that they have a close association with the beetles even though the weight of evidence is against their being primary ambrosia fungi. *Ceratocystis* survives the winter on hibernating ambrosia beetles, growing in yeast-like form in oily secretions on the body surface, but not colonizing the transmission organs. This was reported by Francke-Grosmann (1959) who also found that certain Hyphomycetes and yeasts were carried in the same way and all shared the ability to grow in a yeast phase. Blue-stain organisms can also be transmitted by direct mechanical infection by beetles which do not have an overwintering period.

Perithecia of *Ceratocystis* are particularly evident in older or abandoned tunnels (Doane & Gilliland, 1929; Wilson, 1959). Mathiesen-Käärik (1960b) connects this observation with the nutritional requirements of the fungus; perithecia are formed when available nitrogen is low, a condition which might well be encountered in exhausted galleries.

The literature on yeasts associated with the Scolytoidea is no less extensive than that on blue-stain fungi. It has been fully reviewed by Mrak & Phaff (1948), Shifrine & Phaff (1956), and recently by Callaham & Shifrine (1960), and it is necessary here only to touch on the subject. Almost all bark beetles appear to carry yeasts, and the association between each beetle species and its associated yeast species, usually 1 to 3 in number, is constant (Grosmann, 1930). After introduction into the wood the yeasts grow predominantly in the phloem, and are digested when these cells are eaten. There is also evidence that insect-attractants are produced when yeasts grow in the wood, leading to further bark beetle attack. The commonest yeasts are of the sporogenous genera *Endomycopsis* Dekker, *Hansenula* Syd., *Pichia* Hansen, and species of the anascosporogenous *Candida*. All are of the non-fermenting or weakly fermenting type.

Yeasts are also commonly found with ambrosia beetles. Many of the earlier workers mentioned their presence but considered them of little importance. Webb (1945), however, describing two unnamed species of *Endomycopsis* from tunnels of *Platypus subgranosus*, suggested that they might accelerate the growth of the ambrosia fungus, and Hogan (1948), working with the same species of beetle, thought the yeasts formed part of its food. Wilson (1959) isolated a species of *Pichia* from the tunnels of the Scolytid ambrosia beetle *Corthylus columbianus* in American white oak. This *Pichia* grew mainly in parenchyma cells and there was evidence that it caused a breakdown of tannins in the wood, rendering it more suitable for the growth of other fungi. (Decomposition of pyrogallic tannins by *Pichia* in tanning liquids has been reported by Boidin & Abadie (1954).)

Feeding of ambrosia beetles

Ambrosia fungus growing in the tunnels is constantly browsed by the insects and undoubtedly forms the chief part of their food, but the extent to which the diet is supplemented by wood is uncertain. Wood fragments are always present in the gut of feeding adults, and usually of feeding larvae. Browne (1961b) reports that the first two larval stages of

Doliopygus dubius (Platypodidae) do not eat wood but that from the third instar an increasing amount is eaten. In *Trypodendron lineatum* where the larva lives in the niche, all the wood excavated in enlarging the niche is swallowed and passed through the gut. The larger part of the food, however, consists of the fungal palisade which lines a portion of the wall of the niche and which is constantly renewed as it is browsed (Hadorn, 1933).

It is uncertain whether swallowed wood is digested. Cachan (1957) showed that the starch content of wood which had passed through the gut was diminished. The bark beetle *Phloeosinus bicolor* digests not only the contents of the wood cells but also hemicelluloses A and B of the cell wall (Parkin, 1940). In later experiments (Parkin, unpublished) the digestive enzymes of the ambrosia beetle *Platypus cylindrus* were found to be similar to those of this bark beetle indicating that hemicellulose digestion may occur in ambrosia beetles.

After pupation, teneral adults feed for a time on the ambrosia fungus, then cease feeding and void their digestive tracts (Hadorn, 1933; Chapman, 1958; Roberts, 1961). After arrival at the new log they do not eat until the ambrosia fungus has established itself, and autolysis of the flight muscles occurs during the period when they are boring without feeding. As soon as the fungus is established in the tunnel, the beetle resumes feeding.

INVESTIGATIONS ON THE OAK PINHOLE BORER, *PLATYPUS CYLINDRUS*

Various aspects of the biology of *Platypus cylindrus* have been described by Groschke (1954), Husson (1955) and Baker (1956) but little is known about its ambrosia fungi. The account given here is based on observations made between 1955 and 1958 in south-east England.

(1) *Life history*

Beetles were found on the wing on warm days from June to September, and this can be regarded as the functional emergence period. Although it was shown that sporadic emergence of mature beetles occurs throughout the remainder of the year, such beetles do not survive.

The beetles appear to be attracted to suitable oak logs and stumps which are in a fresh 'green' condition, but the mechanisms underlying attraction are little understood (Bletchly, 1961). After alighting upon the bark the behaviour of the two sexes differs. As usual among the Platypodidae (Hogan, 1948; Jover, 1952; Browne, 1961*b*), the male beetle initiates the boring and excavates the first few centimetres of the tunnel. The male will not accept a female until the tunnel is long enough to accommodate both beetles. A female beetle after alighting moves randomly over the surface until within 5–10 cm. of the entrance shaft of a male, whereupon its movements become directed and it approaches and enters the tunnel. Both insects then momentarily emerge from the tunnel for copulation which follows the elaborate ritual described by Jover (1952).

The female then re-enters the tunnel first, resuming the boring begun by the male who takes no further part in the construction of the gallery, but remains in the proximal part of the tunnel and ejects bore-dust to the outside of the log. The tunnel runs radially inwards until it approaches the boundary of the sapwood and heartwood, then turns and follows a sinuous course roughly following this boundary (Pl. 1, fig. 1). At this stage, when the tunnel is 10–15 cm. long, the first eggs are laid. The female beetle later constructs branch tunnels in the heartwood.

If a gallery is started in June, July or the first half of August, the first clutch of eggs is laid about 4 weeks after entry of the female, but in tunnels started in late August or September, eggs appear either 4–8 weeks later or not until the following spring. They hatch after 2–6 weeks into flattened ovoid larvae (described by Strohmeyer, 1906). The number of larval instars was not determined with certainty, but was either 4 or 5, and larvae moulted to the final instar after about 4 weeks. This is a stage of long duration during which larvae increase their weight by a factor of up to 5 without further increase in the size of the head capsule. Some of the larvae usually pupate in the spring following attack, and in summer emerge as mature adults through the entrance made by the parents. However, when the first eggs are laid late in the year of attack, or during the following spring, the emergence of the first new adults may be delayed until the second summer after attack.

Egg-laying continues at irregular intervals all the year round throughout the 2–3-year life of the female, spermatozoa from the single copulation being stored in the chitinous spermatheca. As the gallery is extended, the number of developing oocytes in the ovarioles, and hence the number of eggs in each clutch, increases. Each of the four ovarioles contains 1–4 developing oocytes in the first months of egg-laying, 6–8 in the second year when the gallery system is longer, and up to 15 in the more extensive systems. A similar correlation between length of gallery and egg-laying potential of the parent female has been reported by Roberts (1961) for *Trachyostus ghanaensis* and Browne (1961*b*) for *Doliopygus dubius*. It appears that a balance is maintained between the available tunnel space and hence food supply, and the population of developing larvae.

Larvae in the first two instars have feebly chitinized mouthparts and the mandibles act merely as scrapers. In older larvae and particularly those in the last instar, these organs are heavily sclerotized and the apposed edges of the mandibles are gouge-like. Larvae were observed to bore only in the last instar, making tunnels indistinguishable from those of the beetle. The amount of tunnelling varies, but is slower than that of the adults. In addition to the elongation of the tunnel system, each last-instar larva constructs its own pupal cradle—a short pocket at right angles to the main gallery. The newly hatched adults feed on fungus but do not bore wood before emerging. All parts of the tunnel are kept free from loose bore dust which is expelled through the tunnel entrance by the male if present, but by the female when the male has been lost or experimentally removed. Removal of the male does not cause the disorganization of the colony which Cachan (1957) reported in *Platyscapulus auricomus*.

The gut of *Platypus cylindrus* is similar to that of Scolytids such as *Trypodendron lineatum* or *Xylosandrus* (*Xyleborus*) *morstatti* which have been described in detail by Hadorn (1933) and Lhoste & Roche (1959) respectively. In the adult, the capacious thin-walled crop leads into a narrow gizzard (proventriculus) lined with eight rows of longitudinal chitinous plates. The midgut is long, consisting of an anterior dilatable ventriculus and a narrow posterior portion leading in turn into a distensible hindgut (colon and rectum). There are no diverticula. The larval gut differs in that no gizzard is present, and the anterior midgut is relatively longer than that of the adult and is capable of even greater distension. The adult gut, therefore, is adapted for cutting and grinding (comminuting) relatively large particles, whereas that of the larva is not.

(2) *The associated fungi of* Platypus cylindrus

Platypus cylindrus is a typical ambrosia beetle in its feeding habits. Fungi which it introduces into its tunnel produce the usual 'ambrosial' lining on which larvae and adults feed. Isolations were made from wood chips, ambrosial lining and, after fractional sterilization (Francke-Grosmann, 1956a), from beetles emerged from tunnels.

Two per cent malt agar proved suitable for all except single chlamydospore isolations, for which clarified Czapek + 0·3 % yeast-extract agar was used.

Cultures were maintained on 2 % malt and on malt, yeast-extract, glucose, peptone agar (MYGPA). A prune-juice + yeast-extract agar

(PYA) was used to induce chlamydospore formation and morphology was studied on Duncan slide cultures.

For study of the growth of fungi on oak, sterile heartwood blocks, prepared from a freshly felled tree, were surface flamed, and placed on cotton-wool at the bottom of boiling tubes. Twenty ml. of distilled water or of 2 % peptone solution was poured aseptically over each block and allowed to soak into the cotton-wool. Contamination-free blocks were inoculated after 1 week.

Auxanographic growth tests were carried out by the method of Lodder & Kreger-van Rij (1952).

(i) Sporothrix

From tunnels. The fungus which appears to be the predominant member of the tunnel flora was first recognized with certainty in artificial culture, when the germination of single chlamydospores was observed on clarified Czapek + yeast-extract agar. Germination occurred at 24° in 10–20 days after implantation but only 10 % germinated. When scrapings of tunnel lining were used as inocula, this fungus never appeared as a sole isolate, and indeed its presence in the mixed cultures resulting from these inoculations was not detected during the earlier stages of this work. After the single chlamydospore isolations had indicated the characteristics of the *Sporothrix*, it was subsequently demonstrated by repeated subculturing from the mixed cultures of tunnel scrapings. Frequently, however, the vigorous and much faster growth of the staining fungi also associated with the beetle prevented its isolation, particularly when subculturing from implanted wood chips; on only three occasions was it recovered from surface sterilized wood chips which had constituted part of the tunnel wall, and never from chips taken even a short distance from the tunnel.

From insects. Culturing from the beetles presented the same problems as culturing from pieces of tunnel scrapings. Although the fractional sterilization period did result in almost complete elimination of moulds, it did not kill staining fungi which grew and swamped *Sporothrix* before it could germinate. Pure cultures of *Sporothrix* were obtained only from the elytra of three females and of one male. However, microscopic examination of scrapings from the body surface of males and females frequently revealed chlamydospores on the head (among hairs on the flat frontal region and around the mouthparts), the thorax (in the folds between meso- and metanotum, and among the long sternal hairs of the coxae) and the elytra (on the hairs of the declivity and on the undersurface of the elytra).

Chlamydospores were also seen in preparations of rectal contents of half of the beetles which were microscopically examined, occurring in both sexes. *Sporothrix* could not be cultured from the gut; the streaks made from rectal contents developed profuse growths of *Candida* and bacteria; those from fore- and midgut were sterile or produced sparse bacterial colonies.

The fungus in culture. The fungus grew slowly on all agar media used, the mean radial growth on 2 % malt agar being 0·2 cm./24 hr. at 24°. Cultures, particularly those on MYGPA, gave off a sour, fruity odour. The most satisfactory medium for morphological studies was PYA, on which spore formation including chlamydospores was most vigorous. The appearance of an early culture strongly resembles that of a yeast, being white, sodden and appressed but mounded in the centre. The mycelium mass is rubbery and extremely tenacious and can be peeled from the agar as a single pellicle. In later cultures the colour darkens in the centre, sometimes in concentric rings, through brown and olivaceous-brown to black. The hyphae are branched, septate and when young are hyaline. Old hyphae are dark, and thick-walled with short cells, which become almost isodiametric and packed together into a stroma. The staining in culture is due to the dark hyphae, not to exudates in the medium. Conidia are always hyaline, forming typical *Sporothrix* heads; successive conidia have slightly different points of origin on the conidiophore. In young cultures on PYA, conidia are also produced in succession from a common growing point as figured by Carrión & Silva (1955) for *Sporotrichum gougerotii*. Conidia are also borne as in *Pullularia*—laterally on conical prominences on the sides of a fertile hypha. The conidia are apiculate or triangular in outline, particularly when packed in heads, and are of the wet type readily sliming down to form spore masses.

Chlamydospores which resemble those of *Sporotrichum schenckii* are formed terminally on differentiated hyphae 3–4 times the diameter of those that bear conidia. They tend to form short 'monilial' chains but are readily dislodged. Chlamydospores are round to pear-shaped, 10–16 μ in diameter including a hyaline, refractile wall 2–3 μ thick (Pl. 1, fig. 2). Sterile hyphae with dense cytoplasmic contents grow abundantly among the conidiophores, and on malt agar are packed together to form a thick felt. Yeast-like or oidial forms frequently occur, especially in young cultures.

The fungus grew on sterile oak blocks only when additional peptone was present. The wood eventually became black-stained on the surface where the cells were packed with dark hyphae.

Identification of the fungus. This strain has strong morphological similarities with *Sporotrichum schenckii* and *S. gougerotii* as figured by Dodge (1935) and Carrión & Silva (1955). The latter authors proposed placing *S. gougerotii* in the Dematiaceae under *Cladosporium* since it produces dark-coloured hyphae and its spores arise from a common meristematic point. Although the *Platypus* ambrosia fungus darkens in old culture, conidiophores and conidia are always hyaline and there seem good grounds for retaining it in the Moniliaceae. Spores are occasionally budded successively from a common meristematic point, but are more typically found as separately developing individuals at slightly different sites of origin. Smith & Elphick of the Common-wealth Mycological Institute suggest (personal communication) that the fungus be regarded as a new species falling into the genus *Sporothrix* Hektoen & Perkins, which includes the fungus commonly known as *Sporotrichum schenckii* (Hektoen & Perkins) Matruchot. The taxonomy of *Sporothrix* and *Sporotrichum* Link ex Fr. is confused and awaits monographic treatment and it would be unwise at this stage to speculate further on the nomenclature of this ambrosia fungus.

(ii) Endomycopsis *sp.*

From tunnels. This yeast was isolated from more than half of the scrape inoculations, from those spore suspension streaks which were not overgrown by moulds and from wood chips which either included some tunnel wall or were immediately adjacent to it, but never from wood remote from the tunnels. It was present in very new tunnels, where the male alone was boring, and until the time that the first batch of eggs was laid it was the most frequently isolated fungus.

From insects. The sites on the body which carried chlamydospores of *Sporothrix* also yielded cultures of *Endomycopsis*, but the yeast was not seen in nor cultured from the gut.

Growth in culture. The yeast grows well on 2 % malt or MYGPA producing a creamy-white to buff colony of a tough rubbery consistency. Mycelium and pseudomycelium are abundant, with blastospores pro-duced terminally and laterally, and budding vigorously after abstriction from the mycelium. Ovate asci containing four hat-shaped ascospores are borne terminally on thicker hyphae (Pl. 1, fig. 3). Cultures have an odour of fermenting apples.

Growth was good on sterilized oak heartwood blocks with or without added peptone; asci and blastospores were formed on and just below the surface of the wood. A slight brown stain was imparted to the wood after 2 months. The yeast was recovered from the blocks by implanting surface slivers of wood in agar.

A series of carbon and nitrogen assimilation auxanographs were prepared from different isolates of this fungus, all of which gave the same results. Glucose, maltose, sucrose and ethanol were assimilated as sole carbon sources; galactose and lactose were not. Splitting of arbutin took place. Interpretation of the nitrogen assimilation tests was difficult. The growth on the basal nitrogen-free medium was so strong that it was not possible to demonstrate any difference in growth when $(NH_4)_2SO_4$ or KNO_3 were added. Despite washing of the agar, and starvation and washing of the inoculating yeast (Barnett & Ingram, 1955), this growth in the basal medium was not eliminated, but it was possible to show that uric acid was assimilated as a source of nitrogen.

Identification of the yeast. Dr Kreger-van Rij of the Yeast Division, Centraalbureau voor Schimmelcultures, Delft, demonstrated feeble fermentation of glucose, assimilation of KNO_3, and lack of growth in the absence of vitamins. She identified the yeast as a new species of *Endomycopsis* bearing some resemblance to *E. bispora* (Beck) Dekker. In Wickerham's (1951) classification it would be regarded as a species of *Hansenula* since it can assimilate KNO_3.

(iii) Candida *sp.*

Isolations from tunnels. This yeast was isolated from scrapings, and from wood chips which included or were adjacent to the tunnel wall. It was not found in very new tunnels, but in gallery systems in their second year or later it was more common than *Endomycopsis*. Often the two yeasts were cultured from the tunnels together, and could be separated only by single spore subculture.

Isolations from beetles. The yeast was taken into culture from the same situations on the body as was *Endomycopsis* but was not found so frequently. Unlike *Endomycopsis* it was cultured from the rectal contents of about half of the beetles examined.

Growth in culture. On 2 % malt agar this species forms typical soft yeasty colonies, dirty-white to cream in colour. Abundant pseudo-mycelium is formed, with blastospores developing in short chains or in small wreaths.

Glucose, sucrose, maltose and ethanol but not galactose and lactose were assimilated and arbutin was hydrolysed. Nitrate and uric acid were not assimilated.

Dr Kreger-van Rij confirmed these assimilation results and showed that glucose and sucrose were fermented, and that vitamins were required for growth in synthetic media. Small hat-shaped spores were only observed on one occasion.

Identification of the fungus. Had the production of ascospores been confirmed the yeast would have been assigned to the genus *Pichia*, *sensu* Phaff, but in their absence Dr Kreger-van Rij preferred to regard it as a new species of *Candida*.

(iv) Cephalosporium *sp.*

Isolations from tunnels. This fungus was more frequently isolated and more readily cultured than the three so far described. It was obtained from tunnel scrapings and from wood chips, including chips from the slightly brown-stained wood which extends for 5–10 mm. along the grain from older tunnels. It was never isolated from wood remote from the galleries. *Cephalosporium* frequently occurred in mixed cultures with the other three ambrosia fungi and with *Ceratocystis* or its conidial stages, but could be isolated by subculture. Very early tunnels did not yield *Cephalosporium*; it occurred from about 4 weeks after initiation of attack and was then present throughout the life of the colony.

Isolations from beetles. The fungus was carried superficially on most hairy parts of the integument, and was isolated from nearly all beetles cultured, but not from the gut.

Growth in culture. The strain grows rapidly on 2 % malt agar, increasing its radius by 1·5 cm./24 hr. at 24° and producing a white flocculent aerial mycelium which becomes sodden as the culture ages, and turns pale-brown, later dark-brown. Browning occurs more rapidly with increasing temperature and is delayed indefinitely at 18° or below. Unlike the darkening of the *Sporothrix* which is due entirely to the hyphae, darkening of *Cephalosporium* is due to staining of the medium as well as of old hyphae.

Conidia are borne in 'Cephalosporium' heads but readily slime down to produce sticky masses. They are very variable in size, and bud after abstriction, often forming oidial groups. In older cultures coremial strands and yeasty mounds are often found.

On oak blocks the strain grows sparsely in the absence of peptone, vigorously in its presence, producing a brown stain apparently caused by hyphae, not by staining of the wood elements.

Identification of the fungus. Although this fungus is similar in many ways to an early stage of *Leptographium lundbergii* Lagerb. & Melin as described by Webb (1945), comparison with authentic cultures of *L. lundbergii* showed large differences. As the *Cephalosporium* type of conidiophore is maintained throughout all stages of culture, this fungus must be allocated to this admittedly unsatisfactory genus.

(v) *Other fungi*

Other species of staining fungi and moulds were often isolated, especially from spore suspensions. The staining fungi were invariably present in older galleries and in the entrance shaft, and were occasionally found when streaks were made with suspensions of tunnel scrapings. They were isolated from wood chips near the tunnel and also from sapwood remote from tunnels, but not from remote heartwood. Fractional sterilization of beetles did not eliminate staining fungi which were isolated from 20 % of beetles examined, particularly from the hairs of the elytral declivity, but not from the gut.

The staining fungi isolated were identified by Savory (personal communication) as *Ceratocystis piceae*, *C. pluriannulata* and *C. pilifera*. In addition, a *Graphium* which could not be induced to form perithecia was regularly found. '*Cephalosporium*' occurred as a conidial stage of *Ceratocystis* but was quite distinct from the 'ambrosial' *Cephalosporium*.

Other fungi which at times were isolated from the tunnels were: *Coryne sarcoides*, several species of *Penicillium*, *Paecilomyces varioti*, *Fusarium* sp., *Stysanus* sp., *Botrytis* sp., *Trichoderma viride*, species of Mucorales and various wood-rotting Basidiomycetes. *Penicillium*, *Paecilomyces* and *T. viride* were found only in tunnels where the colony is declining, and appear incompatible with healthy tunnels. Basidiomycetes are rarely found in the tunnels even when present in wood quite close by, and it appears that some antagonism exists between the tunnel flora and wood-rotting fungi.

Growth of the fungi in the tunnels

Despite the regularity with which *Sporothrix* sp., *Endomycopsis* sp., *Candida* sp. and *Cephalosporium* sp. were isolated, their identity with the organisms comprising the ambrosial layer can be accepted only if their growth in culture can be related to that in the tunnels.

The fungi in the tunnels were examined in thin wood sections. Prior to sectioning, the oak blocks were vacuum-impregnated with lactophenol and left in this medium for 48 hr. Sections were stained in 2 % cotton blue in lactophenol, with gentle heating, then mounted in clear lactophenol.

In new tunnels, corresponding with the results of isolations, *Endomycopsis* was the first fungus to become recognizably established. Its hyphae ramified through the wood for 2–3 mm. around the tunnel, particularly in the ray and parenchyma cells. Asci were formed both within wood elements and projecting into the tunnel. Mycelial growth was more pronounced than in culture, and conidia were abundant in the tunnels. The pseudomycelium so evident in culture was not encountered. This early growth of *Endomycopsis* is very sparse compared with the later growth of *Sporothrix* and could easily be overlooked if stained preparations were not made.

It has already been mentioned that loose borings are ejected from the tunnel; the greater part of the wood which is excavated is dealt with in this way. A small proportion is, however, swallowed by the adults and these pieces of wood were always seen on examination of the crops of actively boring beetles. During their passage through the gizzard the wood-pieces are comminuted and examination of the contents of the midgut showed that a few small cells, particularly the parenchyma remain entire, but fibres, vessels and tracheids are shattered. The comminuted wood after passage through the gut is passed out in a string of moist faeces which are smeared quite evenly over the walls of the tunnel by the movement of the insects.

About 1 month after establishment of the gallery, normally coinciding with the time of first egg-laying, the growth of *Sporothrix* first becomes evident. The layer of finely divided wood is hardly noticeable at this stage, but germinating chlamydospores can be found within it. Hyphae quickly spread through the wood around the tunnel, particularly the ray and parenchyma cells. Sterile hyphae of the dense-cytoplasmic type seen in culture form a palisade projecting into the lumen of the tunnels; later, conidia are formed on thinner conidiophores interspersed among the sterile hyphae. Conidia are apiculate but not as markedly triangular in outline as those in culture. Chlamydospores are produced terminally on thicker conidiophores and occasionally form short monilioid chains. This growth of *Sporothrix* does not replace the *Endomycopsis*; the two species coexist in the palisade layer and in the surrounding wood cells and asci of *Endomycopsis* are seen in considerable numbers throughout the life of the tunnel. The greater part of the growth, however, is *Sporothrix*. In the primary tunnel the fully developed palisade layer occurs only transiently in patches here and there (Pl. 2, fig. 4), since it is browsed as it forms, and all hyphae and spores which project into the tunnel lumen are eaten. The browsing exposes the wood elements of the tunnel wall which are packed with hyphae. Regeneration then occurs with fresh hyphae ramifying through the new layer of comminuted wood which builds up on the tunnel wall. Conidia and chlamydospores of *Sporothrix* are produced throughout the layer and in places asci of *Endomycopsis* can be seen, particularly near parenchyma. The regular palisade structure is lacking. This irregular type of lining appears to be quite suitable for food; it is the most commonly occurring type and is extensively browsed, both fungus and wood particles being ingested. It is quickly renewed as the wood particles are replaced by others from the faeces of the insects, and the fungi continually ramify through this new layer. Mature chlamydospores pass intact through the gut of adults and larvae. The whole growth is sodden, and all spores are of the slimy type.

As this process of browsing and renewal proceeds, the basal layer of the lining, where old hyphae project from the wood, takes on the form of a stroma of cuboidal cells resembling that seen in the old cultures of *Sporothrix*. The stroma is intimately bound up with comminuted wood particles. It is from this basal layer that the continuously renewed growth into the lumen arises.

Chlamydospores are produced abundantly in some parts of the tunnel and sparsely in others (Pl. 2, fig. 5). In browsed or regenerated regions, germinating detached chlamydospores are seen, sometimes in great numbers (Pl. 2, fig. 6); these may have passed through the gut or may have been carried mechanically by moving insects. Germinating chlamydospores were observed in faeces collected from insects removed from the tunnel. Their part in the economy of the tunnel is not clear, but their germination and growth in browsed areas may help to regenerate the lining. It seems more likely that chlamydospores inoculate newly constructed galleries, and are of importance in the transmission of the fungus by emerging beetles.

Once an initial palisade layer has been browsed, subsequent growth is usually irregular, but occasional regeneration from the stroma occurs in sporodochium-like patches especially when active boring is not in progress. The palisade structure is at its most flourishing in branch tunnels penetrating into the heartwood. This type of lining could not be correlated with a particular stage of the insect, but it probably provides a more concentrated food source than the irregular lining, and may have some significance in the nutrition of young larvae or of teneral adults.

Among the growth in the lumen of the tunnel, spores of *Cephalosporium* can be recognized, particularly when they are budding. Hyphae of this species cannot be distinguished among the profusion of mycelium of *Sporothrix* and *Endomycopsis*, but can be recognized in wood elements beyond the zone of densely packed *Sporothrix* hyphae. This accords with the observation that *Cephalosporium* could be cultured from the faintly stained wood several millimetres away from a tunnel.

A yeast with blastospores like *Candida* is also recognizable in the irregular type of lining, and more noticeably in parenchyma cells opening into or immediately adjacent to the tunnel. No pseudomycelium formation was seen.

The appearance of the entrance shaft of the tunnel, which lies in the sapwood, is quite different. Neither the palisade nor the irregular type of lining occurs although the presence of a stroma of *Sporothrix* is discernible in the deeper reaches of the shaft. The predominant fungi are *Ceratocystis* spp. and their conidial forms. *Graphium* heads are frequent in the shafts of younger tunnels while perithecia are common in old shafts. *Coryne sarcoides* sometimes causes diffuse stain of the sapwood and sporodochia of this species may project into the entrance shaft. No growth of dry-spored fungi was seen in the shafts of younger tunnels, but in sapwood portions of abandoned tunnels, spore-bearing heads of *Penicillium* were often present.

Gut contents of different stages of Platypus

Larvae. The gut contents of all stages of larvae were similar. Comminuted wood but not larger wood pieces were present throughout the length of the gut. There was no visual evidence of breakdown of the cell-wall material. In the crop, abundant fungal material was present, among which could be recognized hyphae, conidia, oidia and chlamydospores of *Sporotrichum*, *Cephalosporium* spores, and ascospores which resembled those of *Endomycopsis*. Occasional crescentic, and larger hat-shaped spores which were probably from *Ceratocystis* spp. were also distinguishable. Many of the wood fragments were packed with dark-brown or olive-green hyphae and others were filled with tight masses of yeast cells. The hindgut contents included wood particles in the same condition as those in the crop but most of the contained yeasts had been digested. All young hyphae had been digested but older, dark hyphae were apparently unaffected. Spores had disappeared, with the exception of *Sporothrix* chlamydospores, most of which passed through the gut unaltered.

Teneral adults. When adult beetles hatch from the pupae in the pupal cradles they spend a variable period feeding in the parental tunnel. During this time they do not bore, and the gut contents examined were of the larval pattern.

Adults on the wing. The crop and midgut were devoid of food material but the crop was distended by an air-bubble. The hindgut contained varying quantities of brown material among which could be distinguished chlamydospores and, frequently, budding yeasts, but which occasionally appeared devoid of fungus.

Adults constructing tunnels. It has already been described how, during construction of the tunnel, the adults swallow some of the wood borings. These larger pieces of wood were very evident in the adult crop, giving the contents a quite different appearance from those of larval crops. The comminuted particles in the adult mid- and hindgut after passage through the gizzard were of the same size and appearance as those in the larval gut. Due to the action of the gizzard, pieces of dark mycelium in, and attached to the wood cells were chopped into very fine particles. The distribution of young mycelium and spores resembled that in the larva.

DISCUSSION

The ecology of the ambrosia fungi of some Scolytids, especially *Trypodendron* spp., has been extensively explored by Hadorn (1933), Leach *et al.* (1940), Mathiesen-Käärik (1953, 1960*a*, *b*) and Francke-Grosmann (1952, 1956*a*, *b*, 1958, 1959). With the exception of the account by Webb (1945) of *Platypus subgranosus*, information on the ambrosia fungi of Platypodids is fragmentary. It is therefore of interest to see how the pattern of the ambrosial association in *P. cylindrus* agrees with that in the Scolytidae. There is strong evidence that the *Sporothrix* obtained in culture is the principal ambrosia fungus of *P. cylindrus*. It was isolated from single chlamydospores as the main component of the tunnel flora, and its growth in culture corresponds morphologically with this component. This species may be analogous with the *Monilia* spp. of the Scolytidae. Both are difficult to bring into artificial culture, and are easily swamped in their early stages by adventitious moulds. Both produce a fruity odour and become intensely dark in culture. However, *Sporothrix* does not exude an oily stain as does *Monilia*, nor does it require such specialized high protein media. A striking convergence is seen in the habit of the fungi in nature. A stroma is formed from which a palisade of hyaline hyphae projects into the lumen. Other hyphae ramify away from the stroma, through the wood cells. The palisade of *Monilia* typically consists of chains of cells, which are usually referred to as spores, although they may not be true conidia (Hadorn, 1933). That of *Sporothrix* is composed of sterile hyphae, and of conidiophores bearing either true conidia or chlamydospores, but this palisade is frequently disorganized.

It appears that the *Sporothrix* of *Platypus* and the *Monilia* of Scolytids, though not perhaps closely related, occupy similar ecological positions in their respective habitats. *Monilia* is transported in highly differentiated organs in Scolytids and may in general be regarded as being more narrowly specialized in its adaptation to a symbiotic association than *Sporothrix*.

The evidence from *Platypus cylindrus* supports Francke-Grosmann's (1956*a*) suggestion that the *Leptographium lundbergii* which Webb (1945) claimed to be the principal ambrosia fungus of *P. subgranosus* was in fact an associated staining fungus. The *Cephalosporium* sp. of *P. cylindrus* has considerable resemblances to *Leptographium*, and as it is so readily isolated and brought into culture it could be mistaken for the major component of the tunnel flora. Webb made her isolations from wood chips around the tunnel—a situation which with *P. cylindrus* usually yielded *Cephalosporium* and not *Sporothrix*. It is suggested therefore that Webb's *Leptographium* and the *Cephalosporium* described here are ecologically comparable. They are staining fungi closely associated with the beetles but do not constitute the ambrosial palisades.

Endomycopsis has previously been found associated with other species of *Platypus* (Verrall, 1943; Webb, 1945), and the demonstration that in *P. cylindrus* this yeast is an early colonizer of the tunnel, and persists throughout the life of the palisade, indicates that yeasts may play a considerable part in the nutrition of the insects. The second yeast, *Candida* sp., is less obvious in the tunnel than *Endomycopsis* sp. but is eaten by the beetle particularly in particles of wood. It will be noted that these yeasts are of the same genera as those which Callaham & Shifrine (1960) reported to be regularly associated with bark beetles, and are likewise capable of only weak fermentation. Whether they supply essential B vitamins to the insect has not been demonstrated, but the yeasts themselves are not vitamin-autotrophic. Hopton & Woodbine (1960) found that seven species of *Hansenula* which they investigated had fat contents of up to 20 %. It would be interesting to investigate the fat metabolism of the *Endomycopsis* (*Hansenula*) of *P. cylindrus* to see whether this species is fat-producing, and if so whether this would stimulate the growth of the principal ambrosia fungus which by analogy with *Monilia* might be expected to assimilate fat.

The association between *Platypus cylindrus* and species of *Ceratocystis* appears to be of the type found in the Scolytid ambrosia beetles; the blue-stain fungi are transmitted by the insect and grow in the wood near the tunnels but constitute only an incidental part of the flora of inhabited branches. In old branches and particularly in the entrance shaft they may become the dominant species.

The observations on the carriage of ambrosia fungi on the body surface of *Platypus cylindrus* confirm the account of Francke-Grosmann (1956*a*) who did not, however, specify the nature of the spore masses she found among the glandular hairs of this species. However, the demonstration of *Candida* and of chlamydospores of *Sporothrix* in the rectal contents of some of the emerging beetles does not agree with her findings, nor with the later work of Browne (1961b) and Roberts (1961), who showed that the gut was entirely empty in emerging Platypodids, except for uric acid in the rectum. The presence of spores in the gut of *P. cylindrus* requires confirmation. It may be that old chlamydospores remain fortuitously in the rectum after the pre-emergence feeding and are of no significance as a transfer mechanism. However, *Candida* was cultured from the gut of some individuals and so can evidently be present in a viable condition.

The distension of the foregut of emerged adults by an air-bubble may have some significance. Chapman (1958) has shown that the foregut in *Trypodendron lineatum* becomes distended by a bubble only after flight, and recent work by Graham (1961) suggests that its presence acts as a 'trigger' for the release of the settling and attack phase of behaviour.

The presence of wood in the gut of tunnelling adults and larvae calls for comment and leads to speculation on whether the insects' nutritional requirements are met, in part, from this source. A small proportion of the borings produced by the tunnelling adult is ingested and physically disintegrated by the action of the gizzard. The resultant particles can be seen in the mid- and hindgut and in the faeces which are smeared by the movement of the beetle on to the walls of the tunnel. All the wood particles seen in the gut of the larvae are of this comminuted type and it appears that larvae ingest from the tunnel lining the particles which have been smeared there after passage through the adult gut. Last-stage larvae do not appear to swallow the wood which they remove in boring since their gut contents are identical with those of young non-boring larvae. *P. cylindrus* differs in this respect from another Platy-podid, *Doliopygus dubius*, the later stage larvae of which swallow their borings (Browne, 1961b).

In *Platypus cylindrus* this small quantity of wood which is recirculated through the alimentary canals of the insects seems unlikely to provide a significant part of the food requirements. It must be remembered that the greater part of the tunnel system runs through the heartwood which is composed largely of non-living tissue, low in digestible cell content material and assimilable nitrogen. There is no noticeable digestion of the shattered fragments of cell wall and though it is reasonable to assume that any cell contents present are digested, the amount of food obtained in this way must be small. The fragments may well, however, have the mechanical function of acting as a 'carrier' for the fungal material which is browsed by the insect.

It may, on the other hand, be conjectured that the wood fragments are of importance to the ambrosia fungi themselves. The ability to utilize uric acid was demonstrated in *Endomycopsis*; perhaps the wood fragments after passage through the gut serve to distribute the nitrogenous excretions of the insects and make them available to the fungus. The importance of fatty secretions from the insects for the growth of ambrosia fungi, which was demonstrated by Francke-Grosmann (1956*a*), has already been discussed.

Ambrosia fungi, without any gross disorganization of the structure of the wood around the tunnels, are able to maintain their own profuse growth and at the same time to support a colony of developing insects. The nutrition of these fungi is a virtually untouched field of research. Mathiesen-Käärik showed that blue-stain fungi could utilize the pectins of the cell wall, an ability which may be shared by the ambrosia fungi. Fougerousse (1957) reported that ambrosia fungi in culture could cause reduction in the degree of polymerization of cellulose.

It is frequently stated that the fungi concentrate the nitrogenous matter of the wood into the palisade layer, but there is no quantitative supporting evidence. Peklo & Šatava (1950) demonstrated *in vitro* nitrogen fixation by a *Torulopsis* associated with the ambrosia fungus of *Anisandrus dispar*, but it is not known whether this occurs within the confines of the tunnel, or to what extent fixation of atmospheric nitrogen may be generally important in the economy of all ambrosia beetles. There are indeed many unanswered questions about the extraordinarily complex relationship between these wood-inhabiting beetles, the assembly of fungi which they transmit, and the wood which supports the whole community.

ACKNOWLEDGEMENTS

The investigation on *P. cylindrus* was carried out at the Forest Products Research Laboratory while I was seconded from the West African Timber Borer Research Unit. I wish to thank my colleague Mr J. G. Savory for his help and guidance on the mycological problems encountered. The paper is published by permission of the Department of Scientific and Industrial Research.

REFERENCES

BAKER, J. M. (1956). Investigations on the oak pinhole borer, *Platypus cylindrus*. *Rec. annu. Conv. Brit. Wood Preserving Ass.* 1956, p. 92.
BAKER, J. M. (1960). In 'Ambrosia beetle research in West Africa'. *Rep. 7th Commonw. ent. Conf.* 1960, p. 94.

BAKSHI, B. K. (1950). Fungi associated with ambrosia beetles in Great Britain. *Trans. Brit. mycol. Soc.* **33**, 111.

BAKSHI, B. K. (1952). *Oedocephalum lineatum* is a conidial stage of *Fomes annosus*. *Trans. Brit. mycol. Soc.* **35**, 195.

BARNETT, J. A. & INGRAM, M. (1955). Technique in the study of yeast assimilation reactions. *J. appl. Bact.* **18**, 131.

BATRA, L. R. & FRANCKE-GROSMANN, H. (1961). Contributions to our knowledge of ambrosia fungi. I. *Ascoidea hylecoeti* sp.nov. (Ascomycetes). *Amer. J. Bot.* **48**, 453.

BEAL, J. A. & MASSEY, C. L. (1945). Bark beetles and ambrosia beetles: with special reference to species occurring in North Carolina. *Bull. Duke Sch. For.* no. 10.

BEAUVERIE, J. (1910). Les champignons dit Ambrosia. *Ann. Sci. nat. (Bot.)*, **11**, 31.

BECK, O. (1922). Eine neue *Endomyces*-Art. *Endomyces bisporus*. *Ann. mycol.*, *Berl.* **20**, 219.

BEESON, C. F. C. (1917). The life history of *Diapus furtivus*. *Indian For. Rec.* **6**, 1.

BEESON, C. F. C. (1941). *The Ecology and Control of the Forest Insects of India and the Neighbouring Countries*. Dehra Dun: Beeson.

BLETCHLY, J. D. (1961). A review of factors affecting ambrosia beetle attack in trees and felled logs. *Emp. For. Rev.* **40**, 13.

BOIDIN, J. & ABADIE, F. (1954). Les levures des liqueurs tannantes végétales; leur action sur les tanins pyrogalliques. *Bull. Soc. mycol. Fr.* **70**, 353.

BROWN, E. S. (1954*a*). The biology of the coconut pest *Melittomma insulare* and its control in the Seychelles. *Bull. ent. Res.* **45**, 1.

BROWN, E. S. (1954*b*). *Xyleborus morstatti*, a shot hole borer attacking avocado pear in the Seychelles. *Bull. ent. Res.* **45**, 707.

BROWNE, F. G. (1961*a*). The biology of Malayan Scolytidae and Platypodidae. *Malayan Forest Records*, no. 22.

BROWNE, F. G. (1961*b*). Preliminary observations on *Doliopygus dubius*. 4*th Rep. W. Afr. Timber Borer Res. Unit*, no. 15. London: Crown Agents.

BUCHANAN, W. D. (1940). Ambrosia beetle *Xylosandrus germanus* transmits Dutch Elm disease under controlled conditions. *J. econ. Ent.* **33**, 819.

BUCHANAN, W. D. (1941). Experiments with an ambrosia beetle, *Xylosandrus germanus*. *J. econ. Ent.* **34**, 367.

BUCHNER, P. (1953). *Endosymbiose der Tiere mit pflanzlichen Mikroorganismen*. Basle: Birkhäuser.

CACHAN, P. (1957). Les Scolytoidea mycetophages des forêts de Basse Côte d'Ivoire. *Rev. Path. vég.* **36**, 1.

CALLAHAM, R. Z. & SHIFRINE, M. (1960). The yeasts associated with bark beetles. *For. Sci.* **6**, 146.

CARRIÓN, A. & SILVA, M. (1955). A revision of so-called *Sporotrichum gougerotii*. *Arch. Derm. Syph.*, *N.Y.*, **72**, 523.

CHAMBERLIN, W. J. (1939). *The Bark and Timber Beetles of North America*. Corvallis, Oregon: O.S.C. Co-operative Association.

CHAMBERLIN, W. J. (1958). *The Scolytoidea of the Northwest*. Corvallis, Oregon: Oregon State University Press.

CHAPMAN, J. A. (1958). Studies on the physiology of the ambrosia beetle *Trypodendron* in relation to its ecology. *Proc. 10th int. Congr. Ent.* 1956, **4**, 375.

CHAPMAN, J. A. (1961). The arrangement of abdominal ganglia and flight muscle changes in the ambrosia beetle *Platypus wilsoni*. *Canada Dep. For. Bi-m. Progr. Rep.* **17** (6), 3.

CROWSON, R. A. (1960). The phylogeny of coleoptera. *Annu. Rev. Ent.* **5**, 111.

DOANE, R. W. & GILLILAND, O. J. (1929). Three Californian ambrosia beetles. *J. econ. Ent.* **22**, 915.

DODGE, C. W. (1935). *Medical Mycology*. St Louis: Mosby.

FARRIS, S. H. & CHAPMAN, J. A. (1957). A preliminary study of the deposition and early growth of fungus within the galleries of the ambrosia beetle *Trypodendron lineatum*. *Canada Dep. Agric. For. Biol. Div. Bi-m. Progr. Rep.* **13** (6), 3.

FERNANDO, E. F. W. (1959). Storage and transmission of ambrosia fungus in the adult *Xyleborus fornicatus*. *Ann. Mag. nat. Hist.* **2** (20), 478.

FINDLAY, W. P. K. (1959). Sapstain of timber. *For. Abstr.* **20**, 1, 167.

FISCHER, M. (1954). Untersuchungen über den kleinen Holzbohrer (*Xyleborinus saxesini*). *PflSch.* **12**, 137.

FISHER, R. C., THOMPSON, G. H. & WEBB, W. E. (1953–54). Ambrosia beetles in forest and sawmill. *For. Abstr.* **14**, 381; **15**, 3.

FOUGEROUSSE, M. (1957). Les piqûres des grumes de coupe fraîche en Afrique Tropicale. *Bois For. Trop.* **55**, 39.

FRANCKE-GROSMANN, H. (1952). Über die Ambrosiazucht der beiden Kiefern-borkenkäfer *Myelophilus minor* and *Ips acuminatus*. *Medd. SkogsforskInst., Stockh.* **41** (6), 52 pp.

FRANCKE-GROSMANN, H. (1956a). Hautdrüsen als Träger der Pilzsymbiose bei Ambrosia-käfern. *Z. Morph. Ökol. Tiere*, **45**, 275.

FRANCKE-GROSMANN, H. (1956b). Grundlagen der Symbiose bei pilzzüchtenden Holzinsekten. *Verh. dtsch. zool. Ges. Hamburg*, 1956, 112.

FRANCKE-GROSMANN, H. (1958). Über die Ambrosiazucht holzbrütender Ipiden im Hinblick auf das System. *Verh. dtsch. Ges. angew. Ent.* 1957, 139.

FRANCKE-GROSMANN, H. (1959). Beiträge zur Kenntnis der Übertragungsweise von Pflanzenkrankheiten durch Käfer. *Verh. 4. int. PflSch. Kongr.* **1**, 805.

FRANCKE-GROSMANN, H. & SCHEDL, W. (1960). Ein orales Übertragungsorgan der Nährpilze bei *Xyleborus mascarensis*. *Naturwissenschaften*, **47**, 405.

GADD, C. H. (1947). Observations on the life cycle of *Xyleborus fornicatus* in artificial culture. *Ann. appl. Biol.* **34**, 197.

GADD, C. H. & LOOS, C. A. (1947). The ambrosia fungus of *Xyleborus fornicatus*. *Trans. Brit. mycol. Soc.* **31**, 13.

GRAHAM, K. (1961). Air swallowing: a mechanism in photic reversal of the beetle *Trypodendron*. *Nature, Lond.* **191**, 519.

GREGORY, J. L. (1954). Shot-hole borers of cacao. *Rep. 6th Commonw. ent. Conf.* 1954, p. 293.

GROSCHKE, F. (1954). Zur Lebensweise und Bekämpfungsmöglichkeit des Eichen-kernkäfers *Platypus cylindrus*. *Verh. dtsch. Ges. angew. Ent.* 1952, p. 103.

GROSMANN, H. (1930). Beiträge zur Kenntnis der Lebensgemeinschaft zwischen Borkenkäfern und Pilzen. *Z. Parasitenk.* **3**, 56.

HADORN, C. (1933). Recherches sur la morphologie, les stades évolutifs et l'hiver-nage du bostryche liseré (*Xyloterus lineatus*). *Beih. Z. schweiz. Forstver.* no. 11.

HARTIG, T. (1844). Ambrosia des *Bostrychus dispar*. *Allg. Forst- u. Jagdztg*, **13**, 73.

HOGAN, T. W. (1948). The biology of the pin-hole borer—*Platypus subgranosus*. *J. Dep. Agric. Vict.* **46**, 373.

HOPTON, J. W. & WOODBINE, M. (1960). Fat synthesis by yeasts. I. Comparative assessment of *Hansenula* spp. *J. appl. Bact.* **23**, 283.

HUBBARD, H. G. (1897). The ambrosia beetles of the United States. *Bull. U.S. Bur. Ent.* (N.S.), no. 7.

HUSSON, R. (1955). Sur la biologie du Coléoptère xylophage 'Platypus cylindrus'. *Ann. Univ. Sarav.-Scientia* (1955), p. 348.

HUNT, J. (1956). Taxonomy of the genus *Ceratocystis*. *Lloydia*, **19**, 1.

JOVER, H. (1952). Note préliminaire sur la biologie des Platypodidae de Basse-Côte d'Ivoire. *Rev. Path. vég.* **31**, 73.

KALSHOVEN, L. G. E. (1958–59). Studies on the biology of Indonesian Scolytoidea. 4. Data on the habits of Scolytidae. *Tijdschr. Ent.* **101**, 157; **102**, 135.

KALSHOVEN, L. G. E. (1960). Studies on the biology of Indonesian Scolytoidea. 7. Data on the habits of Platypodidae. *Tijdschr. Ent.* **103**, 31.

KOCH, A. (1960). Intracellular symbiosis in insects. *Annu. Rev. Microbiol.* **14**, 121.

LEACH, J. G., HODSON, A. C., CHILTON, ST J. P. & CHRISTENSEN, C. M. (1940). Observations on two ambrosia beetles and their associated fungi. *Phytopathology*, **30**, 227.

LHOSTE, J. & ROCHE, A. (1959). Contribution à la connaissance de l'anatomie interne de *Xyleborus morstatti*. *Café, cacao, thé*, **3** (2), 76.

LHOSTE, J. & ROCHE, A. (1961). Anatomie comparée des organes transporteurs de champignons chez quelques Scolytoidea. *Proc. 11th int. Congr. Ent.* 1960, **1**, 385.

LODDER, J. & KREGER-VAN RIJ, N. J. W. (1952). *The Yeasts. A Taxonomic Study.* Amsterdam: North Holland Publishing Company.

MATHIESON, A. (1950). Über einige mit Borkenkäfern assoziierte Bläuepilze in Schweden. *Oikos*, **2**, 275.

MATHIESON-KÄÄRIK, A. (1953). Eine Übersicht über die gewöhnlichsten mit Borkenkäfern assoziierten Bläuepilze in Schweden und einige für Schweden neue Bläuepilze. *Medd. SkogforsknInst., Stockh.* **43**, 1.

MATHIESON-KÄÄRIK, A. (1960 a). Studies on the ecology, taxonomy and physiology of Swedish insect-associated blue stain fungi, especially the genus *Ceratocystis*. *Oikos*, **11**, 1.

MATHIESON-KÄÄRIK, A. (1960 b). Growth and sporulation of *Ophiostoma* and some other blueing fungi on synthetic media. *Symb. bot. upsaliens*, **16**, 168 pp.

MRAK, E. M. & PHAFF, H. J. (1948). Yeasts. *Annu. Rev. Microbiol.* **2**, 1.

MULLER, H. R. A. (1933). Ambrosia fungi of tropical Scolytidae in pure culture. *Versl. Afd. Ned.-O.-Ind. ned. ent. Ver.* **1**, 105, 125.

NEGER, F. W. (1908 a). Die pilzzüchtenden Bostrychiden. *Naturw. Z. Forst- u. Landw.* **6**, 274.

NEGER, F. W. (1908 b). Die Pilzkulturen der Nutzholzborkenkäfer. *Zbl. Bakt.* **20**, 279.

NEGER, F. W. (1909). Ambrosiapilze. II. Die Ambrosia der Holzbohrkäfer. *Ber. dtsch. bot. Ges.* **27**, 372.

NEGER, F. W. (1911). Zur Übertragung des Ambrosiapilzes von *Xyleborus dispar*. *Naturw. Z. Forst.- u. Landw.* **9**, 223.

PARKIN, E. A. (1940). The digestive enzymes of some wood-boring beetle larvae. *J. exp. Biol.* **17**, 364.

PEKLO, J. & ŠATAVA, J. (1950). Fixation of free nitrogen by insects. *Experientia*, **6**, 190.

RENNERFELT, E. (1950). Über den Zusammenhang zwischen dem Verblauen des Holzes und den Insekten. *Oikos*, **2**, 120.

ROBERTS, H. (1961). The adult anatomy of *Trachyostus ghanaensis*, a W. African beetle, and its relationship to changes in adult behaviour. *4th Rep. W. Afr. Timber Borer Res. Unit*, p. 31. London: Crown Agents.

ROCHE, A. & LHOSTE, J. (1960). Descriptions d'organes adaptés à la dissémination des champignons chez les Scolytoidea. *C.R. Acad. Sci., Paris*, **250**, 2056.

RUDINSKY, J. A. (1962). Ecology of Scolytidae. *Annu. Rev. Ent.* **7**, 327.

SCHEDL, K. E. (1958). Breeding habits of arboricole insects in Central Africa. *Proc. 10th int. Congr. Ent.* 1956, **1**, 183.

SCHMIDBERGER, J. (1836). Naturgeschichte des Apfelborkenkäfers *Apate dispar*. (Cited by Fisher, Thompson & Webb, 1953–54.)

SCHNEIDER, I. (1959). Untersuchungen über die Aktivitätsgrenzen von pflanzlichen und tierischen Importholzschädlingen. *Mitt. dtsch. Ges. Holzforsch.* **46**, 8.

SCHNEIDER-ORELLI, O. (1913). Untersuchungen über den pilzzüchtenden Obstbaumborkenkäfer *Xyleborus* (*Anisandrus*) *dispar* und seinen Nährpilz. *Zbl. Bakt.* (II), **38**, 25.

PLATE 1

Fig. 1

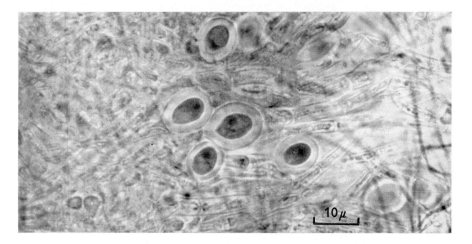

Fig. 2

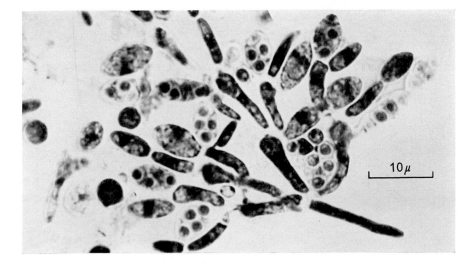

Fig. 3

(*Facing page* 264)

PLATE 2

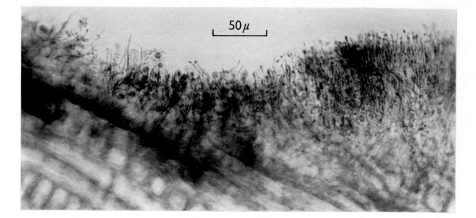

Fig. 4

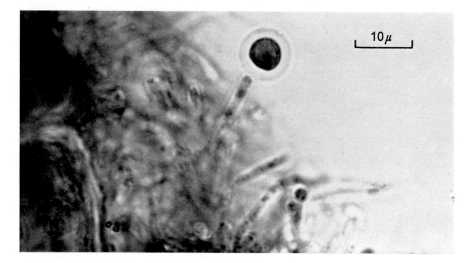

Fig. 5

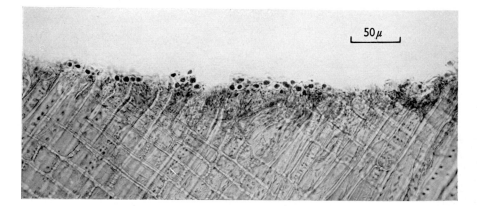

Fig. 6

SHIFRINE, M. & PHAFF, H. J. (1956). The association of yeasts with certain bark beetles. *Mycologia*, **48**, 41.

SMITH, J. H. (1935). The pinhole borer of North Queensland cabinet woods. *Bull. Div. Ent. Pl. Path. Qd* (N.S.), no. 12.

STROHMEYER, H. (1906). Neue Untersuchungen über Biologie, Schädlichkeit und Vorkommen des Eichenkernkäfers *Platypus cylindrus*. *Naturw. Z. Forst- u. Landw.* **4**, 329, 409, 506.

STROHMEYER, H. (1911). Die biologische Bedeutung sekundärer Geschlechtscharaktere am Kopfe weiblicher Platypodiden. *Ent. Bl.* **7**, 103.

TROTTER, A. (1934). Il fungo-Ambrosia delle gallerie di un Xyleborino di ceylon. *Ann. Ist. sup. agr. Portici*, Ser. 3; vi, 256.

VERRALL, A. F. (1943). Fungi associated with certain ambrosia beetles. *J. agric. Res.* **66**, 135.

VITAL, A. F. (1951). An interesting case of symbiosis between *Stysanus microsporus* and *Platypus difficilis* in *Eschweilera luschnathii*. *Bol. Sec. Agric. Pernambuco*, **18**, 177. (in *Rev. appl. Mycol.* **31**, 488.)

WEBB, S. (1945). Australian ambrosia fungi. *Proc. Roy. Soc. Vict.* **57**, 57.

WICKERHAM, L. J. (1951). Taxonomy of yeasts. *U.S. Dept. Agric. Tech. Bull.* no. 1029.

WILSON, C. L. (1959). The Columbian timber beetle and associated fungi in White Oak. *Forest Sci.* **5**, 114.

WOOD, S. L. (1957). Ambrosia beetles of the tribe Xyloterini (Coleoptera: Scolytidae) in North America. *Canad. Ent.* **89**, 337.

WRIGHT, E. (1935). *Trichosporium symbioticum* n.sp., a wood-staining fungus associated with *Scolytus ventralis*. *J. agric. Res.* **50**, 525.

EXPLANATION OF PLATES

PLATE 1

Fig. 1. Tunnel of *Platypus cylindrus* in oak. Primary gallery with early branch gallery into heartwood.

Fig. 2. Old culture of *Sporothrix* sp. on PYA showing chlamydospores.

Fig. 3. Culture of *Endomycopsis* sp. on MYGPA showing asci each containing four ascospores.

PLATE 2

Fig. 4. Transverse section of *P. cylindrus* tunnel showing palisade layer with sterile hyphae, conidia and chlamydospores.

Fig. 5. Transverse section of tunnel showing detail of chlamydospore.

Fig. 6. Transverse section of tunnel with lining in irregular condition. Many chlamydospores present.

SYMBIOTIC ASSOCIATIONS:
THE RUMEN BACTERIA

R. E. HUNGATE

University of California, Davis

Symbiosis as defined by de Bary (1879) is an association in which two different species live together in a close spatial and physiological relationship. If they are reciprocally dependent, the relationship is mutualism. In parasitism, one profits at the expense of the other, and in commensalism, one member benefits without affecting the other.

The true parasite or commensal requires association with the host, but hosts free of symbionts may occur. In contrast, true mutualists must live with the partner. If the mutualism is significant in the lives of the partners, i.e. if it has survival value, the species survive together but not alone. If individuals of two associated species occur in nature independently of each other, their relationship has not evolved to the point of survival advantage and it is questionable that it is mutualism.

Many mutually beneficial relationships exist between species without the close physical association that is the characteristic feature of symbiosis. If close spatial connexion (at least during some essential phase of the life cycle) is not retained as a criterion of symbiosis, loose relationships such as production of oxygen by a plant and its utilization by an animal are included and the term loses both meaning and usefulness.

To evaluate the rumen bacteria as symbionts, their distribution should be known. Unfortunately, there is little information on their occurrence outside the rumen. The bacterium isolated by Hall (1952) from the rabbit caecum is closely related to rumen types; methanogenic bacteria similar to those from the rumen have been isolated from sludge by Smith (private communication), and lactobacilli of the rumen also occur elsewhere. But most species of rumen bacteria have not been reported from other habitats. The rumen provides a highly specialized environment to which bacteria as well as protozoa have become specifically adapted.

THE RUMEN ENVIRONMENT

A feature of the rumen distinguishing it from most natural habitats is its relative constancy. The ruminant is an excellent thermoregulator; the heat of microbial fermentation in the rumen raises its temperature only slightly above that of the ruminant and temperature fluctuations are minor (Krzywanek, 1929). After water intake by the host, the rumen micro-organisms are subjected to a brief increase in the water activity of their environment, but osmotic equilibrium between rumen contents and blood is rapidly re-established (Parthasarathy & Phillipson, 1953).

The ruminant secretes alkaline saliva (containing $NaHCO_3$) and absorbs acid fermentation products, keeping the rumen acidity normally within a pH range of 5·8–6·8. With high intake of readily available carbohydrate (starch or sugar), acid production may lower the pH to 4·0, causing host malaise and a marked change in the flora (Hungate, Dougherty, Bryant & Cello, 1952). In cases of very poor rations, the pH may go above 7·5 (Clark & Lombard, 1951). These extremes are not characteristic of a ruminant on a good quality forage, and usually accompany some nutritional abnormality.

Carbon dioxide is the major component (60–70 %) of the gas in the top of the rumen (Reiset, 1868). The remainder is chiefly methane, with small quantities of nitrogen and other gases. Some air is swallowed and the gas in the top of the rumen occasionally contains a little oxygen. The mass of the rumen digesta is so large in relation to the surface exposed to this gas that the fraction of the digesta affected by oxygen is negligible. The rumen contents readily absorb small concentrations of oxygen. The oxidation-reduction potential, measured with a platinum electrode in the presence of benzylviologen, is about $-0·35$ V., at which the oxygen concentration is extremely low.

As a source of microbial food, the rumen always contains a considerable quantity of watery digesta, containing organic matter to the extent of 10–18 %. Domestic ruminants are sometimes fed only once daily, with consequent fluctuation of available food, but most are fed at least twice daily, and animals on pasture graze about 8 hr. in each 24 (Harker, Taylor & Rollinson, 1954). Rumination (cud-chewing) may occur 8 hr. per day, increasing the food available to the micro-organisms (Dehority & Johnson, 1961) and levelling out the fluctuations due to irregular intake. Other factors such as food storage in the protozoa and the long time required for complete digestion of the fibre, operate to keep a reasonably constant concentration of carbohydrate food in the rumen, at least compared to most other natural microbial habitats.

Because of the constancy of the environment, it might be expected that the rumen would be occupied by a simple bacterial flora dominated by a few especially well-adapted species. Actually, it is the exact reverse; the flora is extremely complex. Part of this complexity can be accounted for by the diversity of chemical components in the plant materials consumed. Uniformly favourable physical factors permit competition by more kinds of organisms than could develop under more rigorous conditions. The rumen habitat has its rigours but they lie in the biotic rather than the physical components.

CULTIVATION OF THE RUMEN BACTERIA

Cultivation of the rumen bacteria *in vitro* has been achieved by simulation of the rumen environment. Since the temperature, moisture, acidity and anaerobiosis vary within narrow limits, simulation has consisted simply in keeping within these limits.

Pure carbon dioxide was used, with 0·5 % sodium bicarbonate, to provide a pH of about 6·7. Selection of an initial value near the upper limit of the rumen range increases the quantity of acid products that can be accommodated within the pH range suitable for growth. Too high a pH diminishes the rate of cellulose digestion.

Gas-bubbling capillary tubes and other devices exclude air during handling of the medium, and the oxidation-reduction potential is kept low by adding small concentrations of cysteine, sodium sulphide, or other reducing agents. An osmotic pressure near to that of the rumen results automatically from the buffer system, plus balanced inorganic ions (Hungate, 1950).

In general, liquid enrichments have been avoided for reasons outlined elsewhere (Hungate, 1962). Culture counts are obtained from agar roll tubes inoculated with serial dilutions of rumen contents, and pure cultures can be subcultured from selected colonies. If necessary, anaerobiosis can be preserved during transfer by passing a gentle stream of carbon dioxide into open tubes. The rubber-stoppered roll tube method possesses several additional advantages: tubes can be examined without loss of anaerobiosis, colonies close together in the thin agar can be picked separately, the ratio of medium to gas phase is large, permitting use of small concentrations of oxygen-absorbing reagents in the medium, and no special glassware, anaerobic jars, or incubators are necessary. A cylinder of pure carbon dioxide and a 45° water-bath to hold the tubes of melted agar are the only special equipment needed.

Rumen fluid is included in the medium to supply nutrients peculiar to the habitat (Hungate, 1947; King & Smith, 1955). Thirty per cent (v/v) rumen fluid provides essential nutrients without introducing inhibitory concentrations of fermentation products.

Two types of media have been used: (1) habitat-simulating media which support the growth of a wide variety of bacteria, and (2) niche-simulating media which provide conditions favourable to the growth of a particular kind of organism. Each has its place in an ecological analysis (Hungate, 1962) and each has contributed to the understanding of rumen bacteria.

The most difficult habitat factor to simulate is the food which normally consists of coarse particulate insoluble material. Conceivably it could be finely ground and incorporated in the medium, but such an agar would be opaque and unsuitable for detecting small colonies. In practice, a mixture of carbohydrates (Bryant & Burkey, 1953; Bryant & Robinson, 1961 b) or a water extract of hay (Hungate, 1957) has been employed in rumen-simulating media in lieu of hay and grain. For niche-simulating media, an attempt is made to simulate all rumen factors except the one selecting for the niche, usually the characteristic substrate, for example lactate (Johns, 1948), starch (Hamlin & Hungate, 1956), cellulose (Hungate, 1947), carbon dioxide plus hydrogen (Smith & Hungate, 1958), or protein (Blackburn & Hobson, 1960 b). As knowledge of the nutritional requirements of individual kinds of bacteria accumulates (Bryant & Doetsch, 1955; Bryant & Robinson, 1961 a), better selective media for niche inhabitants will be devised.

The host rumen functions of continuously removing waste are not easily duplicated *in vitro*; in consequence, it is necessary in laboratory cultures to use smaller concentrations of substrate than are characteristic of the rumen. For short-time experiments, substrate concentrations comparable to those of the rumen can be studied (Carroll & Hungate, 1954), but such systems deviate rapidly from rumen conditions and results after more than 1 or 2 hr. of incubation may be misleading (Warner, 1956).

Such methods have permitted the isolation of numerous bacteria (Bryant, 1959) and have been successful also in growing anaerobic bacteria from other habitats (Hall, 1952; Mylroie & Hungate, 1954; Maki, 1954). However, the total viable count for animals on a hay ration is still only 5–20 % of the direct count. Whether this is due to inclusion of many non-viable cells in the direct count, or to inadequacies in the culture method, has not been determined. For animals on a grain ration the viable count approaches the direct count more closely.

KINDS OF RUMEN BACTERIA

Cellulose digesters

Historically, the rumen bacteria attracting most widespread interest were those capable of digesting cellulose; these seemed most likely mutualistic. Due to the difficulty encountered in early attempts at culture, many misleading and erroneous impressions of instability in cellulolysis and other characteristics developed and the cellulose digesters were regarded as special and peculiar.

The most reliable clues to the nature of the rumen cellulolytic bacteria came from the direct microscopic studies of Henneberg (1922) and Baker (Baker & Martin, 1938; Baker, 1942). Before this, the only identified cellulolytic bacteria from rumen contents were the spore-formers (Ankersmit, 1905; Pochon, 1935), but the sparseness of this group in the rumen precludes their being of importance. Henneberg (1922) described a *Micrococcus ruminantium* and *Streptococcus jodophilus*, and thought they were important cellulose decomposers in the rumen. He used iodophily as one test for cellulolytic bacteria, assuming that the cells would be unable to form iodophilic reserve carbohydrates unless they could digest cellulose; formation of 'Frassbetten' in the fibre was the second criterion. Baker & Martin (1938) concluded that 'the cytoclastic micro-organisms include vibrionic, coccoid, and giant forms'.

Important rumen cellulolytic bacteria were not cultured until the environmental factors of the rumen were simulated *in vitro*. In many of their characteristics, these bacteria resemble non-cellulolytic types.

The first rumen cellulolytic bacterium to be obtained was a Gram-negative non-motile rod (Hungate, 1947), later described as *Bacteroides succinogenes* (Hungate, 1950). Alone among the rumen cellulolytic bacteria it possesses close affinities with the myxobacteria, and is only distinguished from the recently described *Cytophaga succinicans* (Anderson & Ordal, 1961) by its obligate anaerobiosis. It moves through 1 % agar and digests cellulose in the immediate vicinity of migrating cells in a manner resembling the aerobic cytophagas. Succinate is the chief fermentation product and an approximately equal amount of carbon dioxide is fixed. Acetic acid is the other chief product of fermentation with some formic acid produced by certain strains.

Pure cultures of rumen cellulolytic cocci forming yellow or white colonies were also obtained (Hungate, 1947; Sijpesteijn, 1948). They vary widely in their characteristics; strains isolated at different times and places are rarely identical. Those producing yellow pigment and succinic acid as an important fermentation product are placed in the species *Ruminococcus flavefaciens* (Sijpesteijn, 1948). Those with little pigment and producing little succinic acid are assigned to *R. albus* (Hungate, 1957).

Butyrivibrio fibrisolvens is the other important cellulolytic bacterium of the rumen. It was first isolated in the author's laboratory (Hungate,

1950) by Bryant, who later named and described it (Bryant & Small, 1956a). It is a polarly flagellated curved rod which produces butyric acid, among other products, and is abundant in the rumen of animals on most rations. By employing cellulose as the selective substrate (Hungate, 1957), *Butyrivibrio* strains attacking cellulose can be isolated. If a non-selective medium is used (Bryant & Small, 1956a) most of the isolated strains are non-cellulolytic. Although only a small minority of the strains digests cellulose, the genus is so abundant in the rumen that it constitutes an important fraction of the cellulolytic bacteria.

The speed and completeness with which *Butyrivibrio* strains attack cellulose are quite variable. Strains vary also in other characteristics and are rarely identical when isolated at different times and from different localities. Margherita (1962) found common antigens among the butyrivibrios, indicating interrelationships within the group, but strains from different places were not serologically identical. The serological variability was less marked than in the strains of cellulolytic cocci studied by Elizabeth Hall (unpublished experiments at Pullman, Washington), in which no cross-agglutination was observed.

Clostridium lochheadii, C. longisporum (Hungate, 1957) and *Cillobacterium ruminantium* (Bryant, Small, Bouma & Robinson, 1958) are less important strains of cellulolytic bacteria isolated from rumen contents. In a few cases, *C. lochheadii* has been a conspicuous cellulolytic component (Hungate, 1957), but in most ruminants this species is rarely encountered.

Physiology of cellulose digestion. Many of the cellulose-decomposing bacteria in the rumen attack cellulose rapidly and completely. The belief that they attack it slowly and with difficulty probably arose through use of culture media containing more cellulose than could be decomposed within the narrow environmental limits favourable to growth (Hungate, 1950). Cellulose digestion ceased because the accumulation of fermentation products had rendered the medium unsuitable for growth. With 0·5 % bicarbonate and an atmosphere of carbon dioxide in the medium, *Bacteroides succinogenes* and *Ruminococcus* rapidly decompose 0·1 % cellulose supplied as finely divided filter-paper or cotton. With concentrations of cellulose greater than a few tenths of 1 %, the initial attack is rapid but digestion slows down in the later stages; if more cellulose is supplied than can be fermented, it is left undigested or is digested slowly after growth has ceased.

Sugars have been demonstrated as products of the rumen cellulolytic bacteria, except *Bacteroides succinogenes*. Either cellobiose or glucose or both are formed in small amounts in old cultures in which fermentation has ceased. Extracellular intermediates also form during growth. The bacterial colonies in cellulose agar become surrounded by a clear zone which steadily increases in diameter as the colony grows and as the cellulase diffuses outwards.

Relationship between cellulose digesters and other rumen bacteria. The extent to which the rumen cellulolytic bacteria are able to recover the sugar which their enzymes release in the rumen is not known, but the following observations may be relevant. In serial dilutions of rumen contents into tubes of cellulose agar, colonies of *Ruminococcus* and *Butyrivibrio* (as disclosed by clearing of the cellulose) are observed only in the higher dilution tubes. The cellulose never disappears in the low dilutions unless *Bacteroides* or *Clostridium* are also present. Pure cultures of *Ruminococcus* and *Butyrivibrio* similarly diluted into agar medium behave differently; they digest cellulose in all tubes, regardless of dilution. If the concentration of cellulose is not too great, it rapidly disappears in the lower dilutions and ultimately is completely digested also in the high dilutions.

In agar cultures, cellulose breakdown occurs at a slight distance from the colony and the sugar produced must diffuse some distance to reach the cells elaborating the enzyme. With pure cultures, this presents no problem and growth continues even though the margin of undigested cellulose recedes several centimetres from the colony; all the sugar ultimately diffuses back to the cellulolytic cells. With mixed cultures, other microbes can compete for the released sugar, particularly when they are numerous, and when the zone of clearing becomes larger. In low dilutions, the 'contaminants' being very near the cellulose digesters, use so much of the sugar that the amount diffusing back to the cellulase elaborator no longer supports further enzyme production. In consequence, the clearing of the cellulose does not enlarge enough to be seen.

In liquid cultures, *Butyrivibrio* and *Ruminococcus* are able to grow and digest cellulose in the presence of numerous other bacteria, possibly because they can maintain a preferential position in relation to the cellulose.

The histological observations of Henneberg (1922) and Baker & Martin (1938) suggest that the cellulolytic bacteria cling to the plant fibres.

In contrast to *Butyrivibrio* and *Ruminococcus*, *Bacteroides succinogenes* can digest cellulose to completion in low dilutions of rumen contents in solid medium. As previously mentioned, *B. succinogenes* does not form a colony in cellulose agar, but spreads and lies in a ring at the edge of the undigested cellulose. The fact that this species, able to move through the agar, can digest cellulose in the presence of numerous other bacteria, emphasizes the importance of mobility in cellulose digestion.

Clostridium lochheadii can also digest cellulose in low dilutions of rumen contents in agar though how it does so is not clear. Filaments of this species penetrate the medium and form a loose poorly defined colony surrounded by a zone where all cellulose is digested and no cells are visible. *C. lochheadii* gains access to the surface of the thin agar medium, spreads rapidly over it, and digests the cellulose below and just ahead of the advancing growth. Perhaps this species secretes so much cellulase that the sugars released are adequate both for it and for the accompanying non-cellulolytic forms.

One of the puzzles of rumen ecology has been the apparent scarcity of cellulose-digesting bacteria in comparison to the total population. They compose only 1–5 % of the microflora. Since the cellulose digested in the rumen constitutes as much as one-third of the total fermented substrate, the cellulolytic bacteria might reasonably be expected to constitute about one-third of the total population. The discrepancy may be due to inability of the cellulolytic bacteria to harvest the major fraction of the sugars formed by their cellulase; their attachment to substrate may be insufficient to exclude completely their non-cellulolytic competitors.

Failure to isolate all the important cellulolytic bacteria of the rumen could also explain the discrepancy between their apparent numbers and the proportion of cellulose in the substrate. The missing cells could be either: (1) similar to the known types, but aggregated on particles of plant material and therefore not counted as single cells, or (2) they could be unrepresented because they require conditions not provided *in vitro*.

Comminution of the solids fraction of rumen contents with a Waring blendor increases the total count by a factor of about 10, but does not cause a significant change in the percentage of cellulolytic cells, indicating that a relatively greater clumping of known types of cellulolytic cells cannot account for their low recovery in cultures (Storz & Hungate, 1962).

In experiments with sheep (Weller, Gray & Pilgrim, 1958), diaminopimelic acid was used as an index for bacteria, and the amount in rumen solids and liquid determined. The quantity of bacteria associated with the rumen solids and liquids was about equal. This agrees with the plate counts by other investigators, and suggests again that there is no large population of undetected rumen bacteria clinging to the fibres.

Digestion by pure cultures of the water-insoluble components of hay.
The breakdown of wood or cotton cellulose by pure cultures does not
necessarily indicate an ability to digest the cellulose in the natural feed.
When, however, pure cultures of cellulolytic bacteria were inoculated
into media containing sterilized forage (Hungate, 1957), there was a
significant loss in weight partly due to the digestion of cellulose, indicating that the bacteria attacking the filter-paper or cotton cellulose can
also attack the 'native' cellulose in the forage.

Many pure cultures of *Ruminococcus* use only cellulose, xylan, cellobiose and one or two other carbohydrates. Their action in the rumen
would appear to be restricted to the water-insoluble feed components.
Bacteroides succinogenes similarly utilizes only a few sugars. *Butyrivibrio*, on the contrary, can utilize a great many mono-, di-, and polysaccharides, and appears to possess advantages over the other cellulolytic bacteria, using sugars when present and fibre after sugars are
exhausted. However, in the author's experience, the cellulolytic species
of *Butyrivibrio* are not usually more abundant in the rumen than are
other cellulolytic types, particularly *Ruminococcus*. These relationships
may vary greatly with different rations and geographic areas. *Bacteroides succinogenes* is not found in high numbers in alfalfa hay-fed
cattle in Washington and California. A very high count was obtained
from an animal in Texas (Hungate, 1947), and Bryant has found them
abundant in Maryland. The reasons for these variations are not known.

Starch digesters

Some of the rumen cellulolytic bacteria also digest starch. *Butyrivibrio*
species are particularly active in this respect and on starch agar media
form very large colonies within 24 hr., part of which is capsular
material. *Clostridium lochheadii* digests starch, and some strains of
Bacteroides succinogenes use it (Hungate, 1947). *C. lochheadii* has not
been reported from the rumen of animals fed rations high in starch but
Butyrivibrio (chiefly non-cellulolytic strains) and *B. succinogenes* may
constitute 27 and 2 %, respectively, of the bacteria in such animals
(Bryant, Robinson & Lindahl, 1961). This probably reflects their ability
to use starch. However, since non-amylolytic *Ruminococcus* cells were
as abundant as *B. succinogenes* it cannot be concluded that the latter
was necessarily present because of its starch-hydrolysing capacities.

The non-cellulolytic rumen bacteria able to decompose starch include
Streptococcus bovis, *Bacteroides amylophilus*, *Succinimonas amylolytica*
and some strains of *Selenomonas ruminantium* and *Bacteroides ruminicola*.

Streptococcus bovis is one of the fastest growing bacteria of the rumen. It digests starch rapidly but ordinarily does not compose a large fraction of the rumen flora, even in animals on a high grain ration. Its capacity for rapid growth is expressed when an animal on a hay ration is suddenly given large quantities of grain (Hungate *et al.* 1952; Krogh, 1959). Within a few hours the numbers of *S. bovis* increase from a count of 10^7/ml. to more than 10^9/ml., large quantities of lactic acid accumulate, and the pH drops as low as 4·0. The high acidity causes atony of the rumen musculature and in many cases the animal dies within 24 hr.

The high acidity also kills *Streptococcus bovis*, and an aciduric flora develops in which *Lactobacillus* species predominate. Unless the early bloom of *S. bovis* is detected, *Lactobacillus* may appear to be the cause of the indigestion, and it can be responsible for the continuing acidity, but in most cases *S. bovis* is the important initial agent. As the animal adapts to the high grain ration, *Lactobacillus* is abundant but after adaptation it does not predominate unless high acidity continues.

Individual animals differ in the amount of readily available carbohydrate they can consume without developing unfavourable rumen acidity. Variation in the capacity to secrete sodium bicarbonate may explain these individual differences.

Streptococcus bovis is not usually more abundant in well-adapted grain-fed animals than in hay-fed stock, but during the early stages of gradual adaptation it may be more numerous. Its failure to maintain high numbers in adapted animals may be due to lack of accessory nutrients, but evidence on this point is lacking. Its low numbers in animals on a hay ration, about 10^7/ml., are ascribed to lack of food, though consumption of these bacteria by protozoa (Gutierrez & Davis, 1959) could also be responsible. *S. bovis* can use about a dozen simple carbohydrates, so is not limited to the starch in hay, but it must compete with other bacteria and the holotrich protozoa for the sugars and may not capture a large share. Perhaps it is not successful in competing for the sugar from cellulose.

Bacteroides amylophilus (Hamlin & Hungate, 1956) differs markedly from *Streptococcus bovis* in attacking only maltose and starch; glucose is not fermented. It was isolated initially by employing a rumen fluid starch-agar medium. In some animals it occurred in large numbers but was not abundant in others. The variation may have been due to the differences in rations but too few samples were taken to permit generalization. Blackburn & Hobson (1962) have found *B. amylophilus* in sheep on a partial grain ration in Scotland.

The fermentation products are essentially similar to those of *Bac-*

teroides succinogenes, with carbon dioxide fixed into succinic acid as the chief product, and with lesser amounts of acetic and formic acid produced.

Succinimonas amylolytica (Bryant, Small, Bouma & Chu, 1958) resembles *Bacteroides amylophilus* except that it is motile and can ferment glucose. It is about as restricted in the number of sugars it attacks.

Bacteroides ruminicola (Bryant, Small, Bouma & Chu, 1958) attacks many sugars and forms products similar to other rumen species of *Bacteroides*.

Some strains of *Selenomonas ruminantium* (Bryant, 1956) attack starch. The fermentation products are similar to those of *Bacteroides ruminicola*, except that propionic acid is formed and not all strains form succinate. Presumably a succinic decarboxylase is present but if this explains the propionic acid, the enzyme must occur in very small concentrations since succinic acid is also an end product. Both *B. ruminicola* and *S. ruminantium* attack a wide variety of carbohydrates. Some strains of *B. ruminicola* ferment xylan and pectin whereas some *Selenomonas* strains ferment mannitol and glycerol. *Selenomonas* has a characteristic tuft of flagella in the centre of the concave side of the crescent-shaped cell. The cells in rumen contents are quite variable in size, many of them as large as small protozoa (Lessel & Breed, 1954).

Hemicellulose digesters; xylan and pectin

All the cellulolytic strains capable of vigorous attack on the fibre in hay digest both the fraction of hay soluble in 2 % sulphuric acid and that soluble in 70 %, i.e. the hemicellulose and cellulose fractions. The extent of digestion of the two components is about the same. It might be expected that bacteria capable of digesting the hemicellulose but not the cellulose in hay would occur in the rumen, but to date no such examples have been reported.

Strains of *Ruminococcus flavefaciens* and *R. albus* which digest the fibre in alfalfa hay (Hungate, 1957) can ferment purified xylan. For six cellulolytic strains of *Butyrivibrio fibrisolvens* decomposing alfalfa hay, it has been shown that an average of 39 % of the cellulose and 30 % of the hemicellulose were digested. Purified pectin and xylan are readily fermented. Both *Bacteroides succinogenes* and *B. ruminicola* digest pectin.

Eubacterium ruminantium (Bryant, 1959) attacks xylan in pure culture. Its role in the rumen has not been ascertained. It is a weakly Gram-positive rod with a fermentation pattern resembling *Butyrivibrio*.

Pure cultures of two additional strains not mentioned previously, *Lachnospira multiparus* and *Succinivibrio dextrinosolvens* (Bryant &

Small, 1956b) also decompose pectin. It will be interesting to examine the extent to which these strains attack the pectin in hay since they do not ferment the cellulose and xylan which cover the pectin of the middle lamella. If effective, the pectin enzymes of these strains could play a role in dissolving the cementing substance binding the plant cells, thereby assisting the comminution of the digesta. This could be particularly important with lush green forages which disintegrate in the rumen with almost no assistance from the cud-chewing activities of the animal.

Lachnospira multiparus, as the specific epithet implies, produces many fermentation products, carbon dioxide, hydrogen, formic, acetic, and lactic acids, and ethanol. It forms chains of cells which spread on the agar or penetrate it as filaments to give a woolly appearance to the colony, an appearance seen in few other rumen bacteria except certain strains of *Butyrivibrio*. Only a few sugars are attacked, including glucose, fructose, sucrose, cellobiose, esculin and salicin. It is occasionally the most abundant single organism in agar dilutions inoculated with the washed solids fraction of rumen contents from a cow fed alfalfa hay (Storz & Hungate, 1962). Its growth habit may permit it to penetrate to the site of the middle lamella where the pectic enzymes could yield a greater return.

Succinivibrio dextrinosolvens attacks a few more sugars than does *Lachnospira* and exhibits the acetic-formic-succinic type of fermentation. Its known characteristics do not suggest any obvious role in the rumen.

CONVERSIONS OF INTERMEDIATE PRODUCTS OF THE CARBOHYDRATE FERMENTATION

The bacteria which attack the components of forage do not produce all the known end products of rumen fermentation, e.g. methane, and many of them produce in pure culture materials which do not appear as end products in the rumen, e.g. lactate, succinate, formate, hydrogen gas, and ethanol. Conversions by a secondary flora intervene between the attack on the primary substrate and the formation of some final products.

The methane bacteria

Methanobacterium ruminantium (Smith & Hungate, 1958) has been isolated from dilutions of rumen contents as high as 2×10^8. It ferments hydrogen and carbon dioxide to methane and can use limited quantities of formate. The manometric rate per cell for conversion of hydrogen and carbon dioxide to methane by pure cultures, multiplied by the culture count for rumen methane bacteria, gives a value approximately equal to the rate of methane production in the rumen (Smith, private communication). If *M. ruminantium* is entirely responsible for the rumen methane one would expect the pure cultures to be capable of a somewhat higher rate per cell than that needed to account for the rumen methane production. The rate at which hydrogen diffuses into the liquid of the manometric vessel may limit the *in vitro* methane production of the pure culture. In the rumen, with hydrogen produced throughout the fermenting mass, its diffusion to the methanogenic cells would not be as limiting.

The methane bacteria are extremely difficult to grow in pure culture and require the stringent exclusion of oxygen in a medium reduced to a potential of about -0.35 V. for growth to be initiated and maintained. For the inexperienced investigator it is very difficult to open a culture and prevent entrance of inhibitory amounts of oxygen.

Lactate fermentation. Lactic acid is produced by many pure cultures of rumen bacteria but the fraction of substrate carbon appearing as lactate is small except with *Streptococcus bovis* and *Cillobacterium ruminantium*. The lactate concentration in the rumen is ordinarily less than 0·001 % (Balch & Rowland, 1957; Jayasuriya & Hungate, 1959).

Lactate is fermented by *Veillonella alcalescens* (syn. *V. gazogenes, Micrococcus alcalescens, M. lactilyticus*) found in the rumen of sheep in numbers as high as 7×10^6 (Johns, 1951). It also ferments some of the dicarboxylic 4-carbon acids but does not attack sugars. Acetic and propionic acids, carbon dioxide and hydrogen are formed. Succinate is rapidly decarboxylated to carbon dioxide and propionic acid by whole cells.

Peptostreptococcus elsdenii (Gutierrez, Davis, Lindahl & Warwick, 1959) was first isolated by Elsden, Gilchrist, Lewis & Volcani (1951) who designated it the LC (large coccus) organism and showed (Elsden, Volcani, Gilchrist & Lewis, 1956) that it fermented lactate to acetate, propionate, butyrate, and valerate, in addition to carbon dioxide and hydrogen. On sugar some formic acid also appeared. The proportions of propionate and valerate were less on glucose and fructose, and much caproic acid was formed. Succinate was not decarboxylated to propionate by whole cells. These authors found also that liquid lactate enrichments invariably gave rise to *Veillonella alcalescens* instead of the LC organism. But if starch was used as the enrichment substrate *Streptococcus bovis* grew in large numbers and was followed by LC with no significant development of *V. alcalescens*. This suggests interesting interactions between these species.

Gutierrez *et al.* (1959) were able to isolate *Peptostreptococcus elsdenii* by employing the roll-tube technique and inoculating dilutions of rumen contents directly into peptone yeast-extract lactate agar medium. They found a 35-fold increase in the concentration of *Streptococcus bovis* and a 20-fold increase in *P. elsdenii* in animals changing from hay to a bloat-provoking ration high in barley. This agrees with the enrichment results of Elsden *et al.* (1956).

The predominant bacteria in two steers well adapted to the same high barley, bloat-provoking diet have been studied by Bryant *et al.* (1961). They observed microscopically that *Peptostreptococcus elsdenii* was present and that streptococci similar to *Streptococcus bovis* were occasionally quite numerous. The numbers of these bacteria did not correlate with the intensity of bloat. The animals had been fed the high barley diet for a long time and probably represented a more stable condition than the transitional stage studied by the Gutierrez group.

Strains of *Selenomonas ruminantium* attacking lactate have been described by Bryant (1956) but only part of the isolates possess this feature.

At Pullman, Washington, propionibacteria were found in great numbers in cattle on an alfalfa hay ration (Gutierrez, 1953). The strains resembled *Propionibacterium* (*Corynebacterium*) *acnes*. Similar types were abundant in the alfalfa hay. This species has not been encountered by other workers and was perhaps numerous in the Pullman animals because of the very large numbers of cells introduced with the feed. They appeared to grow to some extent in the rumen, but their absence in most animals indicates that they cannot maintain themselves without continuous inoculation.

The turnover of lactate in rumen contents has been measured by Jayasuriya & Hungate (1959), using lactate-2-^{14}C. In hay-fed cattle the turnover is very small, accounting for less than 1 % of the food fermented. This makes it unimportant as a precursor of propionic acid. An average of 81 % of the label was recovered in acetic acid, 14 % in propionic and 4 % in butyric acid. Baldwin, Wood & Emery (1962) obtained a similar distribution and demonstrated that a non-randomizing route, such as the acrylate pathway in *Clostridium propionicum*, was involved in the production of propionate from lactate. These results show that the small amount of propionate from lactate is probably not formed by typical *Propionibacterium* species, since their metabolism is characterized by a symmetrical 4-carbon dicarboxylic acid intermediate.

In animals on the bloat-provoking high-barley ration of Gutierrez *et al.* (1959) the turnover of lactate was faster, sufficient to account for a maximum of one-sixth of the food utilized (Jayasuriya & Hungate, 1959). Even with this greater rate, only 13 % of the ^{14}C of lactate-2-^{14}C was recovered in the propionate.

Occasionally animals will show an unusually high concentration of propionic acid in the rumen (Balch & Rowland, 1957; Phillipson, 1952). It would be interesting in these cases to determine the size and turnover of the lactate pool and the metabolic pathway, and to identify the bacteria forming the propionic acid.

Succinate. Succinate was suggested as a precursor of propionate in the rumen by the early excellent studies of Johns (1948) in which succinate was shown to be readily decarboxylated to propionic acid by *Veillonella alcalescens*. (Washed cell suspensions of bacteria obtained directly from the rumen also rapidly performed this conversion.)

The pool size and turnover rate of succinate in rumen contents has recently been studied in our laboratory by Blackburn. He found that the extracellular succinate pool in rumen contents of a cow on an alfalfa hay ration was about 0·003 μmol./ml., and that the turnover was rapid. Approximately three-quarters of the label in 2,3-^{14}C-succinate was recovered in propionate, and with this fraction of the pool and its turnover, the contribution of intermediate succinate to the propionic acid was calculated. The results showed that a large fraction of the propionic acid in the rumen arose from succinate which equilibrated with the added labelled molecules. This pool would include all extracellular succinate in the rumen, i.e. that given off as an end product. Succinate produced within a cell and there converted to propionic acid would not be detected by this method. Thus, turnover of intracellular succinate could account for the rest of the propionate formed.

The one-quarter of the label from 2,3-^{14}C-succinate not recovered in propionate was not found in the other volatile fatty acids. It was probably used in cell synthesis, but experimental evidence is lacking.

The organisms decarboxylating the added succinate have not been identified. Since the free energy in the decarboxylation is small, the reaction may be trivial in the economy of the cells, there being little difference between propionic and succinic acids as end products. The chief advantage of decarboxylation could be to prevent a net carbon dioxide fixation, a feature which could possess survival value for bacteria growing in habitats low in carbon dioxide, but which would have little value in the rumen. Perhaps this explains the occurrence of succinic acid as an important fermentation product of so many rumen bacteria.

Attempts to detect colonies causing increased alkalinity in succinate media containing an indicator have led to the isolation by Rouf in the author's laboratory of strains of bacteria which appeared to be *Veillonella alcalescens*. They occurred in low numbers in the rumen contents of a cow fed on alfalfa hay. This result, as well as those of Johns (1948), suggest that the succinate produced by some rumen bacteria may be decarboxylated by others forming propionic acid.

Formate. Formate is produced by *Bacteroides*, *Butyrivibrio*, *Ruminococcus*, *Lachnospira*, and *Succinivibrio*. It is not usually present in sufficient concentration in the rumen to show up in routine analyses, though Gray, Pilgrim, Rodda & Weller (1952) reported that it occasionally constituted as much as 5 % of the total volatile fatty acid, and Annison (1954) found small amounts. Fairly large concentrations of formate have been found in our laboratory in caribou rumen contents fixed with formaldehyde and shipped from Alaska through the courtesy of Dr Pruitt of the University of Alaska. The formic acid could have been a contaminant in the formaldehyde fixative, though the concentrations were quite different in different rumen samples. It would be of interest to study this point further, since formic acid has also been found in the caecal contents of the elephant (Hungate, Phillips, MacGregor, Hungate & Buechner, 1959).

Variability in rumen formate concentration can be due to formate in forage. It constitutes as much as 0·114 % of hay (Claren, 1942; Matsumoto, 1961) and can be found in rumen contents soon after hay is consumed. It does not remain there long, but is rapidly converted (Claren, 1942; Beijer, 1952). Later studies (Carroll & Hungate, 1955) showed that rumen contents convert the carbon in ^{14}C-labelled formate to carbon dioxide. Whole rumen contents produced both methane and carbon dioxide from formate. Presumably hydrogen was formed as an intermediate and converted to methane, since washed cell suspensions of rumen bacteria rapidly convert formate to carbon dioxide and hydrogen (Doetsch, Robinson & Shaw, 1953). In washed cell experiments the oxidation-reduction potential is probably not low enough to permit methane production. The rate of conversion of formate added to rumen contents is much greater than its theoretical maximum rate of production (Carroll & Hungate, 1955), making it doubtful that formate ever accumulates in the rumen due to microbial activities.

Preliminary experiments in the author's laboratory have shown that the formate pool in the rumen turns over very rapidly, about 12 times per minute, indicating an active metabolism of this intermediate.

Hydrogen. There have been many reports of hydrogen in the gas above rumen contents but the careful studies of Lugg (1938) seemed to disprove its occurrence. The classical methods for analysis of rumen gas have included combustion with oxygen, measurement of the decreased volume, and determination of the amount of carbon dioxide. Hydrogen was calculated as one-half of the difference between oxygen consumed and carbon dioxide found. Methane is difficult to combust completely and any incompleteness of combustion would be calculated as hydrogen. Also, analytical error in measurement of oxygen or carbon dioxide could appear as hydrogen.

The recent chromatographic techniques are much more precise than the older methods of gas analysis. We have found that samples of rumen gas contain about 0·05 % hydrogen. Since the host cells cannot utilize hydrogen, the partial pressure in the gas above the rumen contents must correspond fairly closely to the partial pressure of dissolved hydrogen. Rumen contents rapidly absorb hydrogen (Carroll & Hungate, 1955), 1 g. of bovine contents consuming 1·4 μmoles of added H_2 per minute. In view of this capacity for utilization, the presence of hydrogen in the rumen indicates that it is continuously produced. The capacity for utilization exceeds that for production and in consequence the steady-state concentration of hydrogen is low; enzyme systems converting it are not saturated.

Since in the rumen only $\frac{1}{6}$ μmole of methane is produced per minute/g. of contents, requiring $\frac{2}{3}$ μmole of H_2 for its production, it is evident that the capacity of rumen contents to convert hydrogen to methane is more than adequate to account for the rate at which methane is formed. The rate of utilization of added hydrogen by rumen contents is not the total rate since it does not include that arising as a fermentation intermediate. The total rate with added hydrogen would be $1·4 + 0·67$ or 2·07 μmoles/g. minute.

Ethanol. The production of ethanol by pure cultures of *Butyrivibrio, Lachnospira, Ruminococcus, Borrelia* and *Clostridium* suggests that it should occur as an intermediate in the rumen. There are no reports of ethanol in rumen contents so that it does not accumulate as an end product. Regier (1956) determined the rate at which ethanol-1-^{14}C turned over in rumen contents incubated *in vitro* and found it extremely slow, 0·43–1·12 times per hour. The label from the converted ethanol was found exclusively in acetic acid, and she was unable to demonstrate a measurable pool of ethanol. The author has found that ethanol added to rumen contents is metabolized at a very slow rate. These findings indicate that ethanol is relatively unimportant as an intermediate in the rumen.

An acid culture medium usually favours alcohol production by micro-organisms. A low pH during the later stages of fermentation by pure cultures might cause alcohol to form whereas in the better regulated rumen it would not appear. From this standpoint it would be of interest to test for ethanol in grain-fed animals in which the rumen contents stay fairly acid.

Another explanation for the absence of ethanol is that the methane bacteria keep the concentration of hydrogen so low that the equilibrium in the direction of ethanol is insufficient to form appreciable quantities. In this connexion it is of interest that all the pure cultures which produce ethanol also form hydrogen. The reduction of acetyl-CoA to ethanol with NADH would use a high-energy compound available for other purposes if the hydrogen were channelled elsewhere. The reduction of carbon dioxide to methane with hydrogen has a large negative free energy change. Catalysis of this process by the methane bacteria keeps the hydrogen concentration low and may cause the reaction,

$$NADH \rightarrow H + NAD,$$

to proceed rapidly and divert the hydrogen from the ethanol. The methane fermentation may be looked upon as an energy sink into which the hydrogen from all rumen organisms drains, allowing them a higher yield of $\sim P$ and generating additional $\sim P$ in the reaction

$$CO_2 + 4H_4 \rightarrow CH_4 + 2H_2O.$$

Morphologically identifiable bacteria

A discussion of rumen bacteria would be incomplete without mention of forms with a distinctive morphology recognizable by direct microscopic examination. The most striking is *Lampropedia* (Moir & Masson, 1952), cocci arranged in a rectangular plane with individual cells in orderly rows in two directions. *Lampropedia* is not abundant in the rumen and may not grow there since the cultured strains of this species appear to be obligately aerobic. Perhaps the rumen *Lampropedia* are derived from the feed.

Sarcina bakeri (Mann, Masson & Oxford, 1954) is a very large coccus occurring singly, in pairs, or in tetrads. It grows well in proteinaceous media without carbohydrates, is euryoxic, and can ferment sugar with production of acid but no gas. The very large size of the cells (4μ diameter) distinguishes them. They have been seen by Baker, Nasr, Morrice & Bruce (1950) in the rumen and in guinea-pig caecal contents. In the rumen they are not abundant. On several occasions we have observed that low dilutions of rumen contents contained colonies of very large cocci resembling *S. bakeri*.

Spirochetes of the genus *Borrelia* are often seen in rumen contents. They usually can be cultured from the margin of clear zones of cellulose digestion in dilutions of rumen contents into cellulose agar medium (Bryant, 1952), or they may occur in sufficient numbers to appear as isolated colonies in high dilutions in sugar media. They vary in morphology from extremely slender loose spirals to fairly thick ones which taper toward both ends. These organisms are able to move through agar. Colonies can be detected by the hazy appearance of the edge and a tendency to become more diffuse with age, until in old cultures no colonies are visible.

Occasionally, chains of large cocci, not *Peptostreptococcus elsdenii*, are seen in rumen contents. The cells are somewhat distant from each other and in stained smears are connected by a thin strand of material. No organisms with this morphology have been grown in pure culture.

Additional species found in the rumen are *Oscillospira* (Moir, 1951) and an organism observed by Quin (1943), initially thought to be a yeast but later shown to be motile (Westhuizen, Oxford & Quin, 1950).

Cells which could be large bacteria or small protozoa have been described from the sheep rumen by Moir & Masson (1952). Some of these may be related to *Selenomonas* but others do not resemble any known cultivated microbes.

Additional new bacteria will undoubtedly be found in the rumen. However, for each of the many known rumen conversions of forage, at least one bacterial type able to accomplish the conversion in pure culture has been found. Digesters of cellulose, starch, pectin and xylan have been identified. Producers of the chief fermentation products of the rumen, acetic, propionic and butyric acids, carbon dioxide and methane, are known, as are converters of intermediates from which energy can be derived. This makes tenable the hypothesis that most of the important kinds of rumen bacteria have been isolated, though more evidence will be needed before this will be known.

Other aspects of the rumen metabolism of carbohydrate

One of the striking features of carbohydrate metabolism in the rumen is the relative constancy of the ratios in which the volatile fatty acids occur. This is modified by changes in the ration, but with a number of different rations the percentage composition of the volatile fatty acids falls within the range; acetic 47–60 %, propionic 18–23 %, and butyric 19–29 % (Card & Shultz, 1953). Although the ratios of the rumen volatile fatty acids at equilibrium do not necessarily represent the ratios in which they are produced, being a function of both production and absorption, the constancy in the steady state does suggest a constancy in the ratios in which the acids are formed. An interesting but unexplained recent finding is that fine grinding of the feed usually causes an increase in the relative concentration of propionate as compared to acetate.

A change in the proportions in which the volatile fatty acids are produced causes a change in the ratios of fermentation carbon dioxide to methane. Hydrogen is used in the conversion of carbohydrate to propionate and is formed in conversions to acetate and butyrate by known biochemical pathways. In the rumen any excess hydrogen is available for reduction of CO_2 to CH_4, or of triose to propionic acid. The greater the proportion of propionic acid formed, the less the proportion of methane. This inverse correlation between methane and propionic acid has been detected in the animal (Hungate, Mah & Simesen, 1961).

These modifications in the ratios of the volatile fatty acids are important in ruminant nutrition, since the volatile fatty acids are absorbed and utilized (Barcroft, McAnally & Phillipson, 1944; Tappeiner, 1884). Acetate is generally regarded as a source of energy, useful in cold climates but deleterious in hot regions. It does not support a net synthesis of carbohydrate in the ruminant (Kleiber, Smith, Black, Brown & Tolbert, 1952). Propionate is readily converted to carbohydrate; butyric acid also appears to be a source of carbohydrate, though the biochemical pathway for butyrate conversion has not been elucidated. There is no evidence that ruminant tissues metabolize methane. It is excreted through the lungs or by eructation.

In view of the anaerobic conversion of acetate, propionate and butyrate to methane in sludge fermentations and in enrichment cultures, the lack of such a conversion in the rumen appears exceptional. Beijer (1952) found that rumen volatile fatty acids were not converted to methane in the rumen. It may be that the methane conversions of these substrates are too slow to keep pace with rumen metabolism. This does not seem likely. The turnover time for rumen contents is not more than once per day, a turnover which would require an organism to divide only once every 17 hr. in order to avoid being washed out of the rumen. This is a slow division rate and it seems probable that most methanogenic bacteria could achieve it.

Another possibility is that the rumen does not provide the nutrients needed by numerous groups of methanogenic bacteria. In enrichments started with rumen contents, Opperman, Nelson & Brown (1957) were able to obtain methanogenic bacteria attacking acetic acid. This suggests that the bacteria were present in rumen contents and that they could grow with the nutrients available. However, in long-term enrichment cultures the nutritive characteristics of the medium could change markedly. If acetivoric methanogenic bacteria are in rumen contents, some factors holding them in check must operate.

A third possibility is that antagonistic relationships prevent growth of methane bacteria attacking the volatile fatty acids in the rumen. In general, antibiotics in nature have no demonstrable effect, making nutritional hypotheses more attractive. Whatever the explanation, the mutualistic usefulness of the rumen fermentation, in so far as carbohydrate is concerned, depends on the circumstance that the methane conversion does not extend to the fermentation products of value to the host.

Several investigators have noted that *Desulfovibrio desulfuricans* does not occur commonly in the rumen. Some sulphate is found in plant materials (Lewis, 1954) and considerable quantities of sulphite are sometimes added to silage as a preservative. Yet in animals fed such silage the digesta has no odour of hydrogen sulphide. *Desulfovibrio* seems to be almost entirely absent from the rumen of normal animals (Lewis, 1954). The report from the author's laboratory (Gutierrez, 1953) that *Desulfovibrio* had been observed in rumen contents was probably in error. The identification was based on the curved shape of the cells and the motility, together with blackening of the medium when ferrous salts were included. However, cysteine was used as a reducing agent and with the subsequent discovery that *Selenomonas* produces hydrogen sulphide from cysteine and is a curved motile rod, it seems probable that the colonies classified as *Desulfovibrio* were actually *Selenomonas*. Evidence that this explanation could be correct was obtained by P. H. Smith in the author's laboratory after the possible error was brought to our attention and *Selenomonas* had been described.

The experiments of Sapiro, Hoflund, Clark & Quin (1949) and Lewis (1951) showed that nitrate was reduced to nitrite in the rumen of sheep. When forage contains large quantities of nitrate, the nitrite formed may be absorbed into the blood stream and convert haemoglobin to methaemoglobin, in some cases causing death of the animal. Small amounts of nitrite are further reduced in the rumen (Pfander, Garner, Ellis & Muhrer, 1957).

NITROGEN CONVERSIONS IN THE RUMEN

Beginning with the early work of McDonald (1948) on the conversions of protein it has been found almost universally that most of the protein in the feed is broken down to ammonia. Production of ammonia from protein by various pure cultures has been studied by Bryant (1961); strains of *Bacteroides ruminicola, Selenomonas ruminantium, Peptostreptococcus elsdenii* and *Eubacterium ruminantium* were particularly active. Some *Butyrivibrio* strains showed slight production but others were inactive. A rumen strain of *Lactobacillus fermenti* forms ammonia from arginine (Perry & Briggs, 1957).

Rumen ammonia is synthesized into microbial nitrogenous components, with any excess absorbed from the rumen and converted to urea. Ammonia is the preferred nitrogenous substrate of *Ruminococcus* strains and *Bacteroides succinogenes* (Bryant & Robinson, 1961 *a*), *Butyrivibrio* (Gill & King, 1958), *Lactobacillus bifidus* (Phillipson, 1959), and *Eubacterium ruminantium* (Bryant, 1961). Many additional types grow on ammonia if other sources of nitrogen are not available (Bryant, 1961).

Ingested proteins are rapidly digested (McDonald, 1948; Blackburn & Hobson, 1960 *a*) in the rumen. The concentration of amino acids in rumen contents is very low. The precise extent to which the nitrogen in feed protein is reassimilated as amino acids or deaminated is unknown. The fact that many pure cultures can use both ammonia and amino acids is evidence that both can be assimilated in the rumen. Measurements of ammonia and amino acid turnover rates in rumen contents will be of much interest. It might be expected that when nitrogen is in short supply a greater proportion of plant protein would be assimilated as amino acids.

There have been a number of attempts to identify the bacteria responsible for the digestion of protein in the rumen. Blackburn & Hobson (1962) measured proteolysis by bacteria in sheep on a ration containing casein as an important ingredient. Isolated strains of *Bacteroides amylophilus* digested casein to the extent of 33–84 %. Two strains of *Lachnospira* hydrolysed less than 10 % in 4 days, but some *Butyrivibrio* isolates showed as much as 94 % digestion of the casein. Others were inactive. *Bacteroides ruminicola* digested 4–69 %, *Selenomonas ruminantium* 46–95 %, and unidentified rods and cocci 0–88 % of the casein. The authors conclude that proteolysis is not a function of particular highly adapted species but is a widely distributed characteristic.

The results described by Bryant (1961) agree with this view. The author has observed that *Clostridium lochheadii* very actively digests the casein in litmus milk. Varying degrees of peptonization and liquefaction of the curd by strains of *Lachnospira*, *Butyrivibrio* and *Selenomonas* have also been observed.

Studies on the agents hydrolysing urea have yielded similar results. The capacity to hydrolyse this material is possessed to a small degree by a great many species (Blackburn & Hobson, 1962; Carroll, 1960), with no particularly active strains, though Gibbons & Doetsch (1959) isolated a *Lactobacilliis bifidus* that hydrolysed urea rapidly.

The extent to which substrate proteins in the ration of the ruminant are converted into microbial protein has been studied by Weller *et al.* (1958). These investigators estimated bacterial nitrogen from the quantity of diaminopimelic acid, a component peculiar to schizophytes. This component of the bacterial envelope was early suggested by Synge (Work, 1950) as an index for measuring production of microbial cells in the rumen. It is absent from the protozoa. Of the total rumen nitrogen 63–82 % was microbial, of which 42–61 % was bacterial, and 21 % protozoan nitrogen.

NUTRITIONAL REQUIREMENTS OF THE BACTERIA

During the initial stages of the study and isolation of the rumen bacteria rumen fluid was used as a source of possible growth factors (Hungate, 1947). Some of the isolated cultures required rumen fluid; others grew well with conventional media such as yeast extract and peptones. Of the types growing without rumen fluid *Butyrivibrio* requires only the ingredients of the ruminant feed in order to grow. It grows better on alfalfa hay extract with carbohydrate than on commercial media containing derivatives from animal or other plant materials (Margherita, 1962). Other rumen bacteria growing well without the addition of rumen fluid are *Lachnospira*, *Bacteroides amylophilus*, *Succinivibrio*, *Succinimonas*, *Peptostreptococcus*, and *Clostridium lochheadii*.

Ruminococcus and *Bacteroides succinogenes* were the first organisms shown to require nutrient factors in rumen fluid (Hungate, 1950). Bryant & Doestsch (1954) have identified branched and straight short-chain fatty acids as the important required constituents. These are products of the rumen fermentation of amino acids (El-Shazly, 1952). They are used for the formation of leucine (Allison, Bryant & Doetsch, 1962) and possibly also for the longer-chained fatty acids which make up the component of the bacterial cell envelope. Many of the fatty acids of milk are derived from those of the bacteria in the rumen (Keeney, Katz & Allison, 1962).

Methanobacterium ruminantium requires rumen fluid for growth, but because of the difficulty in growing this organism its nutrition has not been analysed. The rumen fluid requirement may be concerned with the low electro-potential required for the initiation of growth.

Selenomonas and *Borrelia* are other types for which ingredients in rumen fluid are essential.

Many of the strains of *Bacteroides ruminicola* require rumen fluid. Part of this requirement is met by haemin (Bryant, 1961), apparently used in the synthesis of cytochrome in these bacteria.

A number of rumen bacteria require B vitamins (Gill & King, 1958; Ayers, 1958; Bryant & Robinson, 1961 c; Bryant, Robinson & Chu, 1959).

Osborn & Mendel (1918) observed that milk was rich in B vitamins, and later investigations have shown that the rumen is the site of B vitamin synthesis (Goss, 1943; Bechdel, Honeywell, Dutcher & Knutsen, 1928). Almost all of the studies of vitamin synthesis have used whole rumen contents or *in vitro* incubated rumen contents and little has been done to determine the amounts and kinds of vitamins synthesized by individual types of rumen bacteria. Hardie (1952) showed that the synthesis of vitamin B_{12} by *Streptococcus bovis* could account for only a very small fraction of the total formed in the rumen.

GROWTH

In the last analysis, ability to grow and reproduce determines survival. Growth in the rumen is limited by the amount of food available; addition of food to rumen contents always increases metabolic activity. The bacteria most rapidly converting food into cells would appear to possess an advantage. Most of the rumen bacteria do grow rapidly; *Streptococcus bovis*, *Butyrivibrio* and *Lachnospira* are among the fastest. Colonies of *S. bovis* have been distinguished with the naked eye as early as 6 hr. after the inoculation of the agar media. This was the phenomenon which first attracted attention to *S. bovis* as the agent causing acidity in animals receiving sudden large doses of grain in the ration.

Colony size is not necessarily a good index of growth rate because mucoidal material may greatly increase its apparent size; this happens with some types of *Butyrivibrio* colonies.

In agar media inoculated with dilutions of rumen contents many colonies do not appear until several days after incubation. Bryant & Robinson (1961 b) find that most of these resemble the types appearing early during incubation, rather than other, slow-growing forms.

Methanobacterium ruminantium colonies do not appear in agar tubes until several days after inoculation, suggesting either a slower growth rate or a lag period during which the medium or the organisms are conditioned. The methane bacteria do not need a rapid growth rate in order to maintain themselves in the rumen as they do not compete with the bacteria attacking the primary foods. The total rumen fermentation increases after feeding, and then slowly drops to the pre-feeding level. In an extensive series of studies on animals fed morning and evening (Hungate, Phillips, Hungate & MacGregor, 1960), the fastest fermentation rate per gram of contents was rarely as much as double the slowest, even when rates after feeding were compared with rates just before. On the assumption that hydrogen is a constant component of the products formed by attack on the primary substrate, it is probably supplied fairly continuously to the methane bacteria, which would therefore not be expected to develop especially rapidly just after food is consumed by the ruminant.

In the first stages of utilization of ingested forages the sugars are rapidly fermented. Later the fibre-digesting bacteria are chiefly responsible for production of hydrogen, carbon dioxide, and the acid fermentation products. The kinetics of attack on the various components of the fibre have been studied by adding a sample of hay to manometric flasks containing rumen fluid, and following the course of production of fermentation products.

In experiments with alfalfa hay the 35 % fraction soluble in water at 39° is very rapidly utilized. Much of it is stored in the holotrich protozoa. The burst of activity due to these soluble materials ceases within an hour, being followed by a slower rate which continues for a longer period, about 6 hr., and this in turn is followed by a fermentation which continues at a rate which diminishes very slowly with time and is still measurable 36 hr. after the substrate was added. This last period of fermentation is interpreted as that supported by the slow digestion of the more resistant fibrous material in the hay. This interpretation is supported by experiments in which hay fibre treated to remove the lignin is fermented in parallel with untreated fibre. The initial fermentation rate is indistinguishable in the two cultures, but later the holocellulose (delignified fibre) ferments faster than the control and a greater total amount of fermentation products is formed. The initial similarity in rates presumably indicates that in both substrates an identical surface is exposed to the enzymes and that this surface limits enzymatic activity.

It would seem that the delignification process removes materials protecting the inner parts of the fibres from enzyme action so that a greater depth of material can be attacked. The slow decrease in rate of fermentation during the third extended period may be due to a gradual decrease in the surface area. An alternative explanation would be that there are numerous substrates in the fibres exhausted in turn.

Both the rate of growth and the efficiency with which substrate can be converted into cell material are important.

THERMODYNAMICS OF THE RUMEN SYMBIOSIS

The ruminant can be pictured as a plankton feeder which cultivates its plankton under highly productive conditions (in a high-producing animal 16 % (w/v) of substrate converted each 24 hr.), and in which the planktonic cells and the wastes from their carbohydrate fermentation are absorbed and utilized.

For growth, both energy and building materials are required. The amount of energy available to the rumen micro-organisms is small because it is derived only through fermentation, and usually limits their growth. With this energy 8–12 % of the substrate is converted into microbial cells. Were there more energy, the fraction stored as cells could be greater.

Aerobic conditions provide greater energy, and various aerobically growing organisms conserve as much as 60 % of the substrate in the form of cells.

The energy not available to the rumen microbes because of anaerobiosis is available to the aerobic host. The volatile fatty acids can be absorbed and oxidized, and although loss of methane decreases the host's energy yield (Krogh & Schmidt-Jensen, 1920), the quantity is still much greater than that available to the microbes. Thus, in so far as energy alone is concerned, the ruminant can form more cell material per unit of food than can the microbes.

But growth requires also the building materials for cell constituents. This poses no problem for the rumen microbes. The supply of building materials in most forages is adequate for optimal cell synthesis. The aerobic ruminant, with greater available energy, could synthesize much more cell material than the anaerobic microbes, but the digested micro-organisms are its building materials and since they are limited by the energy from fermentation, the growth of the ruminant is similarly limited. This explains why a smaller fraction of the digested food is utilized for growth in the ruminant than in monogastric animals such as the chicken or pig. The efficiency of conversion of food is lower

because cell synthesis is limited by an anaerobic process even though the
ruminant is an aerobe.

An obvious way to increase ruminant growth would be to supply
additional building material. But this is not easily accomplished. If
more protein is fed than is needed for building microbial cells, it will
be fermented for additional energy. Protein fermentation does not,
however, produce the growth materials required by the host.

This emphasizes that many foods are most economically fed to monogastric animals, including man, rather than to cattle, sheep or goats.
The ruminant is nutritionally superior only when its food can be used
exclusively by micro-organisms and not directly by the animal itself;
viz. (1) when the fibrous components of forages are high, and (2) when
non-fibrous carbohydrate feeds are so poor in essential nutrients that
monogastric animals using them would grow less than rumen micro-
organisms. In the first instance the ruminant superiority is due to the
ability of the microbes to digest fibre and in the second to their ability
to synthesize cell material from simple precursors. Practical research in
ruminant nutrition should be increasingly directed toward exploiting
these two features.

On the basis of the above considerations it would be expected that
for a given food the dry weight yield of milk could be greater than of
flesh. Milk is roughly 1/3 protein, 1/3 carbohydrate, and 1/3 fat, the
total product containing about 6 % nitrogen, whereas the flesh of the
growing animal is 16 % nitrogen. For the milking animal the excess
energy can be stored in lactose and fat. It is also clear that if high
efficiency in food utilization is to be realized, much of the gain in body
weight must be fat.

CONCLUSIONS

The question 'Are the rumen bacteria true symbionts?' can be answered
in the affirmative. The host and the microbial population have adapted
to the point of mutual dependence and neither succeeds in nature
without the other. This does not necessarily apply to each individual
kind of rumen micro-organism. Not all individuals in a ruminant
species contain identical kinds of protozoa. It has even been found that
all the protozoa can be removed and excluded without visibly damaging
the host. The same would probably be observed if the ruminant could
similarly be deflorated of individual types of bacteria. The role of individual microbes in the complex can be identified but each organism is
interrelated in too many ways to permit a simple definition of its
relationship to the ruminant. Microbial interrelationships are equally

important for the nutritional success of the host and as far as the ruminant is concerned, its relationship is to the total microbial population. The complicated ruminant stomach has evolved with the integrated evolution of the complex rumen microcosm and each is now dependent on the other.

It is thus the rumen population as a whole that is symbiotic. To a certain extent this population is fixed. It gives rise to volatile fatty acids, methane, and carbon dioxide under diverse conditions of rations and management. To a certain extent the micro-organisms composing it are characteristic and specific for this habitat but they do not occur in constant ratios nor are the individual kinds constant. They vary in minor characteristics at different times and in different places and some are probably components of other related habitats.

But there are enough that are almost universally present for the whole population to be regarded as a true symbiont. The rumen is a unique habitat and in it has evolved a symbiotic population of micro-organisms peculiarly fitted to the host and to each other.

REFERENCES

ALLISON, M. J., BRYANT, M. P. & DOETSCH, R. N. (1962). Studies on the metabolic function of branched-chain volatile fatty acids, growth factors for rumino-cocci. I. Incorporation of isovalerate into leucine. *J. Bact.* **83**, 523.

ANDERSON, R. L. & ORDAL, E. J. (1961). *Cytophaga succinicans* sp.n., a faculta-tively anaerobic aquatic myxobacterium. *J. Bact.* **81**, 130.

ANKERSMIT, P. (1905). Untersuchungen über die Bakterien im Verdauungskanal des Rindes. *Zbl. Bakt.* (1 Abt.), **39**, 687.

ANNISON, E. F. (1954). Some observations on volatile fatty acids in the sheep's rumen. *Biochem. J.* **57**, 400.

AYERS, W. A. (1958). Nutrition and physiology of *Ruminococcus flavefaciens*. *J. Bact.* **76**, 504.

BAKER, F. (1942). Normal rumen microflora and microfauna of cattle. *Nature, Lond.* **149**, 220.

BAKER, F. & MARTIN, R. (1938). Disintegration of cell-wall substances in the gastro-intestinal tract of herbivora. *Nature, Lond.* **141**, 877.

BAKER, F., NASR, H., MORRICE, F. & BRUCE, J. (1950). Bacterial breakdown of structural starches and starch products in the digestive tract of ruminant and non-ruminant mammals. *J. Path. Bact.* **62**, 617.

BALCH, D. A. & ROWLAND, S. J. (1957). Volatile fatty acids and lactic acid in the rumen of dairy cows receiving a variety of diets. *Brit. J. Nutr.* **11**, 288.

BALDWIN, R. L., WOOD, W. A. & EMERY, R. S. (1962). Conversion of lactate-C^{14} to propionate by the rumen microflora. *J. Bact.* **83**, 907.

BARCROFT, J., McANALLY, R. A. & PHILLIPSON, A. T. (1944). Absorption of volatile acids from the alimentary tract of the sheep and other animals. *J. exp. Biol.* **20**, 120.

BARY, A. DE (1879). De la Symbiose. *Rev. intern. Sci.* **3**, 301.

BECHDEL, S. I., HONEYWELL, H. E., DUTCHER, R. A. & KNUTSEN, M. H. (1928). Synthesis of vitamin B in the rumen of the cow. *J. biol. Chem.* **80**, 231.

294

R. E. HUNGATE

BEIJER, W. H. (1952). Methane fermentation in the rumen of cattle. *Nature, Lond.* **170**, 576.

BLACKBURN, T. H. & HOBSON, P. N. (1960*a*). Proteolysis in the sheep rumen by whole and fractionated rumen contents. *J. gen. Microbiol.* **22**, 272.

BLACKBURN, T. H. & HOBSON, P. N. (1960*b*). Isolation of proteolytic bacteria from the sheep rumen. *J. gen. Microbiol.* **22**, 282.

BLACKBURN, T. H. & HOBSON, P. N. (1962). Further studies on the isolation of proteolytic bacteria from the sheep rumen. *J. gen. Microbiol.* **29**, 69.

BRYANT, M. P. (1952). The isolation and characterization of a spirochete from the bovine rumen. *J. Bact.* **64**, 325.

BRYANT, M. P. (1956). The characteristics of strains of *Selenomonas* isolated from bovine rumen contents. *J. Bact.* **72**, 162.

BRYANT, M. P. (1959). Bacterial species of the rumen. *Bact. Rev.* **23**, 125.

BRYANT, M. P. (1961). The nitrogen metabolism of pure cultures of ruminal bacteria. *U.S.D.A. Agric. Res. Serv.*, April, pp. 44–92.

BRYANT, M. P. & BURKEY, L. A. (1953). Cultural methods and some characteristics of some of the more numerous groups of bacteria in the bovine rumen. *J. Dairy Sci.* **36**, 205.

BRYANT, M. P. & DOETSCH, R. N. (1954). Factors necessary for the growth of *Bacteroides succinogenes* in the volatile acid fraction of rumen fluid. *Science*, **120**, 944.

BRYANT, M. P. & DOETSCH, R. N. (1955). Factors necessary for the growth of *Bacteroides succinogenes* in the volatile acid fraction of rumen fluid. *J. Dairy Sci.* **38**, 340.

BRYANT, M. P. & ROBINSON, I. M. (1961*a*). Studies on the nitrogen requirements of some ruminal cellulolytic cocci. *Appl. Microbiol.* **9**, 96.

BRYANT, M. P. & ROBINSON, I. M. (1961*b*). An improved non-selective culture medium for ruminal bacteria and its use in determining diurnal variation in numbers of bacteria in the rumen. *J. Dairy Sci.* **44**, 1446.

BRYANT, M. P. & ROBINSON, I. M. (1961*c*). Some nutritional requirements of the genus *Ruminococcus*. *Appl. Microbiol.* **9**, 91.

BRYANT, M. P., ROBINSON, I. M. & CHU, H. (1959). Observations on the nutrition of *Bacteroides succinogenes*, a ruminal cellulolytic bacterium. *J. Dairy Sci.* **42**, 1831.

BRYANT, M. P., ROBINSON, I. M. & LINDAHL, I. L. (1961). A note on the flora and fauna in the rumen of steers fed a feedlot bloat-provoking ration and the effect of penicillin. *Appl. Microbiol.* **9**, 511.

BRYANT, M. P. & SMALL, N. (1956*a*). The anaerobic monotrichous butyric acid-producing curved rod-shaped bacteria of the rumen. *J. Bact.* **72**, 16.

BRYANT, M. P. & SMALL, N. (1956*b*). Characteristics of two new genera of anaerobic curved rods isolated from the rumen of cattle. *J. Bact.* **72**, 22.

BRYANT, M. P., SMALL, N., BOUMA, C. & CHU, H. (1958). *Bacteroides ruminicola* n.sp., and *Succinimonas amylolytica* the new genus and species. Species of succinic acid-producing anaerobic bacteria of the bovine rumen. *J. Bact.* **76**, 15.

BRYANT, M. P., SMALL, N., BOUMA, C. & ROBINSON, I. M. (1958). Characteristics of ruminal anaerobic cellulolytic cocci and *Cillobacterium cellulosolvens* n.sp. *J. Bact.* **76**, 529.

CARD, C. S. & SHULTZ, L. H. (1953). Effect of the ration on volatile fatty acid production in the rumen. *J. Dairy Sci.* **36**, 599.

CARROLL, E. J. (1960). Urea-utilizing organisms from the rumen. Ph.D. thesis, University of Missouri.

CARROLL, E. J. & HUNGATE, R. E. (1954). The magnitude of the microbial fermentation in the bovine rumen. *Appl. Microbiol.* **2**, 205.

CARROLL, E. J. & HUNGATE, R. E. (1955). Formate dissimilation and methane production in bovine rumen contents. *Arch. Biochem. Biophys.* **56**, 525.

CLAREN, O. B. (1942). Über das Schicksal verfütterter Ameisensäure im Verdauungstrakt des Wiederkäuers. *Z. physiol. Chem.* **276**, 97.

CLARK, R. & LOMBARD, W. A. (1951). Studies on the alimentary tract of merino sheep in South Africa. 22. The effect of the pH of the ruminal contents on rumen motility. *Onderstepoort J. Vet. Res.* **25**, 79.

DEHORITY, B. A. & JOHNSON, R. R. (1961). Effect of particle size upon the *in vitro* cellulose digestibility of forages by rumen bacteria. *J. Dairy Sci.* **44**, 2242.

DOETSCH, R. N., ROBINSON, R. Q. & SHAW, J. C. (1953). Catabolic reactions of mixed suspensions of bovine rumen bacteria. *J. Dairy Sci.* **36**, 825.

ELSDEN, S. R., GILCHRIST, F. M., LEWIS, D. & VOLCANI, B. E. (1951). The formation of fatty acids by a gram-negative coccus. *Biochem. J.* **49**, lxix.

ELSDEN, S. R., VOLCANI, B. E., GILCHRIST, F. M. & LEWIS, D. (1956). Properties of a fatty acid forming organism isolated from the rumen of sheep. *J. Bact.* **72**, 681.

EL-SHAZLY, K. (1952). Degradation of protein in the rumen of the sheep. 2. The action of rumen micro-organisms on amino-acids. *Biochem. J.* **51**, 647.

GIBBONS, R. J. & DOETSCH, R. N. (1959). Physiological study of an obligately anaerobic ureolytic bacterium. *J. Bact.* **77**, 417.

GILL, J. W. & KING, K. W. (1958). Nutritional characteristics of a *Butyrivibrio*. *J. Bact.* **75**, 666.

GOSS, H. (1943). Some peculiarities of ruminant nutrition. *Nutr. Abst. & Rev.* **12**, 531.

GRAY, F. V., PILGRIM, A. F., RODDA, H. J. & WELLER, R. A. (1952). Fermentation in the rumen of sheep. IV. The nature and origin of the volatile fatty acids in the rumen of the sheep. *J. exp. Biol.* **29**, 57.

GUTIERREZ, J. (1953). Numbers and characteristics of lactate-utilizing organisms in the rumen of cattle. *J. Bact.* **66**, 123.

GUTIERREZ, J. & DAVIS, R. E. (1959). Bacterial ingestion by the rumen ciliates *Entodinium* and *Diplodinium*. *J. Protozool.* **6**, 222.

GUTIERREZ, J., DAVIS, R. E., LINDAHL, I. J. & WARWICK, E. J. (1959). Bacterial changes in the rumen during the onset of feed-lot bloat of cattle and characteristics of *Peptostreptococcus elsdenii* n.sp. *Appl. Microbiol.* **7**, 16.

HALL, E. R. (1952). Investigations on the microbiology of cellulose utilization in domestic rabbits. *J. gen. Microbiol.* **7**, 350.

HAMLIN, L. J. & HUNGATE, R. E. (1956). Culture and physiology of a starch-digesting bacterium (*Bacteroides amylophilus* n.sp.) from the bovine rumen. *J. Bact.* **72**, 548.

HARDIE, W. B. (1952). Vitamin B_{12} production by *Streptococcus bovis* from the bovine rumen. M.S. thesis, Washington State University.

HARKER, K. W., TAYLOR, J. I. & ROLLINSON, D. H. L. (1954). Studies on the habits of zebu cattle. I. Preliminary observations on grazing habits. *J. agric. Sci.* **44**, 193.

HENNEBERG, W. (1922). Untersuchungen über die Darmflora des Menschen mit besonderer Berücksichtigung der jodophilen Bakterien im Menschen- und Tierdarm sowie im Kompostdünger. *Zbl. Bakt.* (2 Abt.), **55**, 242.

HUNGATE, R. E. (1947). Studies on cellulose fermentation. III. The culture and isolation of cellulose-decomposing bacteria from the rumen of cattle. *J. Bact.* **53**, 631.

HUNGATE, R. E. (1950). The anaerobic mesophilic cellulolytic bacteria. *Bact. Rev.* **14**, 1.

HUNGATE, R. E. (1957). Micro-organisms in the rumen of cattle fed a constant ration. *Canad. J. Microbiol.* **3**, 289.

HUNGATE, R. E. (1962). Ecology of bacteria. Chap. in vol. IV, *The Bacteria*. Ed. Gunsalus & Stanier. New York: Academic Press.

HUNGATE, R. E., DOUGHERTY, R. W., BRYANT, M. P. & CELLO, R. M. (1952). Microbiological and physiological changes associated with acute indigestion in sheep. *Cornell Vet.* 42, 423.

HUNGATE, R. E., MAH, R. A. & SIMESEN, M. (1961). Rates of production of individual volatile fatty acids in the rumen of lactating cows. *Appl. Microbiol.* 9, 554.

HUNGATE, R. E., PHILLIPS, G. D., HUNGATE, D. P. & MACGREGOR, A. (1960). A comparison of the rumen fermentation in European and zebu cattle. *J. agric. Sci.* 54, 196.

HUNGATE, R. E., PHILLIPS, G. D., MACGREGOR, A., HUNGATE, D. P. & BUECHNER, H. K. (1959). Microbial fermentation in certain mammals. *Science*, 130, 1192.

JAYASURIYA, G. C. N. & HUNGATE, R. E. (1959). Lactate conversions in the bovine rumen. *Arch. Biochem. Biophys.* 82, 274.

JOHNS, A. T. (1948). The production of propionic acid by decarboxylation of succinic acid in a bacterial fermentation. *Biochem. J.* 42, ii.

JOHNS, A. T. (1951). Isolation of a bacterium, producing propionic acid, from the rumen of sheep. *J. gen. Microbiol.* 5, 317.

KEENEY, M., KATZ, I. & ALLISON, M. J. (1962). On the probable origin of some milk fat acids in rumen microbial lipids. *J. Amer. Oil Chemists' Soc.* 39, 198.

KING, K. W. & SMITH, P. H. (1955). Comparisons of two media proposed for the isolation of bacteria from the rumen. *J. Bact.* 70, 726.

KLEIBER, M., SMITH, A. H., BLACK, A. L., BROWN, M. A. & TOLBERT, B. M. (1952). Acetate as a precursor of milk constituents in the intact dairy cow. *J. biol. Chem.* 197, 371.

KROGH, A. & SCHMIDT-JENSEN, H. O. (1920). The fermentation of cellulose in the paunch of the ox and its significance in metabolism experiments. *Biochem. J.* 14, 686.

KROGH, N. (1959). Studies on alterations in the rumen fluid of sheep, especially concerning the microbial composition, when readily available carbohydrates are added to the food. I. Sucrose. *Acta vet. Scand.* 1, 74.

KRZYWANECK, F. W. (1929). Über die Temperatur im Pansen des Schafes. *Pflüg. Arch. ges. Physiol.* 222, 89.

LESSEL, E. F. & BREED, R. S. (1954). *Selenomonas* Boskamp, 1922—a genus that includes species showing an unusual type of flagellation. *Bact. Rev.* 18, 165.

LEWIS, D. (1951). The metabolism of nitrate and nitrite in the sheep. 2. Hydrogen donators in nitrate reduction by rumen micro-organisms *in vitro*. *Biochem. J.* 49, 149.

LEWIS, D. (1954). The reduction of sulphate in the rumen of the sheep. *Biochem. J.* 56, 391.

LUGG, J. W. H. (1938). Identification and measurement of the combustible gases that occur in the gaseous metabolic products of the sheep. *J. agric. Sci.* 28, 688.

McDONALD, I. W. (1948). The extent of conversion of food protein to microbial protein in the rumen of sheep. *J. Physiol.* 107, 21 P.

MAKI, L. R. (1954). Experiments in the microbiology of cellulose decomposition in a municipal sewage plant. *Antonie van Leeuwenhoek*, 20, 185.

MANN, S. O., MASSON, F. M. & OXFORD, A. E. (1954). Facultative anaerobic bacteria from the sheep's rumen. *J. gen. Microbiol.* 10, 142.

MARGHERITA, S. (1962). Serological studies of *Butyrivibrio*. Ph.D. thesis, University of California, Davis.

MATSUMOTO, T. (1961). The influence of feed and feeding upon the ruminal gas formation. IX. Formate dissimilation and gas production in the rumen of the goat. *Tohoku J. agric. Res.* 12, 213.

MOIR, R. J. (1951). The seasonal variation in the ruminal microorganisms of grazing sheep. *Aust. J. agric. Res.* **2**, 322.

MOIR, R. J. & MASSON, M. J. (1952). An illustrated scheme for the microscopic identification of the rumen microorganisms of sheep. *J. Path. Bact.* **64**, 343.

MYLROIE, R. L. & HUNGATE, R. E. (1954). Experiments on the methane bacteria in sludge. *Canad. J. Microbiol.* **1**, 55.

OPPERMAN, R. A., NELSON, W. O. & BROWN, R. E. (1957). *In vitro* studies on methanogenic rumen bacteria. *J. Dairy Sci.* **40**, 779.

OSBORN, T. B. & MENDEL, L. B. (1918). Milk as a source of water-soluble vitamins. *J. biol. Chem.* **34**, 537.

PARTHASARATHY, D. & PHILLIPSON, A. T. (1953). The movement of potassium, sodium, chloride, and water across the rumen epithelium of sheep. *J. Physiol.* **121**, 452.

PERRY, K. D. & BRIGGS, C. A. E. (1957). The normal flora of the bovine rumen. IV. Qualitative studies of lactobacilli from cows and calves. *J. appl. Bact.* **20**, 119.

PFANDER, W. H., GARNER, G. B., ELLIS, W. C. & MUHRER, M. E. (1957). The etiology of 'nitrate poisoning' in sheep. *Bull. Mont. agric. Exp. Sta.* no. 637, 12 pp.

PHILLIPSON, A. T. (1952). The fatty acids present in the rumen of lambs fed on a flaked maize ration. *Brit. J. Nutr.* **6**, 190.

PHILLIPSON, A. T. (1959). Assimilation of ammonia nitrogen by rumen bacteria. *Nature, Lond.* **183**, 402.

POCHON, J. (1935). Role d'une bactérie cellulolytique de la panse, *Plectridium cellulolyticum*, dans la digestion de la cellulose chez les ruminants. *Ann. Inst. Pasteur*, **55**, 676.

QUIN, J. I. (1943). Studies on the alimentary tract of merino sheep in South Africa. VIII. The pathogenesis of acute tympanites (bloat). *Onderstepoort J. vet. Sci.* **18**, 113.

REGIER, C. (1956). The conversion of ethanol in the bovine rumen. M.S. thesis, Washington State University.

REISET, J. (1868). Étude des gaz produits pendant la météorization des ruminants. Application à la thérapeutique vétérinaire. *C.R. Acad. Sci., Paris*, **66**, 176.

SAPIRO, M. L., HOFLUND, S., CLARK, R. & QUIN, J. I. (1949). The fate of nitrate in ruminal ingesta as studied *in vitro*. *Onderstepoort. J. vet. Sci.* **22**, 357.

SIJPESTEIJN, A. K. (1948). Cellulose-decomposing bacteria from the rumen of cattle. Thesis, Leiden.

SMITH, P. H. & HUNGATE, R. E. (1958). Isolation and characterization of *Methanobacterium ruminantium*, n.sp. *J. Bact.* **75**, 713.

STORZ, H. & HUNGATE, R. E. (1962). Experiments on the attachment of cellulolytic bacteria to solid particles of rumen contents. (In the Press.)

TAPPEINER, H. VON (1884). Untersuchungen über die Gärung der Cellulose, insbesondere über deren Lösung im Darmkanal. *Z. Biol.* **20**, 52.

WARNER, A. C. I. (1956). Criteria for establishing the validity of *in vitro* studies with rumen microorganisms in so-called artificial rumen systems. *J. gen. Microbiol.* **14**, 732.

WELLER, R. A., GRAY, F. V. & PILGRIM, A. F. (1958). The conversion of plant nitrogen to microbial nitrogen in the rumen of the sheep. *Brit. J. Nutr.* **12**, 421.

WESTHUIZEN, G. C. A. VAN DER, OXFORD, A. E. & QUIN, J. I. (1950). Studies on the alimentary tract of merino sheep in South Africa. XVI. On the identity of *Schizosaccharomyces ovis*. Part I. Some yeast-like organisms isolated from the rumen contents of sheep fed on a lucerne diet. *Onderstepoort J. vet. Sci.* **24**, 119.

WORK, E. (1950). Chromatographic investigations of amino-acids from micro-organisms. II. Isolation of two unknown substances from *Corynebacterium diphtheriae. Biochim. biophys. Acta*, **5**, 204.

THE GROWTH AND METABOLISM
OF RUMEN CILIATE PROTOZOA

G. S. COLEMAN

Biochemistry Department, Agricultural Research Council
Institute of Animal Physiology, Babraham, Cambridge

The purpose of this chapter is to discuss the growth and metabolism of rumen ciliates *in vivo* and *in vitro* and the relationship of these organisms to the host and to rumen bacteria. Of necessity the evidence is fragmentary and the conclusions drawn are tentative because ruminants can live and grow in the absence of rumen ciliates and it has not yet proved possible to culture axenic rumen protozoa *in vitro*. Although flagellates occur in the rumen little is known about them and only the ciliate protozoa will be discussed here. Where possible these will be considered in two groups, the holotrichs and the oligotrichs. The latter group has recently been revised by Corliss (1959) who places these rumen protozoa in the order Entodiniomorphida (Oligotrich *sensu lato*) and reserves the order Oligotrichida for protozoa such as *Halteria*. This revision will be followed here.

Rumen protozoa were first described by Gruby & Delafond (1843). The earliest attempts to culture them *in vitro* and the early theories on the role of these organisms in the rumen were summarized by Becker, Schultz & Emmerson (1929) and Becker (1932). The literature prior to 1955 has already been reviewed in detail by Oxford (1955) and Hungate (1955), and will be considered here only where it is essential to the understanding of more recent studies.

MAINTENANCE OF ENTODINIOMORPHID PROTOZOA
IN VITRO

Over the last forty years several attempts to grow rumen Entodiniomorphid protozoa *in vitro* have been made but only two have met with apparent success. Although Margolin (1930) maintained these protozoa alive for more than 3 weeks on a medium which contained cellulose and a hay infusion, Hungate (1942) was the first to maintain cattle ciliates alive *in vitro* for long periods. He used a 0·6 % NaCl solution containing small amounts of phosphate, calcium and magnesium plus a little cellulose, contained in stoppered flasks and equilibrated with

95 % N_2 + 5 % CO_2 in the gas phase. Under these conditions several species of Entodiniomorphid protozoa, principally diplodinia, could be grown from a 1·5 % inoculum of rumen contents at a cell population of 100–1000 per ml., dividing every 24–48 hr. Although Hungate (1942, 1943) produced evidence that diplodinia separated from the bulk of extracellular bacteria contained a cellulase and a cellobiase, no further work was published on these protozoa at that time.

In 1953 Sugden attempted to repeat Hungate's work using Entodiniomorphid protozoa from British sheep but could not maintain *Metadinium medium* or *Entodinium caudatum* alive for more than 14 days, although similar growth conditions were used.

Kandatsu & Takahashi (1955 a, b, 1956) maintained *Entodinium* spp. dividing *in vitro* for over 30 days on a medium of salts, hay extract, rumen bacteria, fresh clover leaves and vitamin B_{12}, and found that neither of the last two constituents was effective without the other.

Oxford (1958), working in New Zealand, attempted to grow *in vitro* *Epidinium ecaudatum*, the predominant Entodiniomorphid protozoon in cows fed on red clover. The method used was based on that of Hungate (1942) except that clover starch and some green plant material were added as source of food, and bacteria were suppressed by the addition of penicillin and neomycin. Nevertheless, the protozoa died after 8 days.

The second successful attempt to grow Entodiniomorphid protozoa indefinitely *in vitro* was that of Coleman (1958, 1960 b). In all previous studies the medium contained high concentrations of NaCl and was based on Ringer solution or saliva (McDougall, 1948). This was replaced by a potassium phosphate buffer containing small amounts of calcium, magnesium and other ions (Coleman, 1958). The food materials were rice-starch grains and dried grass, both of which were added without sterilization. The principal protozoon (over 99 %) grown in these experiments was *Entodinium caudatum*, but its rate of growth, the best growth medium and both the species and numbers of other protozoa present depended on the length of time the protozoa had been growing *in vitro*. Although Entodiniomorphid protozoa direct from the rumen could be maintained for short periods in salts, starch and grass medium, the addition of fresh rumen fluid, from which the protozoa had been removed, and chloramphenicol (but not penicillin, streptomycin or aureomycin) was essential initially for growth and division. The long-term maintenance of these protozoa was very laborious and in the early stages it was essential to replace the growth medium every day. Subsequently it was only necessary to add rice-starch grains and dried grass each day, with replacement or dilution of the medium every 2–4 days. After the organisms had been some months in culture the fresh rumen fluid in the medium could be replaced by autoclaved rumen fluid and after 2 years this could be omitted completely although the number of protozoa was thereby reduced. The morphology of *E. caudatum* grown in the presence of autoclaved rumen fluid for $2\frac{1}{2}$ years gradually changed, and at the time of writing the characteristic long tail spine had shortened until it was indistinguishable from the other spines. This change may, however, represent a gradual change in the predominant organism rather than a change in morphology because the cultures were not grown from a single cell. Pl. 1, figs. 1 and 2, show the difference between *E. caudatum* growing in the sheep's rumen and the same protozoon after $2\frac{1}{2}$ years culture *in vitro*. Even after $2\frac{1}{2}$ years *in vitro* the cultures still contained a few (less than 0·5 %) *Metadinium medium*. For consistent growth of the protozoa it was essential to use rice starch and dried grass that had not been heated, and attempts to use these materials sterilized at temperatures above 100° before addition to the medium had some deleterious effect on the protozoa. It is possible that the cause of Sugden's failure to grow *E. caudatum* was the use of hay, etc., heated at 140° for 1 hr. before use.

In 1959 Gutierrez reported briefly in an abstract that he had grown *Epidinium ecaudatum* (Crawley) *in vitro* dividing twice each day. This represents the fastest division rate obtained for a rumen ciliate in culture but, as far as the author is aware, the observation has never been substantiated.

It appears probable that *Entodinium caudatum* can only be grown successfully *in vitro* if an exact balance between protozoal and bacterial growth is maintained. The only carbohydrate in the media used was rice-starch grains which were given only in sufficient quantity to ensure their almost complete engulfment by the protozoa within a short time. If excess was given, in an attempt to avoid the necessity for feeding the protozoa each day, then bacterial growth occurred on the extra starch and the protozoa died. Coleman (1962) showed that the bacteria-free protozoa can probably utilize soluble starch, which is much more rapidly attacked by bacteria than intact starch grains, and some sugars, but these materials did not maintain growing protozoa and their addition to the medium produced heavy bacterial growth. Similarly, the effect of normal bacteriological culture-medium constituents could not be evaluated as their addition to the medium increased the rate of bacterial growth.

THE PREPARATION AND MAINTENANCE OF
AXENIC ENTODINIOMORPHID PROTOZOA

The Entodiniomorphid protozoal cultures described above contain up to 10,000 times as many bacteria as protozoa. Many attempts have been made to prepare bacteria-free cultures, so far without success. One difficulty in these studies is the necessity of finding media to grow all the bacteria present initially so that the success of treatments designed to remove the bacteria can be verified. As the media available for the growth of rumen bacteria are continually being improved, it is possible that some rumen bacteria have not yet been grown *in vitro* and therefore any conclusions drawn from studies of apparently bacteria-free protozoa may not be final.

Gutierrez & Davis (1959) showed that *Entodinium* and *Diplodinium* spp. contained bacteria within their gastric sacs and that these bacteria could be liberated by crushing the protozoa. The preparation of bacteria-free protozoa therefore involves the use of antibiotics to remove intracellular as well as extracellular bacteria and must be controlled by the estimation of the number of bacteria present in broken as well as intact protozoa.

Oxford (1958, 1959) used penicillin (up to 50 μg./ml.) plus neomycin (up to 30 μg./ml.) or terramycin (up to 100 μg./ml.) in an attempt to prepare and maintain *in vitro* axenic epidinia taken directly from the rumen. By the use of the first two antibiotics he obtained protozoa free from extracellular but not intracellular bacteria, as determined by the use of three different bacteriological culture media. Abou Akkada & Howard (1960) claimed to have prepared axenic *Entodinium caudatum* directly from the rumen by incubation with chloramphenicol. However, their proof that the bacteria were absent rested on the use of a single growth medium that contained soluble starch as source of carbohydrate. Although these protozoa probably contained no starch-utilizing bacteria and may conceivably have been bacteria-free, a more rigorous proof using a variety of culture media is necessary. Other workers, e.g. Williams, Davis, Doetsch & Gutierrez (1961) have added antibiotics during short-term experiments to control bacterial numbers (no counts were made).

Coleman (1962) prepared *Entodinium caudatum* suspensions containing only one viable bacterium (intracellular or extracellular) per ten protozoa by 3-day incubation of protozoa grown *in vitro* (Coleman, 1960b) with high levels of penicillin, neomycin, streptomycin and dihydrostreptomycin, with rice starch as source of carbohydrate, under carefully controlled conditions. Unfortunately the number of protozoa declined steadily and all were dead after a further 3–4 days. The best survival under these conditions was obtained when cysteine, carbon dioxide-bicarbonate and Seitz-filtered rumen fluid were added and the culture was incubated under anaerobic conditions. A wide range of other compounds and omission of the antibiotics had no effect on survival. In these experiments a number of culture media used in medical, food and rumen bacteriology were employed in attempts to

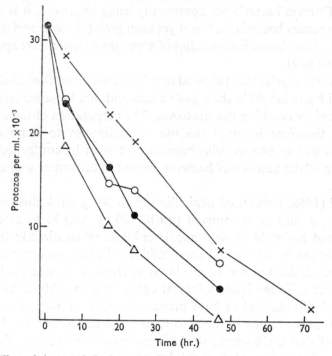

Fig. 1. Effect of three carbohydrates on the survival of starved bacteria-free *Entodinium caudatum*. △—△, No addition; ×—×, soluble starch; ●—●, glucose; ○—○, maltose.

grow all the bacteria present. The results quoted above on the culture of *E. caudatum* at a very low level of contamination were obtained in experiments employing three media which gave the highest counts in the preliminary tests on rumen fluid, protozoal cultures and antibiotic-treated protozoa.

A stimulatory effect of soluble starch, maltose, glucose and some other sugars on the maintenance of bacteria-free protozoa could be demonstrated only if the protozoa were starved before use and the rice starch in the incubation medium used for removal of bacteria was replaced by a low

level of soluble starch (Fig. 1). The effect of these carbohydrates was also
dependent on their concentration (Fig. 2), suggesting that entodinia may
be relatively impermeable to these materials. The finding that only some
sugars were effective suggests a metabolic rather than a physical effect
on the protozoa, but it has not been possible to demonstrate any disappearance of glucose from the large quantities necessary to obtain a
significant effect on survival.

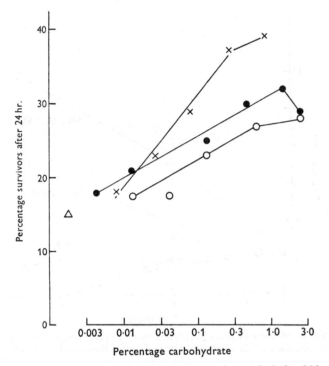

Fig. 2. Effect of concentration of three carbohydrates on the survival after 24 hr. of starved
bacteria-free *Entodinium caudatum*. △, No addition; ×—×, soluble starch; ●—●,
glucose; ○—○, maltose.

Oxford (1959) showed conclusively that starved epidinia would
utilize glucose, and to a lesser extent fructose, to produce gas and
storage polysaccharide in a similar manner to the holotrichs, and that
protozoa which were dying for lack of food could be maintained alive
by glucose.

Of interest is the lack of response of almost bacteria-free entodinia to
a wide range of possible growth stimulatory compounds (Coleman,
1962), while penicillin-treated protozoa incubated in a medium which
still contained many bacteria (Coleman, 1960c, 1962) were stimulated by

addition of yeast extract, grass and autoclaved rumen fluid (Fig. 3). This suggests that the stimulatory effect of these compounds may be mediated indirectly via the bacteria. These penicillin-treated protozoa divided for 3–7 days and survived for 14 days without the daily addition of rice starch and dried grass to the medium, and it seems probable that the penicillin inhibited the growth of the starch-digesting bacteria so that the starch remained for the protozoa to engulf and metabolize.

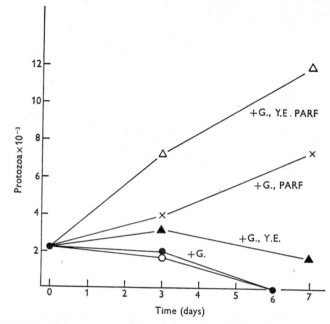

Fig. 3. Effect of addition of grass (G.), yeast extract (Y.E.) and autoclaved rumen fluid (PARF, see Coleman, 1962) on the survival of penicillin-treated *Entodinium caudatum*.

Gutierrez & Davis (1959) approached this problem in a different way and attempted to grow *Entodinia* spp. in the presence of a single bacterial strain. When a culture of *Streptococcus bovis* isolated from crushed entodinia was added to protozoa taken directly from the rumen, and the whole incubated in a salts-grass starch-rice medium, the number of protozoa increased for at least a week. Control tubes without the added bacteria failed to grow. Unfortunately these authors were unable to control the growth of *S. bovis* under these conditions and could not maintain the protozoa for long periods. This approach is of particular importance in view of our inability to grow axenic entodinia and may eventually lead to the culture of protozoa in the presence of only one species of bacterium.

Although there is no evidence that the antibiotics used in the various studies were toxic to the protozoa, it is not possible to decide whether the decline in the protozoa was caused by prolonged exposure to the antibiotics or the disappearance of the bacteria. However, Oxford (1958) and Coleman (1960c) found antibiotics to be essential for even limited survival of their protozoa because they prevented bacteria swamping the protozoa.

THE METABOLISM OF ENTODINIOMORPHID PROTOZOA

'Bacteria-free' protozoa prepared in the various ways described above have been used for studies on the metabolic activities of the protozoa. As these suspensions probably still contain a variable number of viable bacteria, the interpretation of the results in terms of protozoal metabolism must be tentative.

Carbohydrates

As noted above starved epidinia can metabolize glucose to produce gas and storage polysaccharide (Oxford, 1959) and entodinia may metabolize glucose (Coleman, 1962). In contrast, Abou Akkada & Howard (1960) found that intact entodinia did not metabolize soluble starch, cellobiose, maltose, sucrose or glucose, as measured by disappearance of carbohydrate from the medium or stimulation of gas production. Similarly, Williams et al. (1961) reported that intact Ophryoscolex would not utilize cellobiose, galactose, maltose, sucrose or glucose, but in neither case were the protozoa starved. There is general agreement that these protozoa utilize starch grains but it is apparent that they only metabolize sugars when devoid of adequate storage polysaccharide or engulfed starch.

Coleman (unpublished observations) found that starved Entodinium caudatum grown in vitro hydrolysed soluble starch and rice-starch grains both during and after the removal of most of the bacteria (Coleman, 1962). The starch was estimated by the anthrone method after treatment of the whole culture with alkali to destroy the sugars (method of Abou Akkada & Howard, 1960), and glucose by the specific method of Huggett & Nixon (1957). Fig. 4 shows the disappearance of soluble starch, the appearance of glucose and the number of protozoa and bacteria present over a 7-day incubation. As the total carbohydrate in the culture (as measured by anthrone determination before alkali treatment) dropped by less than 5 % during the incubation, there had been extensive breakdown of the soluble starch to glucose and maltose (identified chromatographically) with little further metabolism. Very similar results were obtained with starch grains, except that the amount of starch taken up by the protozoa and the amount of glucose produced was much smaller (about one-tenth). Although these results suggest very strongly that intact entodinia can hydrolyse soluble starch and starch grains to maltose and glucose, it is impossible to be certain

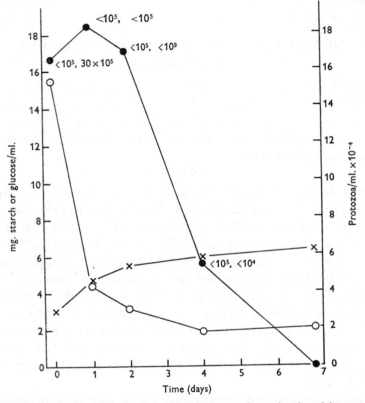

Fig. 4. Breakdown of soluble starch by *Entodinium caudatum* incubated in presence of penicillin, neomycin, streptomycin and dihydrostreptomycin. O—O, Starch; ×—×, glucose; ●—●, protozoa. The numbers at each point are the viable bacteria/ml. as determined on media A and C (Coleman, 1962).

because during the first 2 days some bacteria were present and later there were many disintegrating protozoa.

The products of fermentation of storage starch and of starch grains ingested by entodinia prior to isolation from the rumen are qualitatively similar to those found with other rumen micro-organisms except that very little lactic acid is produced (Abou Akkada & Howard, 1960).

Entodiniomorphid protozoa contain several carbohydrate-degrading enzymes as shown by studies on the soluble fraction obtained after cell breakage. Abou Akkada & Howard (1960) found only amylase, maltase and traces of other hydrolytic enzymes in their chloramphenicol-treated *Entodinium caudatum*. Bailey (1958) found the same enzymes in *Epidinium ecaudatum* treated with penicillin and neomycin. Thomas (1960) also found some invertase in *Entodinium* spp. taken directly from the rumen, but it is difficult to compare these results quantitatively with those of Abou Akkada & Howard (1960). The enzymes found by these workers are probably of protozoal rather than bacterial origin because (*a*) the gentle methods necessary to break protozoa would probably not damage the bacteria, which would be subsequently removed by centrifugation, and (*b*) Abou Akkada & Howard (1960) showed that their *Entodinium caudatum* amylase was different from that of *Streptococcus bovis*, the bacterium found in the gastric sac of their protozoa.

It is still impossible to explain the finding that entodinia which live on starch grains, and which contain active enzymes that produce maltose and glucose from starch, are unable to metabolize these sugars readily.

The only controlled study into the decomposition of pectic substances by entodiniomorphid protozoa was that of Abou Akkada & Howard (1961) who showed that *Entodinium caudatum* extracts contained no pectin-esterase or polygalacturonase activity.

Protein

Warner (1956) suggested, and Blackburn & Hobson (1960) confirmed, that a mixed protozoal fraction, contaminated with bacteria, isolated directly from the rumen, hydrolysed casein more rapidly per unit dry weight than the various bacterial fractions tested. Unfortunately these authors were unable to obtain an active protozoal fraction free from bacteria.

Abou Akkada & Howard (1962) found that chloramphenicol-treated *Entodinium caudatum* rapidly hydrolysed casein to produce peptides and amino acids, but they obtained no evidence for the deamination of the amino acids. It is important that these authors observed that *E. caudatum* readily engulfed and digested stained casein particles. Incubation of the protozoa in buffer produced a steady leakage of cellular nitrogen, principally as ammonia, into the medium, and when casein was added there was negligible conversion of casein nitrogen into cellular nitrogen. Coleman (1962) found that casein would not prolong the life of bacteria-free *E. caudatum*. *Ophryoscolex caudatus* isolated directly from the rumen and incubated with penicillin and dihydrostreptomycin also rapidly hydrolysed several proteins (Williams, Gutierrez & Doetsch, 1960).

Until recently there was no evidence on the biosynthetic abilities of rumen protozoa but Williams *et al.* (1961) have now shown that leucine, valine and alanine are incorporated by *Ophryoscolex caudatus*. Coleman (unpublished observations) found that *Entodinium caudatum* cells grown *in vitro* and washed free of most of their extracellular bacteria, but still containing bacteria in their gastric sacs, readily incorporated ^{14}C-glycine when incubated in the presence of penicillin and neomycin. At the end of the incubation period the protozoa were washed three times on the centrifuge at very low speed with salts containing small quantities

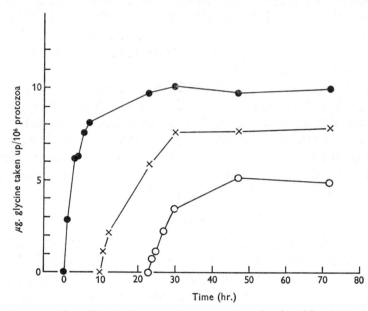

Fig. 5. Incorporation of glycine into *Entodinium caudatum* incubated in presence of penicillin and neomycin. ●—●, Glycine added initially; ×—×, glycine added after 9½ hr.; ○—○, glycine added after 23 hr. Results calculated from uptake of ^{14}C from ^{14}C-glycine of known specific activity.

of ^{12}C-glycine. Fig. 5 shows that under these conditions incorporation of ^{14}C into the protozoa continued for approximately 24 hr. and that protozoa incubated in the absence of glycine for 24 hr. were still capable of incorporating ^{14}C-glycine although the initial rate was slower and the final amount smaller. Unfortunately this does not prove that the glycine was incorporated into the protozoa because the ^{14}C-glycine uptake may have been the result of exchange of the liquid in the gastric sac with external medium containing ^{14}C-glycine, or of incorporation by the extracellular or intracellular bacteria rather than of genuine incorporation into protozoal material.

To determine if the glycine was free in the gastric sac, the protozoa were incubated with various concentrations of ^{14}C-glycine and the incorporation measured. At the lowest concentrations of ^{14}C-glycine (up to 20 μg./ml. in the presence of 2×10^5 protozoa/ml.), over 7 % of the total ^{14}C was taken up in 4 hr. by organisms, the packed cell pad volume of which was less than 1 % of the total, showing that glycine

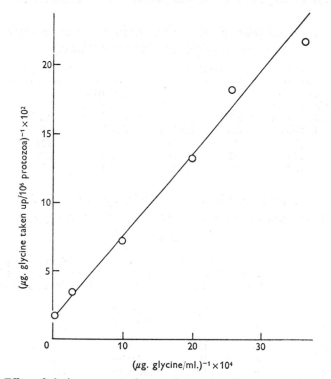

Fig. 6. Effect of glycine concentration on the uptake of glycine. Results calculated from uptake of ^{14}C from ^{14}C-glycine of known specific activity.

was concentrated inside the protozoa. The incorporation of glycine in this experiment, as measured by direct uptake of ^{14}C, increased with increasing glycine concentration, and Fig. 6, which is a double reciprocal plot, shows that over the first 4 hr. the maximum rate of uptake was *c.* 15 μg./hr./10^6 protozoa at infinite glycine concentration.

To test whether uptake was into the bacteria, the protozoal suspensions which had incorporated glycine were broken under conditions that did not break the bacteria. Protozoa were incubated at a density of 3.5×10^5 cells/ml. in salt medium containing 0·75 % soluble starch, 0·025 % L-cysteine, 0·12 % NaHCO$_3$, 0·05 % neomycin sulphate,

1250 u. penicillin G/ml. and 28 μg. glycine (440 counts/min./μg.)/ml. under 95 % $N_2 + 5$ % CO_2. The culture was dispensed in 8 ml. quantities in separate tubes and one tube was harvested for each time point. The protozoa were then broken in a Potter tissue homogenizer and centrifuged to produce supernatant and pellet fractions which were fractionated by a method based on that of Schneider (1945). Table 1 shows that after 1 hr. over 80 % of the ^{14}C was present in the supernatant and that

Table 1. *Distribution of* ^{14}C *in* Entodinium caudatum *during the metabolism of* ^{14}C-glycine (counts/min.)

	Broken cell supernatant			Broken cell pellet		
Time (hr.)	Cold tri- chloroacetic acid-soluble fraction	Nucleic acid	Protein	Cold tri- chloroacetic acid-soluble fraction	Nucleic acid	Protein
1	1710	64	140	95	124	180
5½	2170	168	580	230	232	380
23	1120	160	1440	110	76	700

Table 2. *Incorporation of* ^{14}C-glycine by broken and intact Entodinium caudatum *in the presence and absence of antibiotics*

Protozoa	Anti- biotics	Viable bacteria present per ml.*		^{14}C in protozoa (counts/min.)	^{14}C in protozoa + bacteria (counts/min.)
		Medium A	Medium C		
Initial					
Intact	.	10^5	5×10^5	.	.
Broken	.	10^5	30×10^5	.	.
Final					
Intact	+	10^5	$< 10^5$	196	194
Broken	+	10^5	5×10^5	10	12
Intact	−	10^8	150×10^6	222	226
Broken	−	10^7	300×10^6	144	624

* Counts of viable bacteria were carried out on media A and C respectively (Coleman, 1962).

88 % of that was in the 5 % cold trichloroacetic acid-soluble or small molecular weight fraction. At low external glycine concentrations over 90 % of this ^{14}C was not present as free glycine. This low incorporation of ^{14}C into the pellet after breakage and centrifugation of the protozoa shows that less than 20 % of the total incorporation at this time could have been into the bacteria.

It is likely that the incorporation of ^{14}C into the bacteria was even less. When intact or broken protozoa were incubated at a population density equivalent to 1.4×10^5 cells/ml. in a salt medium (as in the experiment described in Table 1 but with 25 μg. glycine (130 counts/min./μg.)/ml.), intact bacteria took up only 5 % of the labelled carbon taken up by intact protozoa (Table 2). At the end of the experiment the protozoa were washed under conditions such that all bacteria free in the medium would be carried down with them. The amount of ^{14}C incorporated was not increased, showing that the uptake of ^{14}C by bacteria free in the medium was negligible. Omission of antibiotics from the medium increased the incorporation of ^{14}C into intact protozoa by 10–100 %, depending on the experiment. Nearly all this increased ^{14}C was found in the pellet fraction after breakage of the protozoa, suggesting that the increased uptake was into the intracellular bacteria rather than the protozoa (Table 3).

Table 3. *Effect of penicillin and neomycin on the distribution of* ^{14}C *in* Entodinium caudatum *during the metabolism of* ^{14}C-glycine *(counts/min.)*

	Broken cell supernatant			Broken cell pellet		
Antibiotics	Cold trichloroacetic acid-soluble fraction	Nucleic acid	Protein	Cold trichloroacetic acid-soluble fraction	Nucleic acid	Protein
+	620	16	60	10	12	60
−	670	16	40	75	376	690

Medium as in Table 1; density of protozoa $1\cdot6 \times 10^5$ cells/ml.

Lipids

Little is known about the lipid metabolism of Entodiniomorphid protozoa. Wright (1961) found that the addition of penicillin and terramycin to rumen liquor reduced the lipolysis of linseed oil during a 22 hr. incubation and suggested that bacteria rather than protozoa were responsible for the hydrolysis. Trier (1926) observed that ciliates engulfed fat particles but concluded that these were then digested by bacterial action. Wright (1959, 1961) found that penicillin- and neomycin-treated epidinia hydrogenate lipids, and cell-free extracts of epidinia hydrolyse tributyrin but not triolein.

Cellulose

The earlier workers on the digestion of cellulose by rumen ciliates were unable to agree whether or not these organisms digested native cellulose (see Becker, 1932; Westphal, 1934) and there is still some doubt on this point.

Hungate (1942, 1943) was the first to show that *Diplodinium* spp.

could be grown *in vitro* on cellulose (chemically treated cotton-wool) which was digested to form cell glycogen. He also showed that extracts from suspensions of these protozoa washed free of debris and bacteria contained an active cellulase as determined by the appearance of glucose on the addition of cellulose. The optimum pH of this enzyme was, however, lower than that of normal rumen contents.

The large Entodiniomorphid protozoon *Metadinium medium* also ingests cellulose fibres which disintegrate inside the cell (Sugden, 1953), a process associated with the appearance of starch-like material (probably an amylopectin) in the cytoplasm. The presence of cellulose as cotton-wool also prolonged the life of the protozoa. In these studies the actual cellulose present was measured and two-thirds of the cellulose disappeared in 2 days (Oxford, 1955). As pointed out by Oxford (1955), it is important in any studies on cellulose digestion to measure the disappearance of native cellulose and not just the appearance of glucose, which may have arisen from smaller molecular weight compounds associated with the cellulose preparation. Unfortunately at least some of these experiments by Sugden were carried out with HCl-treated cotton-wool and it is not possible to refute the finding of Dogiel (1925, quoted by Becker, 1932) that cellulose elements of plant tissues, which had been ingested by ciliates, were expelled intact through the anus without perceptible change in their morphology.

Although the rate of cellulose breakdown observed by Sugden was probably too large to be caused by associated bacteria, the addition of streptomycin reduced the utilization of the cellulose and the life of the protozoa. Although the streptomycin probably reduced the number of bacteria present, there was no conclusive proof that all the intracellular bacteria were destroyed. It is, therefore, not certain whether this effect of streptomycin was directly on the protozoa or indirectly by causing the death of bacteria or a combination of both.

THE ROLE OF BACTERIA IN THE NUTRITION OF RUMEN CILIATES

As rumen protozoa live in an environment teeming with bacteria it has often been assumed that the protozoa feed on the bacteria. Kofoid & MacLennan (1930) observed bacteria in the gastric sac of entodinia and stated without further proof that the protozoa engulfed bacteria as food. More recently, Appleby, Eadie & Oxford (1956) were unable to demonstrate the presence of viable bacteria in Entodiniomorphid ciliates isolated directly from the rumen. The position was clarified a little when

Gutierrez & Davis (1959) observed 100–150 bacteria in each *Entodinium* taken directly from the rumen and isolated 3–10 viable *Streptococcus bovis* cells from each crushed protozoon. The bacterium isolated from the protozoa was then grown *in vitro* on conventional media and fed to starved entodinia and diplodinia which were observed to engulf the bacterial cells. Although the bacteria observed in Gram-stained smears of entodinia taken directly from the rumen were morphologically similar to the *Streptococcus bovis* isolated from the rumen, very few of the visible bacteria were grown *in vitro*. The possibilities remain therefore that other viable bacteria were present which did not grow on the starch medium used for isolation or that over 90 % of the bacteria in the gastric sac were moribund and being digested. Plate 1, figs. 3 and 4, show Gram-positive cocci present in the washed entodinia grown as described by Coleman (1960*b*).

Although, as a result of microscopic examination, the Entodiniomorphid protozoa have been known for a long time to contain bacteria, similar evidence for the holotrich protozoa was lacking for many years, because the protozoa as isolated contained too much food material for bacteria to be seen. The other evidence was also negative as Sugden & Oxford (1952) found that holotrichs would not engulf *Lactobacillus bulgaricus* and Appleby *et al.* (1956) obtained no evidence for any viable bacteria within the holotrichs taken from the rumen. The first demonstration that bacteria had a role in holotrich nutrition was that of Gutierrez (1958) who starved isotricha and dasytricha isolated from the rumen for 3–4 days and then followed microscopically the ingestion of bacteria from crude rumen contents. The protozoa selectively ingested rods which were at first contained in a vacuole at the base of the gullet. The vacuoles eventually broke off and moved away into the cytoplasm. Four types of bacteria were isolated on a starch medium from these protozoa. When these bacteria were grown *in vitro* and fed to holotrichs, individual bacteria from only three of the strains were engulfed by the protozoa, and bacteria from only one strain prolonged the life of *Isotricha prostoma* for even a limited period. Unfortunately Gutierrez used media selective for the isolation of starch- or cellulose-digesting bacteria and it is possible that other bacteria may be present within the holotrichs.

Gutierrez & Hungate (1957) observed a similar digestion of bacteria by the holotrich *Dasytricha ruminantium* and found that two species of rumen bacteria not apparently isolated from protozoa would maintain the protozoa in a state of division for up to two weeks.

From the above evidence it is apparent that rumen protozoa can selectively engulf certain rumen bacteria, but the metabolism of these bacteria inside the protozoa is still obscure although they may have some growth-promoting effect. In the partially successful experiments on the culture of protozoa with pure strains of bacteria it is possible that the bacteria may act in part by reducing the redox potential beyond that achieved by the cysteine added. *Escherichia coli* has been successfully used by Smith & Hungate (1958) for this purpose in the culture of methane bacteria from the rumen.

THE CULTURE OF HOLOTRICH PROTOZOA

The first partially successful attempt to culture sheep rumen holotrich protozoa *in vitro* was that of Sugden & Oxford (1952) who maintained these organisms in the absence of any antibiotics in constant numbers for 1 month under anaerobic conditions in a salt solution containing 5 % centrifuged but unboiled grass juice. Variations in the concentration of grass juice and the addition of boiled grass juice or Seitz-filtered rumen fluid reduced the protozoal survival time. Some evidence was obtained that a trace metal may be important for the maintenance of these protozoa. Gutierrez (1955) cultured *Isotricha* spp. which divided every 48 hr. on a medium of salts, ground alfalfa, ground wheat and 30 % rumen fluid, but their maintenance involved individual daily transfer of each protozoon to fresh medium and was continued for only a few weeks. Although Sugden & Oxford (1952) and Gutierrez (1955) used different media both found that it was essential to change the media every day.

The attempts of Gutierrez & Hungate (1957) and Gutierrez (1958) to culture holotrichs in the presence of pure strains of bacteria have been considered above.

THE METABOLISM OF HOLOTRICH PROTOZOA

The study of rumen holotrich protozoa in recent years was initiated by the discovery of Oxford (1951) that when glucose, fructose and sucrose were metabolized by these protozoa there was extensive conversion of the sugars into a glucosan (subsequently found to be a starch) within the cell. Masson & Oxford (1951) and Sugden & Oxford (1952) extended these studies and found that several mono-, di-, tri- and polysaccharides were converted into intracellular starch and also prolonged the life *in vitro* of these protozoa by a few days. The presence of the starch makes the protozoa much denser than the surrounding liquid, and protozoa which have metabolized glucose can be separated from rumen liquor by standing for a time in a warm place when they sink to the bottom. The holotrichs can then be washed free of bacteria by repeated sedimentation under gravity and decantation. Heald, Oxford & Sugden (1952) made this the basis of a method for the preparation of holotrichs for mano-metric studies and found that streptomycin reduces the number of residual bacteria without harming the protozoa.

Heald & Oxford (1953) prepared mixed holotrichs using this technique and found that, of the glucose-carbon which was fermented but not converted into cellular starch, approximately one-third appeared as lactic acid, one-third as butyric acid and the remainder as acetic acid and carbon dioxide. Hydrogen was also produced but little or no formic or propionic acid. It is of interest that compared with glucose fermenta-tion by holotrichs, *Entodinium caudatum* produced little lactic acid and relatively more acetic acid and butyric acid during the fermentation of starch.

Gutierrez (1955) separated the smaller *Dasytricha* from the larger *Isotricha* spp. by differential sedimentation of mixed suspensions treated with glucose and determined the fermentation patterns of both groups. Cellobiose and salicin were utilized by dasytricha alone; other-wise both groups fermented the same sugars and, with only small quantitative differences, produced the same fermentation products.

The study of the metabolism of different genera of rumen protozoa was facilitated by the work of Eadie & Oxford (1957) who described a simple method for the removal of the fauna present in the rumen of a sheep and its replacement by a single protozoon species. Howard (1959a) prepared holotrichs from sheep that contained either *Dasytricha* of *Isotricha* spp. and confirmed and extended the work of Gutierrez (1955). He showed that *Dasytricha ruminantium* unlike the *Isotricha* spp. fermented galactose, maltose and cellobiose, and that galactose was con-verted into glucose. Subsequently Howard (1959b) demonstrated the presence of the appropriate carbohydrases in cell-free extracts of these protozoa.

Mould & Thomas (1958) showed that the holotrich starch was an amylopectin and was probably synthesized by the phosphorylase mechanism common to plants and animals.

THE GROWTH AND METABOLISM OF
RUMEN PROTOZOA *IN VIVO*

The growth and metabolism of defaunated animals

The first successful attempt to prepare animals free from rumen protozoa was that of Becker (1929) who treated goats with copper sulphate. Becker & Hsiung (1929) then showed that such goats could only be re-infected with rumen ciliates if they fed on material freshly contaminated with saliva from an infected animal. Comparisons of defaunated goats before and after re-infection with respect to digestibility of various food constituents, nitrogen excretion and changes in body weight failed to show any marked effect of the protozoa (Becker *et al.* 1929). Experiments with growing lambs (Becker & Everett, 1930) showed that animals containing rumen ciliates gained slightly less weight over a 3-month period than defaunated controls. Van der Wath & Myburgh (1941), using the same copper sulphate treatment to defaunate sheep, found the same rate of starch or cellulose digestion in absence or presence of ciliates, and concluded that the protozoa take no part in starch or cellulose digestion.

More recently Pouden & Hibbs (1950) raised two groups of calves which they segregated at birth from all other animals in order to prevent natural infection with rumen ciliates, and then deliberately infected one group by inoculation with cud from an adult animal. They found that there was an insignificant difference in growth rate between the two groups although the uninoculated animals were pot-bellied and had rougher coats.

McNaught, Owen, Henry & Kon (1954) criticized some of Becker's results in the light of later knowledge on the effect of copper, and considered that copper would improve the thrift of the animal by acting as a vermifuge. However, in the critical experiments quoted above, all the animals were treated with copper and then some re-infected with ciliates. It is apparent, therefore, that Holotrich and Entodiniomorphid ciliates are not essential for the growth and maintenance of ruminants, although these protozoa probably play some part in the metabolism of the host. Their role will be considered below.

The growth of ciliates in normal animals

Although most workers concerned with rumen protozoa are familiar with the variations in the rumen fauna as observed under the microscope, there have been relatively few studies on the total number of protozoa or the relative numbers of the different genera in animals on different diets.

There is a regional difference in the genera of protozoa found in the rumen of sheep. Although holotrichs are commonly seen in sheep in Great Britain, there is a suggestion in the literature that they are rarely found in Australian sheep, and this does not appear to be due solely to differences in the diet of the animals. Moir (1951), Moir & Somers (1957) and Purser & Moir (1959) observed few or no holotrichs in their Australian sheep, although Warner (1962) found them in all his animals. Oxford (1958) reported that *Epidinium* was common in New Zealand sheep and *Ophryoscolex* was absent, whereas the reverse was true in Scottish sheep. In contrast Scottish cattle had the pattern of New Zealand sheep. The significance of these differences is unknown.

A major difficulty in quantitative studies of the number of protozoa present is the large day-to-day variation between samples taken from the same sheep on the same diet, and only large differences can be considered significant (Boyne, Eadie & Raitt, 1957; Williams & Moir, 1951). Mowry & Becker (1930) and van der Wath & Myburgh (1941) showed that the number of entodinia depended on the amount of carbohydrate in the diet, a result in agreement with the known requirements of entodinia for growth. Mowry & Becker (1930) also found that protein concentrates and some proteins such as casein, but not dried egg albumin or asparagine, greatly increased the number of protozoa, especially when starch was present.

Purser & Moir (1959), Purser (1961) and Warner (1962) observed a two- to fivefold diurnal variation in the number of entodinia and dasytrichs in sheep fed once daily and found that the pattern of variation depended on the protozoon. The growth rate of entodinia, as determined from the proportion of the protozoa which were dividing at any time, varied from a doubling time of 6 hr. during the period 13–19 hr. after feeding, to a doubling time of 28–58 hr. just before feeding (Warner, 1962). The maximum number of dividing dasytrichs was found 19–22 hr. after feeding.

Two other factors which affect the number of protozoa in the rumen are the frequency of feeding (Moir & Somers, 1956) and the pH of the rumen (Purser & Moir, 1959). On feeding a sheep with oaten chaff, lucerne hay and linseed meal, the pH value of the contents of the rumen dropped from 7·0 to 5·6 in 2 hr. and remained below 6 for a further 4 hr. This low pH value was associated with a drop in the number of dividing protozoa which had doubled in the first hour after feeding. The subsequent rise in the pH value of the rumen contents was associated with an increase in the number of dividing protozoa. The number of protozoa, principally entodinia, was related to the minimum daily pH,

and it is possible that the finding of higher numbers of protozoa present
in sheep fed small amounts of food at more frequent intervals, where
the minimum pH may be higher, is explicable on this basis.

Since the two species of ciliates so far examined have different
patterns of diurnal variation, further work is necessary to determine
whether the marked increase in protozoal division rate, that occurred in
Warner's experiments for a few hours each day in sheep fed once daily,
was related to the increase in the pH value of the rumen contents or to
some other cause.

The metabolism of carbohydrates and proteins by protozoa in vivo

It is now well established that the dietary nitrogen of the ruminant is
largely modified by, and incorporated into, the rumen micro-organisms
before it is digested by the host. Weller, Gray & Pilgrim (1958) used
diaminopimelic acid, which is present in bacteria but not in plants or
protozoa, as a measure of the amount of bacterial nitrogen present in
rumen fractions. By the use of this marker they avoided the need to
prepare the various fractions free from bacteria although it was still
essential to separate the protozoa from plant fibres by physical means.
They concluded that of the rumen nitrogen 46 % was bacterial, 21 %
protozoal, 26 % plant and 7 % soluble. This result agrees well with the
earlier estimates of Hungate (1955) and Schwarz (1925). It is difficult to
obtain an estimate even as reliable as this for the importance of protozoa
in carbohydrate fermentation but Hungate (1955) calculated that 20 %
of the fermentation acids are produced by the protozoa.

McNaught *et al.* (1954) determined for rats the biological value
(percentage of the absorbed food nitrogen retained by the body), true
digestibility and net utilization (biological value × true digestibility/100)
of dried rumen bacteria, dried rumen protozoa and dried brewers yeast.
The biological values of the bacteria and the protozoa were the same,
but the true digestibility of the protozoa was considerably higher than
that of the bacteria. The net utilization of the protozoa was 73 compared
with 60 for the bacteria and yeast. This result suggests that the protozoa
may be more beneficial to the host than the bacteria, and Weller (1957)
has shown that rumen protozoa are richer in isoleucine, leucine, phenyl-
alanine and especially lysine, which are all essential amino acids for the
rat.

Oxford (1955) in a discussion of the results of Becker *et al.* (1929) with
faunated and defaunated goats, pointed out that faunated animals
digested slightly more protein and excreted more nitrogen in their urine
than defaunated controls, especially when the predominant protozoa

were entodinia. He argued that this suggested an increased ammonia production from, for example, protein, due to protozoal activity. Such a loss of combined nitrogen may be detrimental to the host. Some support for this view is provided by the work of Abou Akkada & Howard (1962) who showed that suspensions of *Entodinium caudatum* in buffer lose 1 % of their cellular nitrogen per hour.

A role for rumen protozoa in the digestion of starch was put forward by Becker (1932) reviewing his own work and that of Usuelli (1930). These workers found that when normal faunated sheep were fed on ground barley over 80 % of the starch grains were ingested by the protozoa and in process of conversion into protozoal glycogen in the first 6 hr. Defaunated sheep on a similar diet appeared to be bloated, although otherwise healthy, and it was suggested that the removal of the starch by the protozoa may reduce the rate of CO_2 evolution. *Entodinium caudatum*, a starch-ingesting protozoon, also produces very little lactic acid during the catabolism of ingested starch (Abou Akkada & Howard, 1960), and the ingestion of starch by this organism may reduce the rapid drop in the pH value of the rumen contents observed after a carbohydrate-rich feed.

CONCLUSIONS

Becker and his colleagues 30 years ago came to the conclusion that the number of protozoa depended solely on the amount of food available, and that on a poor diet relatively few were present whereas on a rich diet the ciliates were abundant. If this is true, then the protozoa are probably of least importance on a diet where they could be of most value to the host as a source of essential amino acids and protein. Although these protozoa may contribute up to 20 % of their hosts's nutritional requirements under favourable conditions, no evidence has appeared recently to invalidate Becker's belief that rumen ciliates are commensals and not essential for the nutrition of the host.

The role of the rumen ciliates in the nitrogen metabolism of the host depends not only on their ability to ferment or degrade proteins and on the value of their bodies to the host, but also on their requirements for growth. Oxford (1955) and Chalmers & Synge (1954) discussing the nitrogen requirements of rumen protozoa, assumed that since they were difficult to grow *in vitro*, their nitrogen requirements were very complex, and cited the requirements of other ciliates (Kidder & Dewey, 1951).

I wish to suggest that although the growth-factor requirements of rumen protozoa may be complex, and the optimum physical conditions for growth difficult to obtain in an axenic culture *in vitro*, their nitrogen requirements may be satisfied by ammonia and a few amino acids. Recently the nutritional requirements of several groups of rumen bacteria have been determined and ammonia has been found to be the principal source of nitrogen for *Streptococcus bovis* (Wolin & Weinberg, 1960), some strains of *Ruminococcus, Bacteroides succinogenes* (Bryant & Robinson, 1961), *Butyrivibrio* (Gill & King, 1958) and a sulphate-reducing bacterium (Coleman, 1960a). Although more complicated compounds may have a stimulatory effect (see, for example, Bryant, Small, Bouma & Chu, 1958) there are no examples, as far as the author is aware, of complicated nitrogen requirements for the growth of rumen bacteria. These findings are not unexpected, as the resting level of free α-amino nitrogen is low (Chalmers & Synge, 1954; Annison, 1956) although it rises after a feed of casein. If the protozoa had to assimilate amino acids from the surrounding liquor they would be at a disadvantage in competition with the bacteria because of their larger size and slower metabolism. The protozoa may therefore use ammonia, which is present in relatively large amounts in rumen liquor, as a source of nitrogen. It is, however, possible that as the protozoa engulf bacteria, can hydrolyse protein and presumably can digest the ingested bacteria, the concentration of amino acids in the gastric sac of Entodiniomorphid protozoa or the food vacuoles of holotrichs may be much higher than in the surrounding rumen liquor. These amino acids could then be utilized for growth.

It is hoped that work at present being carried out in the author's laboratory and elsewhere on the metabolism of ^{14}C-amino acids and other compounds will provide some information on the biosynthetic capabilities of rumen protozoa and perhaps provide evidence on their possible nutritional requirements.

REFERENCES

ABOU AKKADA, A. R. & HOWARD, B. H. (1960). The biochemistry of rumen protozoa. 3. The carbohydrate metabolism of *Entodinium*. *Biochem. J.* **76**, 445.

ABOU AKKADA, A. R. & HOWARD, B. H. (1961). The biochemistry of rumen protozoa. 4. The decomposition of pectic substances. *Biochem. J.* **78**, 512.

ABOU AKKADA, A. R. & HOWARD, B. H. (1962). The biochemistry of rumen protozoa. 5. The nitrogen metabolism of *Entodinium*. *Biochem. J.* **82**, 313.

ANNISON, E. F. (1956). Nitrogen metabolism in the sheep: protein digestion in the rumen. *Biochem. J.* **64**, 705.

APPLEBY, J. C., EADIE, J. M. & OXFORD, A. E. (1956). Interrelationships between ciliate protozoa and bacteria in the sheep's rumen. *J. appl. Bact.* **19**, 166.

BAILEY, R. W. (1958). Bloat in cattle. X. The carbohydrates of the cattle rumen ciliate *Epidinium ecaudatum* Crawley isolated from cows fed on red clover (*Trifolium pratense* L.). *N.Z. J. agric. Res.* **1**, 825.

BECKER, E. R. (1929). Methods of rendering the rumen and reticulum of ruminants free from their normal infusorian fauna. *Proc. nat. Acad. Sci., Wash.* **15**, 435.

BECKER, E. R. (1932). The present status of problems relating to the ciliates of ruminants and equidae. *Quart. Rev. Biol.* **7**, 282.

BECKER, E. R. & EVERETT, R. C. (1930). Comparative growths of normal and infusoria-free lambs. *Amer. J. Hyg.* **11**, 362.

BECKER, E. R. & HSIUNG, T. S. (1929). The method by which ruminants acquire their fauna of infusoria, and remarks concerning experiments on the host-specificity of these protozoa. *Proc. nat. Acad. Sci., Wash.* **15**, 684.

BECKER, E. R., SCHULTZ, J. A. & EMMERSON, M. A. (1929). Experiments on the physiological relationships between the stomach infusoria of ruminants and their hosts, with a bibliography. *Iowa State Coll. J. Sci.* **4**, 215.

BLACKBURN, T. H. & HOBSON, P. N. (1960). Proteolysis in the sheep rumen by whole and fractionated contents. *J. gen. Microbiol.* **22**, 272.

BOYNE, A. W., EADIE, J. M. & RAITT, K. (1957). The development and testing of a method of counting rumen ciliate protozoa. *J. gen. Microbiol.* **17**, 414.

BRYANT, M. P. & ROBINSON, I. M. (1961). Some nutritional requirements of the genus *Ruminococcus. Appl. Microbiol.* **9**, 91.

BRYANT, M. P., SMALL, N., BOUMA, C. & CHU, H. (1958). *Bacteroides ruminicola* n.sp. and *Succinimonas amylolytica.* The new genus and species. *J. Bact.* **76**, 15.

CHALMERS, M. J. & SYNGE, R. L. M. (1954). The digestion of protein and nitrogenous compounds in ruminants. *Advanc. Prot. Chem.* **9**, 93.

COLEMAN, G. S. (1958). Maintenance of oligotrich protozoa from the sheep rumen *in vitro. Nature, Lond.* **182**, 1104.

COLEMAN, G. S. (1960a). A sulphate-reducing bacterium from the sheep rumen. *J. gen. Microbiol.* **22**, 423.

COLEMAN, G. S. (1960b). The cultivation of sheep rumen oligotrich protozoa *in vitro. J. gen. Microbiol.* **22**, 555.

COLEMAN, G. S. (1960c). Effect of penicillin on the maintenance of rumen oligotrich protozoa. *Nature, Lond.* **187**, 518.

COLEMAN, G. S. (1962). The preparation and survival of bacteria-free suspensions of *Entodinium caudatum. J. gen. Microbiol.* **28**, 271.

CORLISS, J. O. (1959). An illustrated key to the higher groups of the ciliated protozoa with definition of terms. *J. Protozool.* **6**, 265.

DOGIEL, V. A. (1925). Über die Art der Nahrung und der Nahrungsaufnahme bei den im Darm der Huftiere parasitierenden Infusorien. *Trav. Soc. Nat. St Pétersb.* (*Léningr.*) (*Sect. Zool. and Phys.*), **54**, 67.

EADIE, J. M. & OXFORD, A. E. (1957). A simple and safe procedure for the removal of holotrich ciliates from the rumen of an adult fistulated sheep. *Nature, Lond.* **179**, 485.

GILL, J. W. & KING, K. W. (1958). Nutritional characteristics of a Butyrivibrio. *J. Bact.* **75**, 666.

GRUBY, D. & DELAFOND, O. (1843). Recherches sur des animalcules se développant en grand nombre dans l'estomac et dans les intestins pendant la digestion des animaux herbivores et carnivores. *C.R. Acad. Sci., Paris,* **17**, 1304.

GUTIERREZ, J. (1955). Experiments on the culture and physiology of Holotrichs from the bovine rumen. *Biochem. J.* **60**, 516.

GUTIERREZ, J. (1958). Observations on bacterial feeding by the rumen ciliate *Isotricha prostoma. J. Protozool,* **5**, 122.

GUTIERREZ, J. (1959). Studies on the culture of the rumen ciliate *Epidinium ecaudatum* (Crawley). *J. Protozool.* **6**, Suppl. 21.

GUTIERREZ, J. & DAVIS, R. E. (1959). Bacterial ingestion by the rumen ciliates *Entodinium* and *Diplodinium*. *J. Protozool.* **6**, 222.

GUTIERREZ, J. & HUNGATE, R. E. (1957). Interrelationship between certain bacteria and the rumen ciliate *Dasytricha ruminantium*. *Science*, **126**, 511.

HEALD, P. J., OXFORD, A. E. & SUGDEN, B. (1952). A convenient method for preparing massive suspensions of virtually bacteria-free ciliate protozoa of the genera *Isotricha* and *Dasytricha* for manometric studies. *Nature, Lond.* **169**, 1055.

HEALD, P. J. & OXFORD, A. E. (1953). Fermentation of soluble sugars by anaerobic Holotrich ciliate protozoa of the genera *Isotricha* and *Dasytricha*. *Biochem. J.* **53**, 506.

HOWARD, B. H. (1959 a). Biochemistry of rumen protozoa. 1. Carbohydrate fermentation by *Dasytricha* and *Isotricha*. *Biochem. J.* **71**, 671.

HOWARD, B. H. (1959 b). Biochemistry of rumen protozoa. 2. Some carbohydrases in cell-free extracts of *Dasytricha* and *Isotricha*. *Biochem. J.* **71**, 675.

HUGGETT, A. ST G. & NIXON, D. A. (1957). Use of glucose oxidase, peroxidase and *o*-dianisidine in determination of blood and urinary glucose. *Lancet*, **273**, 368.

HUNGATE, R. E. (1942). The culture of *Eudiplodinium neglectum* with experiments on the digestion of cellulose. *Biol. Bull., Woods Hole*, **83**, 303.

HUNGATE, R. E. (1943). Further experiments on cellulose digestion by protozoa in the rumen of cattle. *Biol. Bull., Woods Hole*, **84**, 157.

HUNGATE, R. E. (1955). Mutualistic intestinal protozoa. *Biochemistry and Physiology of Protozoa*, vol. II, p. 159. Ed. S. H. Hutner and A. Lwoff. New York: Academic Press.

KANDATSU, M. & TAKAHASHI, N. (1955 a). Studies on reticulo-rumen digestion. Part 2. On the artificial culture of some *Entodinia*. I. *J. Agric. Chem. Soc., Japan*, **29**, 833.

KANDATSU, M. & TAKAHASHI, N. (1955 b). Studies on reticulo-rumen digestion. Part 3. On the artificial culture of some *Entodinia*. II. *J. Agric. Chem. Soc., Japan*, **29**, 915.

KANDATSU, M. & TAKAHASHI, N. (1956). Studies on reticulo-rumen digestion. Part 4. On the artificial culture of some *Entodinia*. III. *J. Agric. Chem. Soc., Japan*, **30**, 96.

KIDDER, G. W. & DEWEY, V. C. (1951). The biochemistry of ciliates in pure culture. In *Biochemistry and Physiology of Protozoa*, vol. I, p. 323. Ed. A. Lwoff. New York: Academic Press.

KOFOID, C. A. & MACLENNAN, R. F. (1930). Ciliates from *Bos indicus* Linn. I. The genus *Entodinium* Stein. *Univ. Calif. Publ. Zool.* **33**, 471.

MCDOUGALL, E. I. (1948). Studies on ruminant saliva. 1. Composition and output of sheep's saliva. *Biochem. J.* **43**, 99.

MCNAUGHT, M. L., OWEN, E. C., HENRY, K. M. & KON, S. K. (1954). The utilization of non-protein nitrogen in the bovine rumen. 8. The nutritive value of the proteins of preparations of dried rumen bacteria, rumen protozoa and brewers yeast for rats. *Biochem. J.* **56**, 151.

MARGOLIN, S. (1930). Methods for the cultivation of cattle ciliates. *Biol. Bull., Woods Hole*, **59**, 301.

MASSON, F. M. & OXFORD, A. E. (1951). The action of ciliates of the sheep's rumen upon various water-soluble carbohydrates, including polysaccharides. *J. gen. Microbiol.* **5**, 664.

MOIR, R. J. (1951). The seasonal variation in the ruminal microorganisms of grazing sheep. *Aust. J. agric. Res.* **2**, 322.

MOIR, R. J. & SOMERS, M. (1956). A factor influencing the protozoal population in sheep. *Nature, Lond.* **178**, 1472.

MOIR, R. J. & SOMERS, M. (1957). Ruminal flora studies. VIII. The influence of rate and method of feeding a ration upon its digestibility, upon ruminal function and upon the ruminal population. *Aust. J. agric. Res.* **8**, 253.

MOULD, D. L. & THOMAS, G. J. (1958). The enzymic degradation of starch by holotrich protozoa from the sheep rumen. *Biochem. J.* **69**, 327.

MOWRY, H. A. & BECKER, E. R. (1930). Experiments on the biology of infusoria inhabiting the rumen of goats. *Iowa State Coll. J. Sci.* **5**, 35.

OXFORD, A. E. (1951). The conversion of certain soluble sugars to a glucosan by holotrich ciliates in the rumen of sheep. *J. gen. Microbiol.* **5**, 83.

OXFORD, A. E. (1955). The rumen ciliate protozoa: their chemical composition, metabolism, requirements for maintenance and culture, and physiological significance for the host. *Exp. Parasit.* **4**, 569.

OXFORD, A. E. (1958). Bloat in cattle. IX. Some observations on the culture of the cattle rumen ciliate *Epidinium ecaudatum* Crawley occurring in quantity in cows fed on red clover (*Trifolium pratense* L.). *N.Z. J. agric. Res.* **1**, 809.

OXFORD, A. E. (1959). Bloat in cattle. XV. Further observations concerning the ciliate *Epidinium ecaudatum*, an inhabitant of the rumens of cows liable to legume bloat. *N.Z. J. agric. Res.* **2**, 365.

POUDEN, W. D. & HIBBS, J. W. (1950). The development of calves raised without protozoa and certain other characteristic rumen microorganisms. *J. Dairy Sci.* **33**, 639.

PURSER, D. B. (1961). A diurnal cycle for holotrich protozoa of the rumen. *Nature, Lond.* **190**, 831.

PURSER, D. B. & MOIR, R. J. (1959). Ruminal flora studies in the sheep. IX. The effect of pH on the ciliate population of the rumen *in vivo*. *Aust. J. agric. Res.* **10**, 555.

SCHNEIDER, W. C. (1945). Phosphorus compounds in animal tissues. 1. Extraction and estimation of desoxypentose nucleic acid and of pentose nucleic acid. *J. biol. Chem.* **161**, 293.

SCHWARZ, C. (1925). Die Ernährungphysiologische Bedeutung der Mikroorganismen in dem Vormagen der Wiederkeuer. *Biochem. Z.* **156**, 130.

SMITH, P. H. & HUNGATE, R. E. (1958). Isolation and characteristics of *Methanobacterium ruminantium* n.sp. *J. Bact.* **75**, 713.

SUGDEN, B. (1953). The cultivation and metabolism of oligotrich protozoa from the sheep's rumen. *J. gen. Microbiol.* **9**, 44.

SUGDEN, B. & OXFORD, A. E. (1952). Some cultural studies with holotrich ciliate protozoa from the sheep's rumen. *J. gen. Microbiol.* **7**, 145.

THOMAS, G. J. (1960). Metabolism of soluble carbohydrates of grasses in the rumen of the sheep. *J. agric. Sci.* **54**, 360.

TRIER, H. J. (1926). Der Kohlenhydratstoffwechsel der Panseninfusorien und die Bedeutung der grünen Pflanzenteile für diese Organismen. *Z. vergl. Physiol.* **4**, 305.

USUELLI, F. (1930). Stärkeaufnahme und Glykogenbildung der Panseninfusorien. *Wiss. Arch. Landw.* **3**, 4.

VAN DER WATH, J. G. & MYBURGH, S. J. (1941). Studies on the alimentary tract of merino sheep in South Africa. VI. The role of infusoria in ruminal digestion with some remarks on ruminal bacteria. *Onderstepoort J. Vet. Sci.* **17**, 61.

WARNER, A. C. I. (1956). Proteolysis by rumen microorganisms. *J. gen. Microbiol.* **14**, 749.

WARNER, A. C. I. (1962). Some factors influencing the rumen microbial population. *J. gen. Microbiol.* **28**, 129.

WELLER, R. A. (1957). The amino acid composition of hydrolysates of microbial fractions from the rumen of sheep. *Aust. J. biol. Sci.* **10**, 384.

324 G. S. COLEMAN

WELLER, R. A., GRAY, F. V. & PILGRIM, A. F. (1958). The conversion of plant nitrogen to microbial nitrogen in the rumen of the sheep. *Brit. J. Nutr.* **12**, 421.

WESTPHAL, A. (1934). Studien über Ophryoscoleciden in der Kultur. *Z. Parasitenk.* **7**, 71.

WILLIAMS, P. P., DAVIS, R. E., DOETSCH, R. N. & GUTIERREZ, J. (1961). Physiological studies of the rumen protozoan *Ophryoscolex caudatus* Eberlein. *Appl. Microbiol.* **9**, 405.

WILLIAMS, P. P., GUTIERREZ, J. & DOETSCH, R. N. (1960). Protein degradation and other physiological studies on the rumen ciliate, *Ophryoscolex caudatus* Eberlein. *Bact. Proc.* p. 32.

WILLIAMS, V. J. & MOIR, R. J. (1951). Ruminal flora studies in sheep. III. The influence of different sources of nitrogen upon nitrogen retention and upon total numbers of free microorganisms in the rumen. *Aust. J. Sci. Res.* **4**, 377.

WOLIN, M. J. & WEINBERG, G. (1960). Some factors affecting growth of *Streptococcus bovis* on chemically defined media. *J. Dairy Sci.* **43**, 825.

WRIGHT, D. E. (1959). Hydrogenation of lipids by rumen protozoa. *Nature, Lond.* **184**, 875.

WRIGHT, D. E. (1961). Bloat in cattle. XX. Lipase activity of rumen microorganisms. *N.Z. J. agric. Res.* **4**, 216.

EXPLANATION OF PLATE

Fig. 1. *Entodinium caudatum* present in the sheep's rumen. Phase contrast, × 600.

Fig. 2. *Entodinium caudatum* after 2½ years culture *in vitro* (Coleman, 1960b). Note the absence of the long caudal spine. Phase contrast, × 400.

Fig. 3. Longitudinal section (thickness 5 μ) through *Entodinium caudatum*. Note the Gram-positive cocci in the gastric sac. Gram stain, × 1400.

Fig. 4. *Entodinium caudatum* from a Gram-stained smear. Note the rice-starch grains and their surrounding cocci. × 1300.

PLATE 1

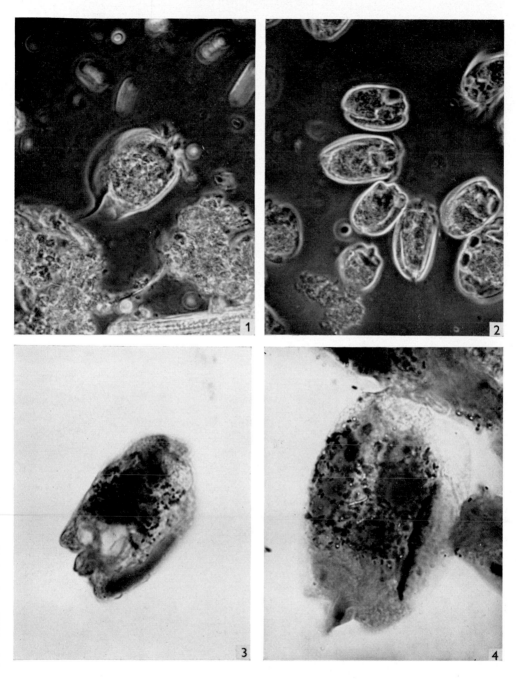

STUDIES ON BACTERIAL ASSOCIATIONS IN GERM-FREE ANIMALS AND ANIMALS WITH DEFINED FLORAS

M. LEV

National Institute for Research in Dairying, Shinfield, Reading

The use of germ-free animals for the study of nutritional, physiological and bacteriological problems has increased in the past few years. All of these studies concern some aspect of bacterial association in the alimentary canal. Germ-free animals offer experimental approaches not obtainable by other means and this justifies the expense and exacting procedure necessary to rear them. Reyniers and Lobund Institute, Gustafsson, Miyakawa and Glimstedt have been pioneers in this field and it is due to their efforts that the technology of producing germ-free animals has reached a point where they are now valuable research tools and no longer laboratory curiosities.

The apparatus used to rear germ-free animals consists essentially of a hermetic compartment to which sterile food and water is supplied through special entry and exit ports which prevent the introduction of bacterial contamination. Stainless steel germ-free rearing units have been used extensively but they are expensive. With the introduction of plastics (Trexler, 1959; Lev, 1962) the price of the apparatus has been greatly reduced, and germ-free animals are now within the scope of many laboratories. Several animal breeders in the U.S.A. (Trexler, 1961) are rearing germ-free animals which may be commercially available in the near future.

Reyniers *et al.* (1949) have proposed a new terminology for germ-free work and use the word 'Gnotobiosis' (known life) to cover not only the germ-free state but also the 'germ-free' animal living in association with certain known, i.e. deliberately inoculated, bacteria or other foreign organisms.

There now exists an extensive literature concerned with germ-free animals, their application to research, and techniques relating to them. These aspects have been dealt with comprehensively in reviews and symposia: Phillips & Smith (1959), *Annals of the New York Academy of Science* (1959), *Recent Progress in Microbiology* (1959), Lev (1961). Research on germ-free animals has given new insight into problems of survival after X-irradiation (McLaughlin *et al.* 1958), the aetiology of dental caries (Orland, 1959; Fitzgerald, Jordon & Stanley, 1961), the origin of antihuman group B agglutinins (Springer, Horton & Forbes, 1959), wound healing (Miyakawa, 1959) and tumorigenesis (Reyniers & Sacksteder, 1959).

The study of the bacteria in the gut or oral cavities of the host animal involves three components—the host, the whole flora and individual species of this flora. The host is affected by many agents other than the gut flora, environment, diet, etc. The bacterial flora is not a constant entity, but is composed of a large number of micro-organisms which in the healthy animal can be considered to be in dynamic equilibrium with the host. This balance may be defined as the normal symbiotic state, and its alteration may produce effects beneficial or detrimental to the host. The effect of a single species on the host is a function of its biochemical activities, the numbers present, the influence of host secretions and the influence of a competitive population of other micro-organisms.

This review will be concerned with (a) direct symbiotic effects involving interactions between two components, often connected with the provision of some growth factor; (b) indirect symbiotic effects involving the synergistic action of several microsymbionts on the host; and (c) effects of bacteria on the physiology of the host.

DIRECT SYMBIOTIC EFFECTS

An example of a direct symbiotic effect is the provision, via bacterial synthesis, of certain vitamins which would otherwise be deficient. As early as 1913 Schottelius raised germ-free chicks, although they grew very poorly compared with conventional animals which grew normally when fed the same sterilized diet. Horton & Hickey (1961) found that germ-free guinea-pigs raised on a particular diet died at 2 weeks of age, whereas the conventional animals grew reasonably well on it. These effects have also been found by other workers and suggest that the flora normally occurring in animals supplies growth factors under natural conditions. If these are not originally present in the diet, the bacterial contribution to the well-being of the animal is of great importance. This was shown by Luckey et al. (1955) who demonstrated that the germ-free, but not the conventional, rat requires biotin which is needed for folic acid metabolism.

By far the most detailed study of the contribution by bacterial flora of the gut to the host's requirement of a specific vitamin has been made by Gustafsson (1959) and Gustafsson et al. (1962) with respect to vitamin K. This vitamin is concerned with the blood-clotting mechanism in animals and is required for the synthesis of prothrombin and perhaps other blood factors. Germ-free rats were fed a vitamin K-deficient diet and showed deficiency symptoms in 5–9 days. On the same diet conventional rats still had a normal blood-clotting mechanism at 24 days. The transfer of the vitamin K-deficient animals to conventional surroundings restored prothrombin times to normal and this was due to the colonization of the gut by certain bacteria synthesizing vitamin K. Gustafsson then compared the ability of various analogues of vitamin K to supply this requirement. Only vitamin K_1 was fully active, menadione phosphate was less active, and menadione (vitamin K_3) and other synthetic forms such as menadione sulphate were inactive. Since these synthetic forms are used in diets it seems probable that bacteria in the gut can modify menadione into the full vitamin which is then utilized by the host. Gustafsson *et al.* (1962) extending this work, found that vitamin K_1 had a full curative effect when administered at a level of 25 μg./kg. of body weight, but ten times as much sodium menadione diphosphate and a hundred times as much menadione were necessary to restore the blood-clotting mechanism in germ-free rats.

Single strains of bacteria isolated from the gut and oral cavities of the rat were introduced into germ-free animals in attempts to correct the deficiency. *Lactobacillus acidophilus*, a 'spore-former', and two species of bacteroides were not active. A strain of *Escherichia coli* and a *Sarcina* sp. restored the prothrombin time to normal after 48 hr.; it has been shown by chemical methods that both these organisms synthesize vitamin K. Gustafsson *et al.* (1962) were therefore the first to demonstrate the correction of an induced deficiency by infecting the host with a single strain of bacterium.

Germ-free animals may be very useful in the discovery of new forms of vitamin K. It has been shown (Lev, 1958, 1959) that rumen strains of the bacterium *Fusiformis nigrescens* have a requirement for this vitamin. *F. nigrescens* was originally isolated from the bovine rumen in association with a species of *Proteus*, and on purification *F. nigrescens* died out. Several forms of the vitamin such as K_1, K_2 and menadione are active in promoting the growth of *F. nigrescens*. However, cells grown on menadione contain no known form of vitamin K, although ether extracts of such cells will support the growth of *F. nigrescens* and other vitamin K requiring bacteria (Lev & Brodie, unpublished results). It would be of great interest to determine whether vitamin K-deficient germ-free rats supplied with menadione and *F. nigrescens* would have a normal prothrombin time, indicating that *F. nigrescens* can convert menadione to a new form of the vitamin which is utilizable by the host. Such direct nutritional effects can be considered as the simplest form of symbiosis.

INDIRECT SYMBIOTIC EFFECTS

Studies of this aspect of microbial interrelationships have been made by Phillips (1957) and Phillips & Wolfe (1959) in relation to the pathogenicity of *Entamoeba histolytica*. Microbial factors were suspected in the aetiology of the disease because alleviation sometimes occurred when bactericidal but not amoebicidal drugs were administered. There was also much variation in the symptoms of the disease, which could not be wholly explained by individual susceptibility or gradation in virulence of the amoeba. Phillips & Wolfe (1959) used germ-free guinea-pigs to elucidate the factors involved. When these animals were inoculated with *E. hystolytica* no lesions developed and the amoeba survived only a few days in the germ-free intestine. However, conventional animals fed the same sterile diet as the germ-free, developed acute disease on inoculation with amoeba. Phillips & Wolfe then implanted various bacteria in germ-free guinea-pigs and tested the effect of amoebae on these animals. Amoebic lesions with extensive ulceration of the caecal wall developed when *Escherichia coli* or *Aerobacter aerogenes* were populating the alimentary tract. *Bacillus subtilis* also had this effect, but an organism which grew very poorly in the germ-free gut had no ability to stimulate the pathogenicity of the amoeba; possibly large numbers are required.

The bacterial factors were further investigated by feeding autoclaved caecal contents of conventional guinea-pigs to germ-free animals which were then inoculated with amoebae. Lesions developed only at the site of inoculation. Finely filtered and autoclaved caecal contents produced no lesions. There is a marked difference in E_h between the caeca of conventional (-367 mV.) and the germ-free guinea-pig (-90 mV.) and so the effect of chemical reducing agents such as thioglycollate and cysteine were next tested. These treatments also resulted in lesions at the site of inoculation with amoebae. Trauma of the surface of the caecum had therefore a pronounced bearing on the ability of the amoebae to infect. The bacterial factors involved appear to be complex, the amoebae depending on the bacteria both to survive in the caecum of the host and to infect the host's tissues. The latter effect of bacteria may be to destroy tissue which is first damaged by the amoebae (Phillips, 1957). These experiments illustrate the ability of two organisms living together to infect the host whereas singly they are completely harmless.

Among symbiotic effects may also be included the case where a component of the gut flora protects the host against other invasive organisms. The normal, i.e. conventional guinea-pig is resistant to infection with *Shigella flexneri* or *Vibrio cholerae*. Derangement of the normal flora by feeding large doses of antibiotics such as streptomycin can result in infection (Freter, 1956; Formal, Dammin, Schneider & La Brec, 1959). This protective effect of components of the gut flora has been studied by Formal *et al.* (1961). Germ-free guinea-pigs inoculated with *S. flexneri* succumbed rapidly to infection, although the disease did not simulate classical dysentery. Heat-killed shigellas had no effect when administered to germ-free guinea-pigs. Similar animals were then implanted with a strain of *Escherichia coli* and subsequently challenged with *S. flexneri*. The animals survived, and when killed, *E. coli* only was isolated from the intestinal contents. In another experiment a lacto-bacillus was implanted in the intestines of germ-free guinea-pigs which were challenged with *S. flexneri* after a week. All the animals succumbed to infection. The nature of the inhibition of *S. flexneri* by *E. coli* is not understood. The *E. coli* used did not produce a colicine or have any other *in vitro* effect on the Shigella, and it was not possible to immunize the germ-free guinea-pigs by subcutaneous inoculation of heat-killed Shigellas. It was, however, found by means of continuous culture techniques that *E. coli* growing in a steady state inhibited the multipli-cation of *S. flexneri* in an anaerobic but not in an aerobic environment. Nevertheless, *E. coli* cannot play a part in the natural resistance of conventional guinea-pigs to infection with *S. flexneri*, because it is not a natural component of the flora. It is therefore obvious that other organisms must be capable of antagonizing *S. flexneri*. Freter (1959) has shown an antagonism of *E. coli* to *S. flexneri* in mice; *Aerobacter aerogenes* was less antagonistic.

Deleterious effects of bacteria on their host animals are also known. Forbes & Park (1959) showed that germ-free chicks grow better than conventionally raised birds and this indicates that bacteria inhabiting the host have a growth-depressing action. The organisms involved in causing growth depression remain in most cases to be discovered. This problem has also been studied from a slightly different aspect, the well-known growth stimulation which occurs when antibiotics are fed to young animals. Although bacteria in the gut were assumed from the first to be concerned in this phenomenon, there is no agreement in the theories put forward in explanation; toxigenic bacteria were thought by some to be eliminated by antibiotics, while others considered that such substances may stimulate vitamin-synthesizing bacteria. The only bacteriological differences found by Lev, Briggs & Coates (1957) to be correlated with the antibiotic growth response in chicks concerned the activities of *Clostridium welchii*. This organism was present in the gut of the young chick and was generally eliminated by the antibiotic in the diet. Even when *C. welchii* was not eliminated, its toxigenicity was reduced, the effect for the host being equivalent to elimination of the organism.

To elucidate further the effect of *C. welchii* in the gut, spores were fed to germ-free chicks with and without penicillin (Lev & Forbes, 1959). Chicks fed *C. welchii* grew less well than the germ-free controls and penicillin corrected this depression. Since an animal with one species of bacterium in its gut is a very artificial experimental subject, a model bacterial flora consisting of a representative of each of the three major groups of bacteria of the gut flora was implanted in the germ-free chicks, viz. *Lactobacillus lactis*, *Streptococcus liquefaciens* and *Escherichia coli*. These had no effect on the growth rate and dietary antibiotics were also without effect. However, when *C. welchii* was added to this flora a growth depression resulted which was relieved in part by the antibiotic in the diet. Thus *C. welchii* is an example of an organism—a component of the normal bacterial flora—which can have a pronounced physiological effect on the host without producing a clinical disease. In the germ-free and defined flora studies, *C. welchii* reached far larger numbers than are found under natural conditions. This indicates that the normal bacterial floral can contribute to the well-being of the host by suppressing a toxigenic organism.

EFFECT OF BACTERIA ON PHYSIOLOGY OF THE HOST

The development of the germ-free animal is abnormal in several respects. All organs of the animal naturally associated with bacteria are reduced in size or capacity. The lymphatic system is poorly developed, and there are fewer plasma cells and secondary nodules in the reticulo endothelial system of germ-free as compared to conventional rats (Thorbecke, 1959). The gut of the germ-free animal is thin and lacks tonus. The reduction in γ-globulin content of the blood of germ-free animals has been described by many workers (e.g. Gustafsson, 1959). Bacteria seem to be necessary in order to effect a normal development and it is possibly due to evolutionary change in close association with their flora that animals, and mammals in particular, depend on certain bacteria for physiological stimuli.

It has been found by several workers from Nuttal & Thierfelder (1896) onwards, that germ-free mammals such as guinea-pigs, rats and rabbits have enlarged caeca. This enlargement may be so great that obstruction of the abdominal organs occurs and death may result. This phenomenon does not occur in conventional animals fed the same sterilized diet as the germ-free. Such enlargement of an internal organ is of obvious importance in the use of germ-free mammals as experimental animals.

Gustafsson (1959) found that the weight of caecum content was five times greater in germ-free rats than in controls. The weight of the caecal wall was 0·5 % of body weight in germ-free and 0·2 % in control rats. Water content of caecum wall was the same in both groups. The effect of certain bacteria on caecum size was also studied. When germ-free rats were contaminated with faeces from conventional animals, the caeca became 'normal' after 4 days, although the rats suffered from diarrhoea and became ill. In a further series of experiments germ-free rats were given *Lactobacillus lactis*, a species of *Proteus* and a coliform organism. Although the coliform and *Proteus*-fed rats had smaller caeca, the caecum size after 8 days remained larger than that of conventional or post-germ-free animals.

Germ-free guinea-pigs were studied by Wynngate, Horton & Forbes (1959) whose illustration of the enlarged caecum is reproduced in Pl. 1, fig. 1, by Phillips, Wolfe & Gordon (1959) who regard the enlarged caecum with its elastic, thin wall as a consistent characteristic of the germ-free guinea-pig, and by Wynngate *et al.* (1959) who found that both the conventional and the germ-free guinea-pig reached the same weight at 4 weeks. The caecum was 22 % of body weight in germ-free and 8·6 % in conventional controls. There was no difference between the weights of the caecal tissues of conventional and germ-free guinea-pigs, indicating that the caecal wall of the germ-free animals was stretched. The water content of the germ-free caecum was higher. An analysis of caecal contents of both groups showed differences which could be interpreted as the effect of the gut flora on ingested foodstuff; the contents of caeca of conventional guinea-pigs contained more fatty acids and more lipids per gram dry weight than germ-free caecal contents. The pH was the same in both groups.

The enlarged caecum phenomenon thus differs in some minor respects between germ-free rats and guinea-pigs. In the germ-free state the caecum lacks the stimulus to evacuate and continuously fills up. The mechanism of evacuation can be considered from two aspects. Caecal or intestinal organisms could supply a nutritional factor without which the caecum fails to develop functionally. In support of this premise, Horton & Hickey (1961) have described a guinea-pig diet which is sterilized by irradiation as opposed to the usual diet (Phillips *et al.* 1959) which is sterilized by autoclaving. Irradiation sterilization is less deleterious to most vitamins than is autoclaving. Horton & Hickey found that their guinea-pigs had a 'normal'-sized caecum, although the wall was thinner and lacked tonus. This would indicate that certain factors were preserved in their diet. On the other hand, bacterial products which may have more of an irritating than a nutritional character may be necessary. These are not included in the diet and cannot be synthesized in the germ-free animals. Skatole and other products of bacterial fermentation (Bokai, 1888) are known to increase gut movements and may be the stimulus required for caecum evacuation.

From the relatively few examples described above, it can be seen that the uses of germ-free animals in the study of gut flora-host relationships has barely begun and many of the quoted problems have been only partially resolved. Germ-free animals provide unique opportunities for investigating isolated symbiotic systems and hence obtaining a clearer picture of the underlying mechanisms. By building from a simple association between one or two organisms in the gut to a complex flora found under natural conditions, symbiotic, and antagonistic mechanisms may be resolved. There is no doubt that the greatest potential for such studies lies in ruminant nutrition, and in the related fields of rumen bacteriology and metabolism.

REFERENCES

Annals of the New York Academy of Science (1959). **78**, 1–400.

BOKAI, A. (1888). Experimentelle Beiträge zur Kenntnis der Darmbewegungen. *Arch. exp. Path. Pharmak,* **24**, 153.

FITZGERALD, R. J., JORDAN, H. V. & STANLEY, H. R. (1961). Experimental caries and gingival pathological changes in the Gnotobiotic rat. *J. Dental Res.* **39**, 923.

FORBES, M. & PARK, J. T. (1959). Growth of germ-free and conventional chicks; effect of diet, dietary penicillin and bacterial environment. *J. Nutr.* **67**, 69.

FORMAL, S. B., DAMMIN, G. J., SCHNEIDER, H. & LA BREC, E. H. (1959). Experimental Shigella infections. II. Characteristics of a fatal enteric infection in guinea-pigs following the subcutaneous inoculation of carbon tetrachloride. *J. Bact.* **78**, 800.

FORMAL, S. B., DAMMIN, G. J., SPRINZ, H., KUNDEL, D., SCHNEIDER, H., HOROWITZ, R. E. & FORBES, M. (1961). Experimental Shigella infections. V. Studies in germ-free guinea-pigs. *J. Bact.* **82**, 284.

FRETER, R. (1956). Experimental enteric Shigella and vibrio infections in mice and guinea-pigs. *J. exp. Med.* **104**, 411.

FRETER, R. (1959). Study of *in vivo* and *in vitro* antagonism between *Shigella flexneri* and normal enteric flora. *Bact. Proc.* p. 97.

GUSTAFSSON, B. E. (1959). *Recent Progress in Microbiology.* VIIth International Congress for Microbiology, Stockholm, 1958, p. 327.

GUSTAFSSON, B. E., DAFT, F. S., McDANIEL, E. G., SMITH, J. C. & FITZGERALD, R. J. (1962). Effect of vitamin K-active compounds or intestinal micro-organisms. *J. Nutr.* (in the Press).

HORTON, R. E. & HICKEY, J. L. S. (1961). Irradiated diets for rearing germ-free guinea-pigs. *Proc. Animal Care Panel*, **11**, 93.

LEV, M. (1958). Apparent requirements for vitamin K of rumen strains of *Fusiformis nigrescens*. *Nature, Lond.* **181**, 203.

LEV, M. (1959). The growth-promoting activity of compounds of the vitamin K group and analogues for a rumen strain of *Fusiformis nigrescens*. *J. gen. Microbiol.* **20**, 697.

LEV, M. (1961). Germ-free animals and their uses in elucidating the action of the gut flora on the host. *J. appl. Bact.* **24**, 307.

LEV, M. (1962). An autoclavable plastic unit for rearing animals under germ-free conditions. *J. appl. Bact.* **25**, 30.

LEV, M., BRIGGS, C. A. E. & COATES, M. E. (1957). The gut flora of the chick. 3. Differences in caecal flora between 'infected', 'uninfected' and penicillin-fed chicks. *Brit. J. Nutr.* **111**, 364.

LEV, M. & FORBES, M. (1959). Growth response to dietary penicillin of germ-free chicks and of chicks with a defined intestinal flora. *Brit. J. Nutr.* **13**, 78.

LUCKEY, T. D., PLEASANTS, J. R., WAGNER, M., GORDON, H. A. & REYNIERS, J. A. (1955). Some observations on vitamin metabolism in germ-free rats. *J. Nutr.* **57**, 169.

McLAUGHLIN, M. M., DACQUISTO, M. P., JACOBUS, D. P., FORBES, M. & PARK, P. E. (1958). The effect of the germ-free state on survival of the ten-day-old chick after X-irradiation. *Rad. Res.* **9**, 147.

MIYAKAWA, M. (1959). Report on germ-free research at the Department of Pathology, University of Nagoya, Japan, and some observations on wound healing, transplantation and foreign body inflammation in the germ-free guinea-pig. *Recent Progress in Microbiology*, p. 299. Ed. G. Tuneval. Oxford: Blackwell Scientific Publications Ltd.

NUTTAL, G. H. F. & THIERFELDER, H. (1896). Thierisches Leben ohne Bakterien im Verdauungskanal. *Z. Physiol. Chem.* **22**, 62.

ORLAND, F. J. (1959). A review of dental research using germ-free animals. *Ann. N.Y. Acad. Sci.* **79**, 285.

PHILLIPS, A. W. & SMITH, J. E. (1959). Germ-free animal techniques. *Adv. appl. Microbiol.* **1**, 141.

PHILLIPS, B. P. (1957). The pathogenic mechanisms in amoebiosis. *Amer. J. Protology*, **8**, 445.

PHILLIPS, B. P. & WOLFE, P. A. (1959). The use of germ-free guinea-pigs in studies on the microbial interrelationships in amoebiosis. *Ann. N.Y. Acad. Sci.* **78**, 308.

PHILLIPS, B. P., WOLFE, P. A. & GORDON, H. A. (1959). Studies on rearing the guinea-pig germ-free. *Ann. N.Y. Acad. Sci.* **78**, 183.

Recent Progress in Microbiology (1959), p. 259. Ed. G. Tuneval. Oxford: Blackwell Scientific Publications Ltd.

REYNIERS, J. & SACKSTEDER, M. R. (1959). Tumerigenesis and the germ-free chick. A preliminary report. *Ann. N.Y. Acad. Sci.* **78**, 328.

REYNIERS, J. A., TREXLER, P. C., ERVIN, R. F., WAGNER, M., LUCKEY, T. D. & GORDON, H. A. (1949). The need for a unified terminology in germ-free life studies. *Lobund Rep.* no. 2, p. 151. University of Notre Dame Press.

SCHOTTELIUS, M. (1913). Die Bedeutung der Darmbakterien für die Ernährung. IV. *Arch. Hyg.* **79**, 289.

SPRINGER, G. F., HORTON, R. E. & FORBES, M. (1959). Origin of anti-human group B agglutinins in germ-free chicks. *Ann. N.Y. Acad. Sci.* **78**, 272.

THORBECKE, G. J. (1959). Some histological and functional aspects of lymphoid tissue in germ-free animals. I. Morphological studies. *Ann. N.Y. Acad. Sci.* **78**, 237.

TREXLER, P. C. (1959). The use of plastics in the design of isolator systems. *Ann. N.Y. Acad. Sci.* **78**, 29.

TREXLER, P. C. (1961). The Gnotobiote. Review and Future. *Biomed. Purview.* **1**, 47.

WYNNGATE, A. E., HORTON, R. E. & FORBES, M. (1959). Biochemical studies on caecal contents of germ-free and conventional guinea-pigs. Final Report Walter Reed Army Institute of Research, Washington, D.C.

EXPLANATION OF PLATE

Fig. 1. Comparative sizes of caeca from germ-free and conventional guinea-pigs (from Wynngate, Horton & Forbes, 1959). Photographed by the Medical Audio-Visual Department, Walter Reed Army Institute of Research, Washington, D.C.

PLATE 1

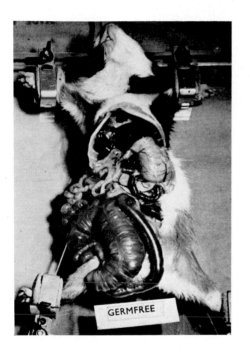

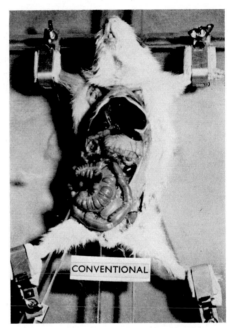

DEFENCE REACTIONS IN ORCHID BULBS

J. NÜESCH*

Institut für spezielle Botanik, Eidgenössische Technische Hochschule, Zürich

INTRODUCTION

Under natural conditions, the germination and development of orchid seeds depends upon infection with mycorrhizal fungi. Bernard (1904) demonstrated this *in vitro* by incubating orchid seeds in an artificial medium with a fungus which he had isolated from orchid roots; the seeds incubated together with the fungus germinated and developed, whereas the controls without the fungus remained sterile. Bernard (1906), and later Burgeff (1936), showed that this ability to induce germination and development was specific for certain fungi, whereas other fungi either destroyed the host or were themselves destroyed by it. Development of the seed is thus dependent upon a specific host-parasite relationship. This is affected by the virulence of the fungus, its properties as a donor of nutritional and growth factors for the young orchid embryo, and the defence reactions of the host itself. Bernard (1909) attempted to elucidate the host-parasite relationship by inoculating young orchid embryos with an avirulent fungus. He found that the hyphae penetrated to a depth of only a few cell layers at the point of attachment of the suspensor, where they were then halted and destroyed, the embryo retaining its potential vitality. When this same embryo was subsequently inoculated with a highly virulent fungus, which would normally have killed the orchid seed, the micro-organism was no longer able to infect. Bernard (1909) referred to this phenomenon as 'pre-immunity'. It seems to be due to defence reactions induced during the first infection. This example would appear to have an important bearing on the symbiotic relationship between micro-organisms and higher plants.

Since orchid embryos are too small to be suitable for such studies we investigated instead the storage organs or bulbs of the Ophrydeae which displayed the requisite properties. In contrast to the roots, which are regularly infected with fungi, the bulbs are far less subject to infections and such as occur are always localized. Moreover, the storage organs very rarely rot, although the very thin cortex of the bulbs offers little mechanical protection. It therefore seemed possible that the bulbs also

* Present address: CIBA Ltd., Basle.

protect themselves against infection by the same type of defence reaction which does not lead to a symbiotic balance, as in seedling development, but instead protects the bulbs from decay.

Bernard (1911), and later Nobécourt (1923, 1928), demonstrated the defence reactions of orchid bulbs *in vitro*. They placed fragments of the bulbs of *Loroglossum hircinum* (L.) Rich. near to a fungus, previously isolated from roots of the same orchid, and grown on an agar plate. They observed that the fungus was unable to invade the bulb tissue and that a zone of inhibition appeared around the bulb fragment. This experiment was confirmed by Gäumann & Jaag (1945), who showed that the bulbs of *Orchis militaris* L. react similarly. Gäumann, Braun & Bazzigher (1950) conducted experiments with a fungus which they isolated from the roots of *O. militaris* and which they identified as *Rhizoctonia repens* Bern. They found that only living tissues are capable of secreting the defensive substances; bulbs killed by heat, cold, or chemical agents were successfully invaded by the fungus. All tissues from the bulbs of *O. militaris* display these defence reactions.

This paper reviews recent work in this field, and gives additional data on the synthesis, nature and activity of the defensive substances which the orchid bulbs are presumed to secrete.

METHODS

(a) Isolation of the root fungi

The fungi were isolated from pieces of externally sterilized root on to malt-agar containing terramycin, as described by Gäumann, Müller, Nüesch & Rimpau (1961).

(b) Assay by the fungus-bulb reaction

The method employed corresponded largely to that described by Gäumann *et al.* (1950). A 5 mm. deep layer of 2% malt agar was poured into Petri dishes and inoculated with the fungus to be tested. After growth was completed, strips (3×0.5 cm.) of the mycelium together with the underlying agar were cut out using a sterile razor blade. The strips, with the mycelium underneath, were now placed on newly prepared malt agar plates. After the mycelium had become established, a sterile fragment of bulb (at least 1 cm.³) was placed about 1 cm. from the strips (depending on the rate of growth of the fungus), and the plates were incubated at 25°. The results were recorded after 3–5 days. Depending on the intensity of the defence reaction, it was possible to distinguish three types of response: (i) a clear zone of inhibition was formed around the bulb; fragment; (ii) fungus growth was halted immediately in front of the bulb; there was no zone of inhibition, but the orchid tissue was not invaded by the fungus; and (iii) the bulb was completely overgrown by the fungus.

The biological activity of the bulb extracts was also assayed by serial dilution tests in which filter-paper discs or cups dipped into the test solution were used instead of bulb fragments.

RESULTS

(1) *Isolation and properties of orchinol*

The agar diffusion test showed that fungistatic substances produced by *Orchis militaris* diffuse out of the bulb tissue into the agar, where they inhibit growth of the mycorrhizal fungus *Rhizoctonia repens*. Fresh storage organs which had been killed by heat, cold, or chemical agents were no longer able to inhibit growth of the same fungus. The inhibitory factor thus seems to be a post-infective defensive substance. We therefore incubated considerable quantities of *Orchis militaris* bulbs, after thorough disinfection, for 10–12 days with *Rhizoctonia repens* in Glaxo flasks at 24°. The bulbs were then extracted and the extract was assayed to determine its content of inhibitory agents. We succeeded in this way in isolating a freshly synthesized antifungal substance (orchinol). This substance, processed and extracted according to the method of Gäumann & Kern (1959) and Gäumann, Nüesch & Rimpau (1960), was identified and assayed by paper chromatography.

Orchinol is neutral and soluble in benzene (Boller *et al.* 1957). Subsequent elucidation of its constitution showed it to be a dihydro-phenanthrene derivative with a melting point of 127° and the formula $C_{16}H_{16}O_3$ (MW = 256). Orchinol was always accompanied by *p*-hydroxy-benzyl alcohol.

Orchinol displays a relatively weak fungistatic activity against various root fungi of orchids and other soil-inhabiting fungi. Its activity is quite unspecific. The minimum inhibiting concentration in the agar diffusion test with 6 mm. filter discs was 10^{-3} to 10^{-4} M. The associated *p*-hydroxy-benzyl alcohol shows no detectable activity and its role has not yet been clarified. Orchinol may be found in many terrestrial orchids. Of twenty-four orchid species incubated with *Rhizoctonia repens*, orchinol was identified with certainty in the bulbs of sixteen (Table 1). The ability to synthesize orchinol was widespread although species within a genus may differ in their capacity to synthesize this substance. In *Serapias*, for example, all three species examined produced orchinol. In the genus *Ophrys*, on the other hand, neither of the species studied

was able to synthesize orchinol. *Loroglossum hircinum* (L.) Rich.
produces *p*-hydroxybenzyl alcohol and very little orchinol, whereas
Loroglossum longibracteatum (Biv.) Moris synthesizes both substances.
   The time required for the synthesis of orchinol was determined by
incubating tissue cylinders (11 mm. in diameter and 12–13 mm. in
length) from *Orchis militaris* bulbs on a 5 day old layer of mycelium
from *Rhizoctonia repens* in Petri dishes (Gäumann & Hohl, 1960). The
cylinders were removed after 12 hr. and after 1, 2, 5, 8 and 12 days and
were cut into 2 mm. thick discs. These discs were extracted in eight
repetitive series and the content of orchinol was determined by the usual
method. The first trace of orchinol was detected after 36 hr. in those parts
of the discs directly in contact with the mycelium, and the highest final
concentration—920 μg./g. tissue, corresponding to nearly 1 % of fresh
weight or to an orchinol concentration of at least 0.5×10^{-2}M—was

Table 1. *Synthesis of orchinol and of* p-*hydroxybenzyl alcohol in the
bulbs of different orchids incubated with* Rhizoctonia repens Bern (*strain
from* Orchis militaris *L.*)

(From Gäumann, Nüesch & Rimpau, 1960)

Host species	Orchinol	*p*-Hydroxy-benzyl alcohol
Aceras anthropophora (L.) R.Br.	+ +	+
Anacamptis pyramidalis (L.) Rich.	+ +	+ +
Charmorchis alpina (L.) Rich.	+ + +	+ +
Coeloglossum viride (L.) Hart.	+ + +	+ +
Gymnadenia albida (L.) Hartm.[1]	+ + +	+ +
G. conopea (L.) R.Br.	?	?
G. odoratissima (L.) Rich.	+	+
Loroglossum hircinum (L.) Rich.[2]	+	+
L. longibracteatum (Biv.) Moris[3]	+ + +	+
Nigritella nigra (L.) Rchb.[4]	+ +	+
Ophrys apifera Huds.	0	0
O. Arachnites (Scop.) Murray[5]	0	0
Orchis coriophora L.	?	+ +
O. latifolia L.	+	+ +
O. maculata L.	0	0
O. mascula L.	+ + +	+ +
O. militaris L.	+ +	+ +
O. morio L.	+ + +	+ + +
O. sambucina L.	+ + +	+ +
O. ustulata L.	0	?
Platanthera bifolia (L.) Rich.	0	0
Serapias lingua L.	+ + +	+ +
S. neglecta Not.	+ +	+ +
S. vomeracea Burm.	+ +	+ +

[1] *Coeloglossum albidum* Hartm.;   [2] *Himantoglossum hircinum* Sprengel;   [3] *Barlia longibracteata* Parlat = *Aceras longibracteata* Rchg. = *Orchis longibracteata* Biv.;   [4] *Nigritella angustifolia* Rich.;   [5] *Ophrys fuciflora* Crantz.

found in such discs. Production of orchinol reaches a maximum at 8 days (see Table 2). The orchinol content diminishes the greater the distance from the point of contact between tissue and fungus.

The way in which the synthesis of orchinol is induced is not yet clear. Orchinol is not found in intact bulbs, but, as noted above, mechanical injury also leads to its formation, though the amount produced is far smaller than in response to infection with a root fungus. The promoting effect of the fungus on synthesis is not specific; a single fungus may induce synthesis in various species of orchid, while several different fungi may activate the synthesis of orchinol in the same orchid. Other species of fungus, e.g. *Rhizoctonia solani* Kuhn, rapidly destroy orchinol and an infection with this fungus also leads to the destruction of the bulbs. The same effect may be seen with very fast growing fungi, such as *Fusarium solani* (Mart) App. et Wr., which invade the mechanically unprotected tissue so quickly that there may not be sufficient time for the synthesis of the defensive substances. If the intact tissues are mechanically destroyed or deprived of air the synthesis of orchinol is likewise inhibited.

Table 2. *The concentration of orchinol (μ/g.) in the tissue-cylinders from the bulbs of* Orchis militaris *incubated with* Rhizoctonia repens (*strain from* O. militaris)

(From Gäumann & Hohl, 1960)

Time of incubation (days)	2 mm. thick discs numbered from the bottom to the top					
	1	2	3	4	5	6
1	0	0	0	0	0	0
2	28	10	0	0	0	0
5	200	112	15	Traces	0	0
8	920	380	160	50	35	45
12	650	540	460	110	100	80

(2) *Isolation and properties of hircinol*

According to Bernard (1904) and Nobécourt (1923, 1928), the bulbs of *Loroglossum hircinum* also inhibit growth of certain mycorrhizal fungi, although in these bulbs little or no orchinol can be found (see Table 1). From the roots of this orchid a fungus, *Rhizoctonia versicolor* Müller et Nüesch (Gäumann *et al.* 1960), was isolated which induces formation of another fungistatic substance in the bulbs of *Loroglossum hircinum*. This substance is also believed to be a dihydrophenanthrene derivative and is called hircinol. Extraction and assay of hircinol was performed by the technique of Dr J. Urech, CIBA Ltd., Basle (not yet published).

Sterile cylinders of bulb tissues are incubated with the fungus under study (Gäumann *et al.* 1960). After incubation for 5–10 days, the bulb cylinders are homogenized and extracted with ether and acetone; the organic solvents are evaporated under vacuum at 50°, the aqueous residue is re-extracted with petroleum ether, and—after adjusting the pH to 5·0 with HCl and adding 0·07% (w/v) NaCl—is extracted six times with 70% (v/v) ether. The petroleum-ether and ether extracts are mixed and concentrated under vacuum.

Hircinol can be identified and semi-quantitatively evaluated by paper chromatography, using Whatman No. 1 paper (impregnated with 20% formamide) and a cyclo-hexane-butanol (1:2) system (Zaffaroni, 1953). The running time is approximately $2\frac{1}{2}$ hr. at 22°. The chromatogram is dried for 1 hr. at 90°. Hircinol gives a red-violet fluorescence in the UV (2500 Å.), with an absorption maximum at 2740 Å. The spots are planimetrically evaluated and the hircinol contents semi-quantitatively assayed by means of a standard curve. The minimum concentration measurable is 0·625 μg./ml. Hircinol shows the same broad spectrum of activity as orchinol. *In vitro* it is active not only against *Rhizoctonia versicolor*, but also against many root fungi and against other saprophytic and pathogenic fungi, including some pathogenic in man.

Gäumann *et al.* (1961) investigated the mycorrhizal fungi of *Loroglossum hircinum*, but failed to observe any clear specificity. Fungi isolated from the roots varied considerably with the geographical location of the host. Fifteen different species were isolated, of which eight were Ascomycetes and only one, *Rhizoctonia versicolor*, probably a Basidiomycete.

Hircinol, like orchinol, cannot be found in intact tissues of bulbs. Its synthesis, like that of orchinol, is induced simply by injury, but as with orchinol is markedly stimulated by various fungi.

In order to obtain further information on this effect of the fungus, we tried to extract and concentrate the stimulating agent. The mycelium was mechanically homogenized with water in a Potter flask or Turmix and filtered. A fraction was isolated from the aqueous solution by precipitation with acetone (4 parts acetone:1 part solution). When incubated with sterile tissues of bulbs, this precipitate—like the mycelium itself—was found to activate the synthesis of hircinol. The mycelium of *Rhizoctonia solani*, which destroys the bulbs of *Orchis militaris* as well as of *Loroglossum hircinum*, was treated in the same way, but with this fungus no activity was found. Further investigations on the induction of hircinol are in progress.

DISCUSSION

Whereas most green plants display passive and active defence reactions that serve to keep them free from micro-organisms (Gäumann, 1951), orchids growing under natural conditions enter into a specialized 'mycorrhizal' relationship with invading fungi, which is necessary for the development of the young seedling. Knudson (1922, 1925, 1930) showed that *in vitro* orchid seedlings can also develop without a fungus on a medium containing sugar. There is evidence that the mycorrhizal association in orchids is often unspecific. Burgeff (1936) showed that growth of the embryos of a single orchid species can be induced by several different fungi; and Knudson (1925) and Derx (1937) showed that fungi not usually associated with orchids can perform this function. It was demonstrated by Gäumann *et al.* (1961) that the fungi associated with *Loroglossum hircinum* roots belonged to quite different systematic groups. This lack of specificity led Curtis (1939) to suggest that the fungus-orchid relationship is a form of attenuated parasitism. The fungal invasion of the orchid embryos and roots can be regarded as a primarily parasitic process, which, however, becomes stabilized due to the localization of the infection.

Investigation of the defence reaction in orchid bulbs led to the isolation of two chemically related phenanthrenes—orchinol and hircinol—with phenol-like properties. The two fungistatic substances are newly synthesized in the host cells, sensitized by the invading fungus, and after 5–8 days reach concentrations that effectively control further fungal spread. Few similar reactions are known, but Condon & Kuč (1960) have reported the formation of a fungistatic phenol in carrot tissue inoculated with *Ceratocystis fimbriata* Ell. et Kalst. Very little is yet known about the biosynthesis of orchinol and hircinol, but the stimulating substance from the fungus does not seem to be chemically related to the antifungal agent formed by the host. It is unlikely that preformed substances are merely chemically degraded or oxidized during the infection, as such reactions usually take place at once (e.g. mustard oil glycosides to mustard oil, or alliin to allicin), whereas the induction of orchinol and hircinol requires about 36 hr. Moreover, neither are formed if the cells have been completely destroyed by mechanical means.

Both the induction and the action-spectrum of the synthesized antifungal substances are relatively unspecific. This does not necessarily preclude their importance as factors in the establishment of mycorrhizal association in the orchids. Evidence has been cited that, for a manifestation of their characteristic effects on seedling growth, such associations also do not necessarily depend on a high degree of specificity. Moreover some fungi such as *Rhizoctonia solani* or *Fusarium solani* may always remain fully active parasites because they destroy the orchinol, or the host tissue, before a proper defence reaction can occur. It therefore seems possible that induced antifungal agents like orchinol and hircinol may have some significance in the establishment of a stable mycorrhizal condition in orchids.

REFERENCES

BERNARD, N. (1904). Recherches expérimentales sur les Orchidées. *Rev. gén. de Bot.* **16**, 405.

BERNARD, N. (1906). Symbiose d'orchidées et de divers champignons endophytes. *C.R. Acad. Sci., Paris*, **142**, 52.

BERNARD, N. (1909). L'évolution dans la symbiose. Les orchidées et leurs champignons commensaux. *Ann. sci. nat. (Bot.)*, **9**, 1.

BERNARD, N. (1911). Sur la fonction fungicide des bulbes d'Ophrydées. *Ann. sci. nat. (Bot.)*, **14**, 221.

BOLLER, A., CORRODI, H., GÄUMANN, E., HARDEGGER, E., KERN, H. & WINTERHALTER-WILD, N. (1957). Über induzierte Abwehrstoffe bei Orchideen. I. *Helv. chim. Acta*, **40**, 1062.

BURGEFF, H. (1936). *Samenkeimung der Orchideen*, Jena: Verlag Gustav Fischer.

CONDON, P. & KUČ, J. (1960). Isolation of a fungitoxic compound from carrot root tissue inoculated with *Ceratocystis fimbriata. Phytopathology*, **50**, 267.

CURTIS, J. T. (1939). The relation of specificity of orchid mycorrhizal fungi to the problem of symbiosis. *Amer. J. Bot.* **26**, 390.

DERX, H. G. (1937). Tentative de synthèse d'une symbiose. *Ann. sci. nat. (Bot.)*, sér. 10, **19**, 155.

GÄUMANN, E. (1951). *Pflanzliche Infektionslehre*, 2 Auflage. Basel: Verlag Birkhäuser.

GÄUMANN, E., BRAUN, R. & BAZZIGHER, G. (1950). Über induzierte Abwehrreaktionen bei Orchideen. *Phytopath. Z.* **17**, 36.

GÄUMANN, E. & HOHL, H. R. (1960). Weitere Untersuchungen über die chemischen Abwehrreaktionen der Orchideen. *Phytopath. Z.* **38**, 93.

GÄUMANN, E. & JAAG, O. (1945). Über induzierte Abwehrreaktionen bei Pflanzen. *Experientia*, **1**, 21.

GÄUMANN, E. & KERN, H. (1959). Über die Isolierung und den chemischen Nachweis des Orchinols. *Phytopath. Z.* **35**, 347.

GÄUMANN, E., MÜLLER, E., NÜESCH, J. & RIMPAU, R. H. (1961). Über die Wurzelpilze von *Loroglossum hircinum* (L.) Rich. *Phytopath. Z.* **41**, 89.

GÄUMANN, E., NÜESCH, J. & RIMPAU, R. H. (1960). Weitere Untersuchungen über die chemischen Abwehrreaktionen der Orchideen. *Phytopath. Z.* **38**, 274.

KNUDSON, L. (1922). Non-symbiotic germination of orchid seeds. *Bot. Gaz.* **73**, 1.

KNUDSON, L. (1925). Physiological studies of the symbiotic germination of orchid seeds. *Bot. Gaz.* **79**, 345.

KNUDSON, L. (1930). Flower production by orchids grown non-symbiotically. *Bot. Gaz.* **89**, 192.

NOBÉCOURT, P. (1923). Sur la production d'anticorps par les tubercules des Ophrydées. *C.R. Acad. Sci., Paris,* **177**, 1055.

NOBÉCOURT, P. (1928). *Contribution à l'étude de l'immunité chez les végétaux.* Thesis, Lyon, 2nd ed.

ZAFFARONI, A. (1953). Corticosteroid microanalysis. *Rec. Progr. Horm. Res.* **8**, 51.

INDEX

Acarospora, 40, 41, 42, 44, 46, 166
acetic acid, produced by protozoa, 209, 315;
by rumen bacteria, 270, 276, 277, 278, 279, 282, 284, 285; used by algae, 185, 189, by insects, 209
acetyl-CoA, 283
acetyl phosphate, 96, 97, 99
Achromobacter, nitrogen fixation by, 93
Actinomyces alni, 77
adenosine diphosphate (ADP), 9; triphosphate (ATP), 104
Aeolidiella, algae in, 175, 177
Aerobacter, in germ-free animals, 328; manganese in, 113; nitrogen fixation by, 93
Agrobacterium, and vesicular-arbuscular mycorrhiza, 158
air-bubble, in crop of ambrosia beetles, 257, 260
alanine, deaminated by glutamic dehydrogenase, 9; as growth factor for lichen fungi, 41
Alcaligenes, and vesicular-arbuscular mycorrhiza, 158
aldehyde oxidase, 110
algae, symbiotic in other algae, 175, in invertebrates, 171–99, in lichens, 4, 5, 35, 37–46
algae, blue-green, nitrogen fixation by, 33, 38, 92, 93, 96, 104
Allium, vesicular-arbuscular mycorrhiza of, 148, 155, 158, 160
Alnus, root nodules of, 72–86 *passim*, 115, 117
Amanita, 125; growth of, *in vitro*, 131
ambrosia beetles and their fungi, 232–65
Ambrosiamyces, 239
amides, in nitrogen fixation in root nodules, 85
amino acids, exuded by roots, 64, 126, 127; in metabolism of rumen bacteria, 288, of rumen protozoa, 307, 308, 320; in nutrition of aposymbiotic cockroach, 213, of mycorrhizal fungi, 130, 131; in the rumen, 287; synthesized in extracts of nitrogen-fixing organisms, 101, 105, in root nodules, 56, 115; used by symbiotic algae, 187
ammonia, as intermediate in nitrogen fixation, 102, 105–9, 115, 116; in metabolism of symbiotic invertebrates and algae, 187; produced by rumen bacteria, 287–8; as source of nitrogen for rumen

organisms, 287, 320; *see also* nitrogen (combined)
amylase, 307
amylopectin, in rumen protozoa, 314–5, 319
Anabaena, extracts of, 104; nitrogen fixation by, 92, 93
anaerobiosis, for culture *in vitro* of protozoa from insects, 209, of rumen bacteria, 268, 278, of rumen protozoa, 302, 314; in thermodynamics of rumen bacteria, 291–2
Anemonia, algae in, 173
aneurin (thiamine), 41, 42, 65, 128, 141, 190, 210; *see also* B vitamins.
Anisandrus, 235, 236, 237, 238, 242, 243, 261
Anobiid beetles, intestinal structure of, 201; metabolism of symbionts of, 209–10
Anodonta, parasitic alga in, 179
anthocyanin pigment, in root nodules, 74
antibiotics, for culture of aposymbiotic insects, 212–3, 218–9; of rumen protozoa, 299, 301, 305, 307, 311; fed to young animals, 329–30; inhibit cell-wall formation, 219–20; produced by: coelenterates, 190; legume seedlings, 64; lichens, 33; orchid bulbs, 335–43; soil organisms, 64; Zooxanthellae, 190; *see also* individual antibiotics
antifungal agents, in orchid bulbs, 335–53
antigens, and antibodies, 9; produced by lysogenized bacteria, 8, 20
ants, symbionts in, 202, 203, 204
aphids, transmission of symbionts of, 206; yeasts in, 201
aposymbiosis in arthropods, 200–31 *passim*
appressoria, 150, 158
arabinose, exuded by roots, 127
Arachis, entry of root nodule bacteria into, 54
arbuscules, 147–9
Artemia, nutrition of, 188
arthropods, symbiosis and aposymbiosis in, 200–31
Ascomycetes, mycorrhizal genera of, 125, 340
asparagine, in root nodules, 116
aspartic acid, stimulant for growth of mycorrhizal fungi, 129, 130
Astragalus, root nodule bacteria of, 61
atmosphere pollution of, sensitivity of lichens to, 5, 34, 45
ATP (adenosine triphosphate), 104
auxin, produced by mycorrhizal fungi, 141

nitrate, inhibits infection of roots by
nodule bacteria, 52; reduced by plant,
83–4, by rumen organisms, 286
nitrite, 109, 115, 286
nitrogen (combined), content of, in different
parts of lichens, 36, 37; excreted into
medium by *Cl. pasteurianum*, 105; in
metabolism of symbiotic cockroaches and
bacteria, 222; of invertebrates and algae,
187; inhibits: formation of perithecia by
ambrosia fungi, 245, infection of roots by
vesicular-arbuscular mycorrhiza, 158–9,
165, nodulation, 80, 166; proportions of
bacterial, plant, and protozoal, in rumen,
288, 318; sources of, for *Endogone*, 156,
for lichen symbionts, 41, for rumen
protozoa, 319–20, for yeasts, 251–2;
transfer of, from nodules to plant, 85–6;
see also ammonia, etc.
nitrogen fixation, biochemistry of, 92–124;
by blue-green algae, 38, 92, 93, 96, 104;
by flagellate symbionts of arthropods
(doubtful), 208, 209; by lichens, 33, 38,
41; in root nodules, 51–2, 53, 54, 55–6,
81–3, 113–16; by *Torulopsis*, 261; *in vitro*
(by extracts of nitrogen-fixing organisms),
55, 95–104
nitrogen (gas), effect of, on nitrogen-fixing
systems, 56, 99, 100, 110, 111, 114
nitrogenase, 99, 108, 112, 118
nitrous oxide, inhibits nitrogen fixation,
105–7
Nocardia, symbiotic in *Hippophaë*, 77, in
Rhodnius, 210–11
nodules, *see* leaf nodules, root nodules
Nostoc, extracts of, 104; in lichens, 40, 41,
43, 44; nitrogen fixation by, 92, 93, 116
nucleic acid, constituents of, exuded from
roots, 126; production of, in bacteria, 25
nucleotides, exuded from roots, 139
nucleus, diffuse, in symbiotic bacteria, 218;
enlarged, in orchids, 148, in root cells
infected with nodule bacteria, 53, 54, and
with vesicular-arbuscular endophytes,
147–8
nurse cells, of insect ovaries, 205–6
nutrient-absorbing organ, mycorrhiza as, 126
nutrition, of aposymbiotic cockroaches, 213;
of arthropods, and presence of symbionts,
203–4, 208–11; of lichens, 5, 33, 34, 40–1,
45–7, 116; of rumen protozoa, rôle of
bacteria in, 312–14; of symbiotic invertebrates and algae, 184–93; and transmission of bacterial symbionts in cockroaches, 218–22; *see also* diet

Odontoglossum, 148
oil-containing organs, of ambrosia beetles,
241–4

oil droplets, in vesicles of vesicular-arbuscular endophyte, 149
Oocystis, 179
Ophiostoma, 237
Ophridium, 2
Ophrydeae, defence reactions in, 336–43
Ophryoscolex, 305, 307, 308, 317
Ophrys, 338–9
orchids, defence reactions in bulbs of,
335–43; mycorrhizas of, 2, 4–5, 166;
nuclear changes in, 148
orchinol (dihydrophenanthrene), antifungal
agent produced in orchid bulbs, 335, 337,
338, 339, 340, 341, 342
Orchis, 335, 336, 337, 338, 339
organelles, removal of, from cells, 10
Oryzaephilus, 205, 212
Oscillospira, 284
ova, transmission of symbionts in, 59,
177–8, 205, 214, 215, 218–22
oxidation-reduction potential, of caeca of
normal and germ-free animals, 328; for
culture *in vitro* of rumen bacteria, 268,
278, 281, and protozoa, 314; of rumen
contents, 267
oximes, in nitrogen fixation, 105
oxygen, produced by algal symbionts,
184–5; required for nitrogen fixation, 55,
84–5, 102; for synthesis of orchinol,
340; used for defaunation of insects,
208, 224
Oxymirus, symbiotic yeast of, 216

~ P (high-energy phosphorus), 283
Paecilomyces, in tunnels of ambrosia
beetles, 254
pantothenic acid, 129, 210; *see also* B
vitamins
Papilionatae, 65
Paramoecium, algae in, 177–93 *passim*;
alga-free culture of, 182; bacteria-free
culture of, 190, 194
parasitism, and symbiosis, 4–5, 59, 66, 125,
164, 180, 189, 217, 341
parasitoid insects, 3–4
Parcoblattus, symbionts of, 204
pathogen-free mice and rats, 5–7; *see also*
germ-free animals
Paulinella, algae in, 174, 179, 188
Paxillus, 125
PCA (dihydropyridazinone - 5 - carboxylic
acid), as possible intermediate in nitrogen
fixation, 106, 107, 108, 109, 118
peas, root nodule bacteria of, 60; vesicular-
arbuscular mycorrhiza of, 163
pectin, 261, 276–7, 284
pectinase, 158
pectin-esterase, absent from *Entodinium*,
307